THE FOUNDATIONS OF INTUITIONISTIC MATHEMATICS

STUDIES IN LOGIC

AND

THE FOUNDATIONS OF MATHEMATICS

Editors

L. E. J. BROUWER, *Laren (N.H.)*
A. HEYTING, *Amsterdam*
A. ROBINSON, *Los Angelos*
P. SUPPES, *Stanford*

Advisory Editorial Board

Y. BAR-HILLEL, *Jerusalem*
K. L. DE BOUVÈRE, *Amsterdam*
H. HERMES, *Munster i/W.*
J. HINTIKKA, *Helsinki*
A. MOSTOWSKI, *Warszawa*
J. C. SHEPHERDSON, *Bristol*
E. P. SPECKER, *Zürich*

NORTH-HOLLAND PUBLISHING COMPANY
AMSTERDAM

THE FOUNDATIONS OF INTUITIONISTIC MATHEMATICS

ESPECIALLY IN RELATION TO RECURSIVE FUNCTIONS

STEPHEN COLE KLEENE

Cyrus C. MacDuffee Professor of Mathematics
The University of Wisconsin, Madison, Wis.

RICHARD EUGENE VESLEY

Assistant Professor of Mathematics
Case Institute of Technology, Cleveland, Ohio

1965

NORTH-HOLLAND PUBLISHING COMPANY
AMSTERDAM

PRINTED IN THE NETHERLANDS

PREFACE

In his intuitionistic analysis, L. E. J. Brouwer created a theory which diverges from classical mathematics and which in its details has not been widely known or understood. Hence it has seemed to us that this theory especially has presented a challenge to the metamathematical and model-theoretic methods. In Chapter I we formalize it, differently than Heyting did in 1930. This formalization includes a continuation of the formal development of (intuitionistic and classical) elementary number theory, which should interest readers of the senior author's "Introduction to Metamathematics", where the development of formalized mathematics in Part II ended somewhat abruptly. The interpretation, or model-theoretic treatment, is undertaken in Chapter II, using the theory of recursive functions from Part III of "Introduction to Metamathematics". This chapter contains two new realizability interpretations (one briefly announced in another form in 1957), with variations. Vesley's Chapter III formalizes Brouwer's theory of the continuum, and Chapter IV applies both metamathematical and model-theoretic considerations to some moot points in that theory. Nothing is presupposed of the reader except familiarity with some material which can be found in Chapters I–XII of "Introduction to Metamathematics" or elsewhere.

The research now definitively reported in this monograph was begun essentially in 1950, while the senior author held a John Simon Guggenheim Memorial Fellowship granted for the purpose. The final stages in the research and editing fell in a period (from July 1, 1963) during which his research was supported by the National Science Foundation of the United States under Grant GP1621.

The senior author was in residence at the University of Amsterdam from January to June 1950, and is indebted to L. E. J. Brouwer, Arend Heyting, Johan J. de Iongh, and the late David van Dantzig and Evert W. Beth, for orienting him in intuitionistic mathematics, and for many fruitful conversations then and later. He must take the sole responsibility for the direction in which his research led him and for the interpretations he reached.

Richard E. Vesley joined the project as a Ph.D. student. Each author's contributions are identified by chapters; but each has influenced the other's chapters (as noted on pp. 88, 150, 179, and in other details). A proof by Joan Rand Moschovakis enters on p. 68.

We thank Douglas A. Clarke, John Moschovakis, Joan Rand Moschovakis and Georg Kreisel for many helpful suggestions on the manuscript. Joan Rand Moschovakis also read the printer's proof.

We lament the death, during the printing, of Evert W. Beth, who gave us so much encouragement in this undertaking.

June 1964 S. C. KLEENE

TABLE OF CONTENTS

CHAPTER I

A FORMAL SYSTEM OF INTUITIONISTIC ANALYSIS

by S. C. KLEENE

§ 1. Introduction (to the monograph). The constructive tendency in mathematics has been represented, prior to or apart from intuitionism, in the criticism of "classical" analysis by Kronecker around 1880–1890, and in the work of Poincaré 1902, 1905–6, Borel **1898**, **1922** and Lusin **1927**, **1930** (cf. Heyting 1934 or **1955**).[1]

Modern intuitionism, founded by Brouwer, constitutes a vigorous manifestation of the constructive tendency. In his thesis **1907** Brouwer treats of the intuitive genesis of the natural number series and argues the distinctness of mathematics from the language in which mathematics is expressed. In his famous paper 1908 on the untrustworthiness of logical principles, he rejects the traditional belief that the classical logic derived from Aristotle has an a priori validity (cf. IM § 13). His writings, such as **1912**, **1921**, contain many examples proposed as

[1] Dates in gothic numerals (e.g. **1898**) refer to the bibliography of this monograph, and in medieval tailciphers (e.g. 1902) to the bibliography of Kleene "Introduction to Metamathematics" **1952b**. The latter book we cite simply as "IM", and furthermore we cite things in IM simply as they were cited within IM when like designations do not exist in the present monograph.

Some errata in early printings of IM are given in the bibliography below. Also: (a) At the top of p. 51 of IM, the seeming circularity that *not B* is used in explaining *not A* is to be avoided thus. Sameness and distinctness of two natural numbers (or of two finite sequences of symbols) are basic concepts (cf. p. 51 lines 20–24). For any B of the form $m=n$ where m and n are natural numbers, *not B* shall mean that m and n are distinct. The explanation of *not A* in lines 5–8 then serves for any A other than of that form, by taking the B in it to be of that form. Equivalently, since the distinctness of 1 from 0 is given by intuition (so *not* $1=0$ holds), *not A* means that one possesses a method which, from any proof of A, would procure a proof of $1=0$ (cf. lines 8–9). (b) On p. 506, the notation in 10 and 11 overlooks the possibility that t contains x; but the reasoning in that case is just the same. (Similarly on p. 155 of the addendum to Kleene **1962a**.)

confuting classical arguments and results. In **1918–9** he began the construction of an intuitionistic mathematics divergent from the classical; the results are collected in **1924–7**, **1927**. Subsequently he was joined by Heyting **1925**, **1927**, **1927a**, **1929**, **1941**, **1953a**, de Loor **1925**, Belinfante **1929**, Freudenthal **1936**, van Dantzig **1942**, Dijkman **1948**, **1952**, van Rootselaar **1952**, **1960**, van Dalen **1963** and others in continuing the development of intuitionistic mathematics; cf. the bibliographies in Heyting **1955** and **1956**. (For "negationless intuitionistic mathematics" Griss **1944**, **1946–51**, **1955**, cf. Heyting **1956** 8.2.) Besides the approach to intuitionism through Brouwer's original papers (including those already cited, and e.g. **1928a**, **1929**, **1948**, **1952**, **1954**), there are the approaches through Heyting's expositions (1934, re-edited **1955**, and **1956**) and Beth's (**1955**, **1959**).

Meanwhile there arose in the 1930's a different manifestation of the constructive tendency, in the form of a general theory of constructive processes, or algorithms, for calculating functions or deciding predicates (i.e. properties and relations). This theory is commonly called the "theory of (general) recursive functions", after one of the mathematical definitions of a class of number-theoretic functions which has been proposed as coinciding with the intuitively-understood class of the number-theoretic functions which can be calculated effectively by a uniform method. The definition of the *general recursive functions* was given by Gödel in 1934 building on a suggestion of Herbrand (cf. IM § 55). The identification of the general recursive functions with the effectively calculable functions, which we call *Church's thesis*, was first proposed by Church in 1936 (cf. IM p. 300 and § 62, Kleene **1958** § 2). Whether or not one accepts the thesis, the considerations adduced in support of it bear on the significance of the theory. Simultaneously, the class of the *λ-definable functions*, which had been introduced and studied by Church and Kleene beginning in 1932, was proved by them to be the same as the class of the general recursive functions (Church 1936, Kleene 1936a; cf. Church 1941). Turing independently proposed the like thesis for the *computable functions*, introduced by him in 1936–7 and independently in less detail by Post 1936; and in 1937 Turing proved them to be the same as the other two classes of functions just mentioned. Subsequently other equivalent formulations have been introduced, as by Post 1943 and by Markov **1951c**. On this basis an extensive theory has been developing in a variety of directions, e.g. decision problems, hierarchies, degrees of unsolvability, recursive equivalence types, recursive analysis, constructive ordinals, recursive functionals. Among the books are Péter **1951**, Tarski-Mostowski-R. M. Robinson **1953**, Markov **1954**, Davis **1958**, Trahtenbrot **1960**, Dekker-Myhill **1960**, Uspenskiĭ **1960**, Hermes **1961**, Klaua **1961**, Sacks **1963**, Rogers **1964**. We have counted (by October 15, 1963) 150 contributors to the subject.[2] The proceedings of several recent gatherings indicate the scope of the work that is going on (Rosser **1957**, Heyting **1959**, Mostowski **1961***, Nagel-Suppes-Tarski **1962**, Dekker **1962**).

The theory of general recursive functions had antecedents in work of Skolem 1923, Hilbert 1926, Ackermann 1928, Gödel 1931, Péter 1934 dealing with special classes of recursive functions, which has been continued by 35 investigators.[3]

[2] The 20 cited already, and J. W. Addison, S. I. Adyan, Günter Asser, Gilbert Baumslag, N. V. Beljakin, Bernays, William W. Boone, R. E. Bradford, Donald Bratton, J. L. Britton, J. R. Büchi, C. S. Ceĭtin, V. S. Černjavskiĭ, André Chauvin, C. R. Clapham, Douglas A. Clarke, Alan Cobham, J. N. Crossley, V. K. Detlovs, V. H. Dyson, Andrzej Ehrenfeucht, Calvin C. Elgot, Herbert B. Enderton, A. P. Eršov, Walter J. Feeney, Solomon Feferman, Patrick C. Fischer, Frederic B. Fitch, Roland Fraïssé, A. A. Fridman, Richard M. Friedberg, A. Fröhlich, D. C. Furguson, R. O. Gandy, Seymour Ginsburg, Andrzej Grzegorczyk, William P. Hanf, Harrop, Hasenjaeger, Graham Higman, Louis Hodes, P. K. Hooper, Hu Shih-Hua, A. Janiczak, Ju. I. Janov, Jaśkowski, A. S. Kahr, Kalmár, L. A. Kalužnin, Carol Karp, Clement F. Kent, Akiko Kino, Kolmogorov, V. S. Koroljuk, Donald L. Kreider, Kreisel, Kuznecov, Daniel Lacombe, Azriel Lévy, Samuel Linial, Shih-Chao Liu, Loh Chung-Wan, Paul Lorenzen, David C. Luckham, M. Machover, Maehara, A. I. Mal'cev, Werner Markwald, D. A. Martin, B. H. Mayoh, S. Mazur, John McCarthy, Thomas G. McLaughlin, Medvedev, Marvin L. Minsky, Moh Shaw-Kewi, John N. Moschovakis, S. Mrówka, A. A. Mučnik, A. A. Mullin, N. M. Nagornyĭ, Nelson, Anil Nerode, B. H. Neumann, P. S. Novikov, Walter Oberschelp, È. S. Orlovskiĭ, Rohit J. Parikh, Porte, Marian Boykan Pour-El, Hilary Putnam, Michael O. Rabin, Wolfgang Rautenberg, H. G. Rice, W. H. Richter, William E. Ritter, Julia Robinson, Alan Rose, Gene F. Rose, Rosser, N. A. Routledge, Ryll-Nardzewski, Šanin, F. J. Sansone, Schütte, Scott, Claude E. Shannon, Norman Shapiro, J. C. Shepherdson, J. R. Shoenfield, Skolem, Smullyan, Ernst Specker, Spector, Peter H. Starke, Yoshindo Suzuki, M. A. Taĭclin, W. W. Tait, Gaisi Takeuti, Stanley Tennenbaum, Tosiyuke Tugué, J. S. Ullian, Robert L. Vaught, Vladeta Vučović, Wang, Jesse B. Wright, C. D. Wyman, C. E. M. Yates, P. R. Young, I. D. Zaslavskiĭ.

[3] Ackermann, David R. Anderson, Paul Axt, James H. Bennett, Bereczki, Church, John P. Cleave, Paul Csillag, R. J. Fabian, Feferman, Fischer, Joyce Friedman, Gödel, R. L. Goodstein, Grzegorczyk, James R. Guard, Juris Hart-

But, although intuitionism began twenty-five years before the theory of general recursive functions and so contributed to the climate of research in foundations in which the theory of general recursive functions arose (cf. Church 1936 Footnote 10, 1936a p. 102), the latter theory in its details developed initially quite independently of the intuitionistic mathematics. On the other side, in the next twenty-five years after the theory of general recursive functions had appeared, intuitionism under Brouwer continued on its way without taking explicit notice of the theory of general recursive functions.

However it is natural to look for connections between intuitionism and the theory of general recursive functions, which (at least until recently) could be characterized as the two principal manifestations of the constructive tendency in modern mathematics. (This characterization no longer fits as well, with the recent rapid proliferation of notions of constructiveness; cf. the volume "Constructivity in mathematics" ed. by Heyting **1959**.) So Kleene conjectured (in 1939–40) that only number-theoretic functions which are general recursive can be proved to exist in the existing formal system of intuitionistic number theory. This conjecture was confirmed, and other metamathematical results concerning the intuitionistic formal number theory were obtained, in his 1945 with Nelson's 1947 (exposition in IM § 82; also cf. Šanin **1958**, **1958a**, Kleene **1960**). Kleene (in 1941) and Beth **1947** proposed to interpret the "laws" in Brouwer's definition of a *set* or *spread* (**1918–9**, **1919**, **1924–7**, **1954**) to be general recursive functions. Proposals for a thoroughgoing use of recursive functions in interpreting intuitionistic analysis were elaborated by Kleene in 1950a, which is a prospectus for the present monograph; and Kleene in **1957** gave a progress report. (Ideas from intuitionistic analysis were applied in recursive function theory by Kleene in **1955a** pp. 417, 420 or **1955b** p. 203.)

In that prospectus 1950a it was suggested that use of recursive functions might help to make Brouwer's analysis more accessible. The appearance meanwhile of new expositions of intuitionism, especially Heyting **1956**, has diminished the need for such help. In these expositions one can follow the development of intuitionistic mathematics in a quite straightforward manner, and we shall not

mans, Hu, Kalmár, Kalužnin, Kent, Kleene, Kreisel, A. H. Lachlan, R. S. Lehman, Moh, Nelson, Péter, Robert W. Ritchie, Julia Robinson, Raphael M. Robinson, Rogers, H. S. Shapiro, R. E. Stearns, Tait.

duplicate them here except to a limited extent. We do aim in the present monograph, among other things, to clarify the differences between intuitionistic and classical mathematics. These differences consist in part in the intuitionist's merely refraining from affirming classical results, and in part in the intuitionist's affirming results contradictory to classical ones.

§ 2. Status of the formal system. We shall begin by setting up a formal system, in which, so far as we have pursued it, the existing intuitionistic mathematics exclusive of the theory of species of higher order (Brouwer **1918–9**, **1924–7**) can be developed. This investigation is continued in Vesley's "The intuitionistic continuum", constituting Chapter III below.

Formalization may at first seem inappropriate. For one of the Brouwerian tenets is that mathematics should consist in intuitive constructions, and intuitive reasoning on the basis of the meaning of the propositions as about those constructions, rather than in formal deductions from formally stated axioms. However this need not forbid the use of formal processes of deduction to save the labor of going back to the meaning at each step, when the axioms and rules of inference have once been justified in terms of the meaning so that we know we always could go back to it. As is familiar from the historical development of elementary algebra e.g., such formal manipulations add much to the rapidity with which mathematical deductions can be performed. The Brouwerian objection is surely only against formal reasoning that could not be performed in terms of the meaning. (Cf. IM p. 62 last paragraph, or Heyting **1953** pp. 59–60.)

Brouwer took the position on philosophical grounds that the possibilities of construction cannot be confined within the bounds of any given formal system, long before this position received confirmation in the famous proof by Gödel 1931 that formalisms adequate for a certain portion of number theory are incomplete. So it is to be understood from the outset that our formal system for intuitionistic mathematics is not complete, even for the portion of intuitionistic mathematics without species of higher order which is expressible in its symbolism. It, or any intuitionistically correct extension of it, could be extended by the Gödel process (which is valid intuitionistically), i.e. by adding a formally undecidable but true formula (cf. IM §§ 42, 60). We shall indeed frequently think of our formal system as extensible, any extension to satisfy certain general conditions.

There are advantages in describing formally a portion of intuitionistic mathematics, even though that portion be only a fragment. Indeed the intuitionist Heyting did exactly this in 1930, 1930a, in response to a challenge by Mannoury to do so. The formation rules and postulates help the outsider by separating out for him a relatively small number of concepts he must grasp, and of propositions and methods of inference to which he must assent, in order that in principle nothing else should remain problematical for him in the body of the intuitionistic mathematics deducible within the system. Furthermore, metamathematical investigations are made possible, as well as investigations, which we may call *model-theoretic* or *semantical* (cf. IM pp. 63, 175, 176, 501), employing interpretations not necessarily elementary or finitary.

Thus Heyting's formalization 1930, 1930a §§ 5, 6 of the intuitionistic logic led to a variety of metamathematical and semantical investigations and applications, of interest in diverse connections not all intuitionistic, by Glivenko 1929, Kolmogorov **1932**, Gödel 1932, 1932–3, **1933**, Gentzen 1934–5, Žegalkin **1936***, Jaśkowski 1936, Stone **1937–8**, Tarski **1938**, Wajsberg 1938, McKinsey 1939, Garrett Birkhoff **1940**, Kleene 1945 with Nelson 1947, Heyting 1946, McKinsey-Tarski **1946**, 1948, de Iongh 1948, Ohnishi **1953**, Mostowski 1948, Rieger **1949**, Henkin 1950a, Curry 1950, Kleene 1948, 1952, **1952a**, IM (**1952b**), Ridder **1950–1**, Schütte **1962**, Kuroda **1951**, Łukasiewicz **1952**, Maehara **1954**, Umezawa **1955**, **1959**, **1959a**, Nishimura **1960**, Schröter **1956**, **1957**, Porte **1958**, Schmidt **1958**, Skolem **1958**, Leblanc-Belnap **1962**, McCall **1962**, Vesley **1963**, Rasiowa **1951**, **1954**, **1954a**, Rasiowa-Sikorski **1953**, **1954**, **1955**, **1959**, Sikorski **1959**, Pil'čak **1950**, **1952**, G. F. Rose **1953**, Gál-Rosser-Scott **1958**, Medvedev **1962**, Kabakov **1963**, Harrop **1956**, **1960**, Beth **1956**, **1959a**, Dyson-Kreisel **1961**, Scott **1957**, Kreisel-Putnam **1957**, Kreisel **1958a**, **1958b**, **1959a**, **1959c**, **1961**, **1962**, **1962a**, Kleene **1962a**. (Glivenko **1928** partially anticipated Heyting 1930. Kolmogorov **1924–5** first gave, for a related system, a result like Gödel 1932–3, IM p. 495.)

Gentzen in 1934–5 gave a formalization of the intuitionistic logic in which it differs from the classical logic in just one axiom schema, that for "negation elimination"; cf. IM p. 101, and Remark 1 p. 120 reading "intuitionistic" for "same". (Glivenko 1929 had already observed that adjoining the law of the excluded middle to the intuitionistic propositional calculus produces the classical.) This facilitates comparison of the two systems. The formal system of intuitionistic arithmetic and number theory which is obtained by adjoining Peano's axioms and recursion equations for suitable number-theoretic functions to the intuitionistic logic (in Heyting's or Gentzen's formalization) has been studied metamathematically or semantically by Gödel 1932–3, Kleene 1945 with Nelson 1947 (as indicated in § 1 above), Nelson 1949, Kleene 1948, IM, Šanin **1953**, **1954**, **1955**, **1958**, **1958a**, Kleene **1960**, Kreisel **1958***, **1959**, **1959a**, **1959b**, **1959d**, **1962**, **1962d**, **1962e**, **1962f**, Harrop **1956**, **1960**, Gödel **1958**, Kleene **1962a**, T. T. Robinson **1963**. Incidentally, the formal system of intuitionistic arithmetic and number theory does not occur as a subsystem readily separated out from Heyting's full system of intuitionistic mathematics 1930a (with 1930).

At the time this investigation was begun, and up through our progress report **1957**, there were so far as we are aware no other metamathematical or semantical studies in print of the Heyting 1930, 1930a system of intuitionistic mathematics including set theory, or of any other formal system for a like portion of intuitionistic mathematics. Kreisel **1959a**, **1959c**, **1962**, **1962a** refer to our **1957**. In **1958b** he describes another such formal system, and treats some of the same matters as the present monograph. Kleene, after unsatisfactory attempts to use recursive functions in the interpretation of the Heyting 1930a system, came to feel it not well suited for the metamathematical studies proposed. It employs somewhat bizarre primitives, namely symbols "σ" and "τ" where "$\sigma p \tau q$" is to be read "from the species q the element p is chosen" and "$p \tau q$" to be read "the p chosen from q". In any case, such symbolism is an impediment to the comparison of the intuitionistic with a classical system. So in 1950a Kleene proposed the innovation of using simply one-place number-theoretic function variables for Brouwer's "choice sequences". Thus the part of intuitionistic mathematics up through intuitionistic set theory excluding higher-order species, which we call *intuitionistic analysis*, comes to be expressible in the same language as a portion of classical analysis. The differences between intuitionistic and classical analysis are then to be found formally in the different treatment of the function variables under the deductive postulates rather than in formation rules. Kreisel **1958b** also uses one-place function variables, under a restriction (citing on p. 372 our **1957**). Spector **1962** takes into account the present formalism (cited on p. 9), details of which were made available to him in November 1959 and January 1961; and Kreisel **1962d, 1962e, 1962f** cite Spector **1962.**

Our postulates will begin with ones which will be common to (or hold in) both the intuitionistic and classical systems, namely those of Postulate Groups A–C in § 4, of Postulate Group D in § 5 (to which additions may be made later), and Axiom Schema $^{x}26.3$ in § 6. This gives what we call the *basic system*. Strengthening the intuitionistic negation elimination schema 8^{I} IM p. 101 (included in Postulate Group A) to the classical one $8°$ IM p. 82 for the *classical system*, and for the *intuitionistic system* adding a single non-classical postulate $^{x}27.1$ in § 7, will then produce the divergence between the two systems. Of course other additions could be made to the classical system, including further formation rules (cf. Hilbert-Bernays 1939 Supplement IV); but this would take us further afield than we wish to go from our primary subject, the foundations of intuitionistic mathematics.

§ 3. Formation rules. 3.1. We must now assume familiarity with the concept of a formal system. A plethora of references could be given; but, when we can, we shall confine our references for this and related purposes to IM. We now cite IM §§ 15 ff.

3.2. The formal symbols of the system shall be the logical symbols $\supset$ ("implies", or "if ... then ..."), & ("and"), $\vee$ ("or"), $\neg$ ("not"), $\forall$ ("for all") and $\exists$ ("there exists"), number variables a, b, c, ..., x, y, z, ..., function variables $\alpha, \beta, \gamma, \ldots,$ the predicate symbol $=$ ("equals"), certain function symbols $f_0, \ldots, f_p$ where f_0 is 0 ("zero") and f_1 is $'$ ("successor" or "plus one"), Church's λ, commas and parentheses.

The number variables are variables over the natural numbers 0, 1, 2, ... The function variables are variables for one-place number-theoretic functions, i.e. one-place functions from the natural numbers to the natural numbers. We suppose potentially available a countable infinity of variables of each sort.

In IM a special font of type (lower case script "a", "b", "c", ...) was used for the number variables themselves as distinguished from letters (lower case Roman "a", "b", "c", ..., "x", "y", "z") standing as names for number variables. We simplify the notation here by using the latter only, as there is little further need (after the basic discussion of formal systems in IM) for exhibiting specimens of the number variables themselves.[4] We can likewise understand that the lower case Greek letters in the font "α", "β", "γ", ... (Roman) are names for function variables.[5]

When we use letters as names for variables, we have to be clear when different letters name variables required to be distinct and when they may be names for the same variable. The same problem was present in IM, though there in some cases we could avoid it by using specimens of the variables themselves. The problem is complicated by the use in IM and here of metamathematical expressions to name formulas etc. that may contain variables not shown or named within those metamathematical expressions. We shall be explicit in these matters at the outset and in our more formal statements. Elsewhere the simplest blanket rule is that variables not required by the notation to be the same (i.e. not named alike, or at corresponding positions in alike named formulas etc., within a connected context) should be distinct whenever it makes a difference for the discussion in progress whether they are the same or distinct. The passages in which we are explicit about the stipulations should provide ample illustration for the application of this blanket rule.

The function symbols f_i for $i > 1$ will be specified later. It will be somewhat a matter of convenience which ones we introduce. The formation rules regarding them, and some other features of the system, we state first in general terms to leave the way open to other applications besides to the specific system considered below. Under the interpretation, each function symbol f_i ($i = 0, \ldots, p$) shall express a "primitive recursive" function $f_i(a_1, \ldots, a_{k_i}, \alpha_1, \ldots, \alpha_{l_i})$ of a specified number k_i of natural numbers and a specified number l_i of one-place number-theoretic functions ($k_i, l_i \geq 0$; in particular, $k_0 = l_0 = l_1 = 0$, $k_1 = 1$). We could admit here "general recursive" functions of such variables; but we shall not have occasion to introduce any that are not "primitive recursive". For the notion of 'primitive

[4] We shall retain the script letters, capital and lower case, in writing particular formulas of the pure propositional and predicate calculi.

[5] Italic Greek letters $\alpha, \beta, \gamma, \ldots$ will be used here as names for functions; in IM Roman Greek letters were so used, and elsewhere, e.g. Kleene **1955**, **1955a**, **1955b**, italic Greek letters.

('general) ['partial] recursive' function, we refer to IM Part III, especially § 43 (§ 55) [§ 63]. But we are following Kleene 1950a § 2 and **1955** § 1 in allowing one-place function, as well as number, variables; that $f_i(a_1, \ldots, a_{k_i}, \alpha_1, \ldots, \alpha_{l_i})$ with $l_i > 0$ is *primitive (general) [partial] recursive* then means that $f_i(a_1, \ldots, a_{k_i}, \alpha_1, \ldots, \alpha_{l_i})$ as a function of $a_1, \ldots, a_{k_i}$ is primitive (general) [partial] recursive uniformly in $\alpha_1, \ldots, \alpha_{l_i}$ IM § 47 (§ 55) [§ 63]. When we come to specify the function symbols, we shall introduce such special symbols, not always prefixed, as will be convenient; so the "f_i" can be considered as names for the symbols to be introduced later, or for $i = 0, 1$ already introduced.

3.3. We define 'term' and 'functor' inductively, as follows. The terms are to be the formal expressions in the role of nouns expressing natural numbers, and the functors those expressing one-place number-theoretic functions. 1. The number variables a, b, c, ..., x, y, z, ... are *terms*. 2. The function variables $\alpha, \beta, \gamma, \ldots$ are *functors*. 3a. 0 is a *term*. 3. For each i ($i = 0, \ldots, p$), if $\mathrm{t}_1, \ldots, \mathrm{t}_{k_i}$ are *terms*, and $\mathrm{u}_1, \ldots, \mathrm{u}_{l_i}$ are *functors*, $\mathrm{f}_i(\mathrm{t}_1, \ldots, \mathrm{t}_{k_i}, \mathrm{u}_1, \ldots, \mathrm{u}_{l_i})$ is a *term*. (Clause 3a is the case of Clause 3 for $i = 0$.) 4a. $'$ is a *functor*. 4. For each i ($i = 0, \ldots, p$) such that $k_i = 1$ and $l_i = 0$, f_i is a *functor*. (Clause 4a is the case of Clause 4 for $i = 1$.) 5. If u is a *functor*, and t is a *term*, (u)(t) is a *term*. (When u is a functor f_i by Clause 4, Clause 5 duplicates Clause 3, and (u)(t) will be written in whatever manner may be adopted for $\mathrm{f}_i(\mathrm{t})$, e.g. for $i = 1$ as (t)$'$.) 6. If x is a number variable, and s is a *term*, λx(s) is a *functor*. 7. A formal expression is a *term* or *functor* only as required by Clauses 1–6.

Next we define 'formula' inductively as follows. The formulas are to be the formal expressions in the role of sentences, i.e. expressing propositions (but they may contain "free" variables). 8. If s and t are terms, (s)=(t) is a *formula*. 9. If A and B are *formulas*, so are (A) ⊃ (B), (A) & (B) and (A) ∨ (B); if A is a *formula*, so is ¬(A). 10. If x is a number variable, and A is a *formula*, ∀x(A) and ∃x(A) are *formulas*. 11. If α is a function variable, and A is a *formula*, ∀α(A) and ∃α(A) are *formulas*. 12. A formal expression is a *formula* only as required by Clauses 8–11.

In writing terms, functors and formulas, parentheses may be omitted under the familiar conventions when no confusion can result (IM § 17), or they may be changed to braces or brackets. (They can always be omitted in using Clauses 6 and 8.) We adopt the abbreviations "$\neq$", "1", "2", "3", ... IM p. 75, "$\sim$" p. 113.

Instead of including among our formal or primitive symbols just the one predicate symbol $=$, we could have allowed others, each one P_j expressing a primitive recursive predicate $P_j(a_1, \ldots, a_{m_j}, \alpha_1, \ldots, \alpha_{n_j})$. We shall get the same effect by providing, when we want a symbol for such a predicate, that $\mathrm{P}_j(\mathrm{a}_1, \ldots, \mathrm{a}_{m_j}, \alpha_1, \ldots, \alpha_{n_j})$ be an abbreviation for $\mathrm{s}_j=0$ where s_j is a term (containing at most the variables $\mathrm{a}_1, \ldots, \mathrm{a}_{m_j}, \alpha_1, \ldots, \alpha_{n_j}$) which expresses the representing function $p_j(a_1, \ldots, a_{m_j}, \alpha_1, \ldots, \alpha_{n_j})$ of P_j (IM p. 227). For the P_j (as for the f_i) we shall introduce such special symbols, not necessarily prefixed, as will be convenient.

3.4. *Free* and *bound* occurrences of variables in terms, functors and formulas are distinguished in the familiar manner (IM § 18). The operators which bind variables now are the *λ-prefixes* $\lambda\mathrm{x}$ and the *quantifiers* $\forall\mathrm{x}$, $\exists\mathrm{x}$, $\forall\alpha$, $\exists\alpha$ where x is any number variable and α is any function variable. As in IM, our definitions of 'term', 'functor' and 'formula' do not exclude the application of one of these operators to a scope containing bound occurrences of the same variable; then in the resulting expression it is only the occurrences free in the scope which are *bound* by that operator.

The result of *substituting* a term for (the free occurrences of) a number variable, or a functor for (the free occurrences of) a function variable, or of several such substitutions performed simultaneously for several variables, and *freedom* at the substitution positions or of the substitution, are defined as before (IM § 18). We use the same notation as before. Thus if we are to substitute for x in a term, functor or formula (or sometimes simply for emphasis), we may introduce a composite notation "E(x)" for it, after which for any t "E(t)" will denote the result of substituting t for (the free occurrences of) x in E(x).

There will be no conflict between "r(x)" as a composite notation for a term and "r(t)" as expressing the result of a substitution into r(x) on the one hand, and "r(x)" and "r(t)" as obtained by omitting the first pairs of parentheses in writing terms (r)(x) and (r)(t) introduced under Clause 5 in 3.3 on the other. For in the first case the "r" separately from the composite notations "r(x)" and "r(t)" will have no status, while in the second the "r" by itself will already denote a functor (it will be one of "α", "β", "γ", ..., or an f_i, or a letter either explicitly introduced, or indicated by the context, as denoting a functor). Similarly, conflict will not arise between composite notations with several argument places and terms formed under Clause 3 (the f_i identify the latter).

"$P_j(t_1, \ldots, t_{m_j}, u_1, \ldots, u_{n_j})$" (cf. end 3.3) will denote the result of a free substitution into $s_j=0$ after any necessary changes of bound variables in s_j (cf. IM end § 33).

A term, functor or formula is *closed* if it contains no variable free (and following Tarski, a closed formula is a *sentence*), otherwise *open*. A formula is *prime* if it contains no logical symbol.

LEMMA 3.1. *If* $a_1, \ldots, a_k$ *are number variables,* $\alpha_1, \ldots, \alpha_l$ *are function variables,* $t_1, \ldots, t_k$ *are terms,* $u_1, \ldots, u_l$ *are functors, and* $E(a_1, \ldots, a_k, \alpha_1, \ldots, \alpha_l)$ *is a term (functor) [formula], then* $E(t_1, \ldots, t_k, u_1, \ldots, u_l)$ *is a term (functor) [formula].*

PROOF is by induction corresponding to the inductive definitions of 'term', 'functor' and 'formula'.

3.5. A term u(t) introduced by Clause 5 of the definition of 'term' and 'functor' expresses the value of the function expressed by u for the number expressed by t as argument. A functor $\lambda x s$ introduced by Clause 6 expresses the one-place number-theoretic function whose value for a given argument is expressed by s when x expresses that argument, or briefly it expresses s as a function of x (Church 1932, IM p. 34).

LEMMA 3.2. *If the functions* $\psi(a_1, \ldots, a_k, \alpha_1, \ldots, \alpha_l, \beta)$ *and* $\chi(a_1, \ldots, a_k, \alpha_1, \ldots, \alpha_l, x)$ *are primitive (general) [partial] recursive, so is the function* $\psi(a_1, \ldots, a_k, \alpha_1, \ldots, \alpha_l, \lambda x\, \chi(a_1, \ldots, a_k, \alpha_1, \ldots, \alpha_l, x))$. (Kleene **1955** 1.3 [**1956** p. 279].)

PROOF. By IM Lemma I p. 236 with the obvious transitivity of the relation 'primitive recursive in' (Lemma VI p. 344 with Theorem II p. 275) [Lemma VI with Theorem XVII (a) p. 329].

LEMMA 3.3. *Let* s *be a term* (u *be a functor*) [P *be a prime formula*] *containing free no variables other than the number variables* $a_1, \ldots, a_k$ *and the function variables* $\alpha_1, \ldots, \alpha_l$. *Then under the intended interpretation* s (u(x) *where* x *is another number variable*) [P] *expresses, as the ambiguous value* IM *p.* 33, *a primitive recursive function of* $a_1, \ldots, a_k, \alpha_1, \ldots, \alpha_l$ *(function of* $a_1, \ldots, a_k, \alpha_1, \ldots, \alpha_l, x$*) [predicate of* $a_1, \ldots, a_k, \alpha_1, \ldots, \alpha_l$*].*

PROOF is by induction corresponding to the inductive definitions of 'term', 'functor' and 'formula', using Lemma 3.2 for Clause 3. (A functor u not of the form λxs is equivalent under the interpretation to the functor of this form λxu(x).)

REMARK 3.4. Instead of a prime formula P in Lemma 3.3, we can allow one merely containing no quantifiers, or none except bounded number quantifiers. (Cf. IM #D, #E p. 228.)

§ 4. Postulates for predicate calculus, for number theory, and concerning functions (Postulate Groups A—C). **4.1.** In giving the transformation rules or postulates (IM § 19) for the basic formal system or the intuitionistic formal system, we begin with postulates for the intuitionistic predicate calculus; but instead of one sort of variables we employ our two sorts.

GROUP A. Postulates for
the two-sorted intuitionistic predicate calculus.

GROUP A1. Postulates for
the intuitionistic propositional calculus.

For these postulates, A, B, C are any formulas; and the postulates are 1a, 1b, 2, 3, 4a, 4b, 5a, 5b, 6, 7, 8^{I} as in IM top p. 82 except:

8^{I}. $\neg A \supset (A \supset B)$ (as in IM p. 101).

GROUP A2. (Additional) Postulates for
the two-sorted intuitionistic predicate calculus.

The first four postulates 9N, 10N, 11N, 12N appear exactly as 9, 10, 11, 12 in IM middle p. 82 with the x in the stipulations bottom p. 81 now a number variable. For the second four as follows, α is any function variable, A(α) any formula, C any formula not containing α free, and u any functor free for α in A(α).

9F. $\dfrac{C \supset A(\alpha)}{C \supset \forall\alpha A(\alpha).}$ 10F. $\forall\alpha A(\alpha) \supset A(u).$

11F. $A(u) \supset \exists\alpha A(\alpha).$ 12F. $\dfrac{A(\alpha) \supset C}{\exists\alpha A(\alpha) \supset C.}$

The classical formal system arises from the basic system by adopting

8°. $\neg\neg A \supset A$ (as in IM p. 82)

as an additional postulate (or equivalently, since 8^{I} then becomes a derived rule IM p. 101, as an alternative to 8^{I}).

4.2. Next we take over the additional postulates for the usual system of intuitionistic number theory. We specify that the next two f_2, f_3 of our function symbols be $+$, $\cdot$ (with $k_2 = k_3 = 2$, $l_2 = l_3 = 0$).

GROUP B. (Additional) Postulates for intuitionistic number theory.

The postulates are 13–21 as in IM bottom p. 82, with the x now a number variable, and the *a*, *b*, *c* particular number variables.

4.3. Anticipating that in Group D in § 5 we shall postulate the recursion equations for the exponential function (a^b or a exp b), we now let f_4 express this function (with $k_4 = 2$, $l_4 = 0$), and use it in Axiom Schema ˣ2.1 of Group C. (Beyond Group B our postulates do not correspond to ones in IM, and we change to another system of numbering, which will distinguish references within this monograph from ones to IM.)

GROUP C. Postulates concerning functions.

For Axiom Schema ˣ0.1, x is any number variable, r(x) is any term, and t is any term free for x in r(x). For Axiom ˣ1.1, a and b are particular distinct number variables, and α is a particular function variable. For Axiom Schema ˣ2.1, x and y are any distinct number variables, α is any function variable, and A(x, α) is any formula in which x is free for α.

ˣ0.1. $\{\lambda x r(x)\}(t) = r(t)$.
ˣ1.1. $a = b \supset \alpha(a) = \alpha(b)$.
ˣ2.1. $\forall x \exists \alpha A(x, \alpha) \supset \exists \alpha \forall x A(x, \lambda y \alpha(2^x \cdot 3^y))$.

4.4. The intuitionistic and classical systems of predicate calculus are treated in IM (Chapters IV–VII, XIV, XV), mostly simultaneously, those results which are established only for the classical system being marked with "°" (cf. p. 101). This material can be taken over for the purpose of proving formulas now containing no function variables or no number variables. When both sorts of variables appear, we can still use the methods developed there, only exercising care not to violate the definition of formula; the restrictions appear in Postulates 10N, 11N where t must be a term, and in Postulates 10F, 11F where u must be a functor. For example, *79 p. 162 gives now $\vdash \forall x \forall y A(x, y) \supset \forall x A(x, x)$, $\vdash \forall \alpha \forall \beta A(\alpha, \beta) \supset \forall \alpha A(\alpha, \alpha)$, but not $\vdash \forall x \forall \beta A(x, \beta) \supset \forall x A(x, x)$ nor $\vdash \forall \alpha \forall y A(\alpha, y) \supset \forall \alpha A(\alpha, \alpha)$. The consequences of the postulates of Group B, taken with the predicate calculus with number variables, are likewise developed in IM (Chapter VIII and elsewhere). We assume enough familiarity not only with the results of IM, but also with the methods underlying them, to permit us here to adapt them to the development of the present formal system without detailed explanation in cases when the adaptation is straightforward. We must stay attentive to the precautions in the handling of variables (IM beginning § 32 p. 146).

Remark 4.1. Since each prime formula of the present system is an equation s=t between terms, it is immediate by substitution in IM *158 p. 192 that $\vdash P \vee \neg P$ for each prime formula. Using also *150, *151 p. 191 and Remark 1 (b) p. 134, likewise $\vdash E \vee \neg E$ for any formula E built from prime formulas (or formulas for each one P of which $\vdash P \vee \neg P$) using only the propositional connectives ($\supset$, &, $\vee$, $\neg$) and bounded number quantifiers. By Remark 1 (a) p. 134 with Theorem 9 p. 128, all provable formulas of the classical propositional calculus (e.g. the equivalences *56–*61 p. 119) are provable in the intuitionistic system when $\vdash P \vee \neg P$ for each of their distinct components P prime for the propositional calculus p. 112.

4.5. Our formation rules provide = as a primitive symbol only between terms. Now, when u and v are functors, "u=v" shall be an abbreviation for $\forall x(u(x)=v(x))$ where x is any variable not occurring free in u or v.

In the following, x is any number variable, r(x) and s(x) are any terms, b is any number variable free for x in r(x) and not occurring free in r(x) (unless b is x), and α is a function variable.

*0.2. $r(x)=s(x) \vdash^{x} \lambda x r(x)=\lambda x s(x)$. *0.3. $\vdash \lambda x r(x)=\lambda b r(b)$.
*0.4. $\vdash \lambda x \alpha(x)=\alpha$.

Proofs. *0.2. Using Ax. Sch. ^{x}0.1 twice and *101, *102 p. 183 (cf. end § 26 p. 118), $r(x)=s(x) \vdash \{\lambda x r(x)\}(x) = r(x) = s(x) = \{\lambda x s(x)\}(x) \vdash^{x} \forall x[\{\lambda x r(x)\}(x) = \{\lambda x s(x)\}(x)]$, i.e. $\lambda x r(x)=\lambda x s(x)$ (by $\forall$-introd. IM § 23).

*0.3. Similarly $\vdash \{\lambda x r(x)\}(x) = r(x) = \{\lambda b r(b)\}(x)$ (cf. IM Example 9 p. 80), whence by $\forall$-introd. $\vdash \lambda x r(x)=\lambda b r(b)$.

The reflexive, symmetric and transitive properties of equality between functors are immediate by the predicate calculus from the same properties of equality between terms (*100–*102 p. 183). The replacement property of equality (IM Theorem 24 and Corollaries pp. 184–185) requires additional care because of the presence of the λ-operator.

LEMMA 4.2. (Replacement theorem.) *Let* E_R *be a term or functor [formula] containing a specified occurrence of a term {functor}* R *not as the variable of a λ-prefix [λ-prefix or quantifier], let* E_S *be the result of replacing this occurrence by a term {functor}* S, *and let* $x_1, \ldots, x_n$ $[x_1, \ldots, x_n, \alpha_1, \ldots, \alpha_m]$ *be the free variables of* R *or* S *which belong to a λ-prefix [λ-prefix or quantifier] having the specified occurrence of* R *within its scope. Then*

$$R{=}S \vdash^{x_1 \ldots x_n} E_R{=}E_S \qquad [R{=}S \vdash^{x_1 \ldots x_n \alpha_1 \ldots \alpha_m} E_R \sim E_S],$$

provided that each of the equality axioms (IM *p.* 403) *for each function symbol* f_i *occurring in* E_R *with the specified occurrence of* R *in its scope is provable.* (The proviso will be met for the way we select the axioms of Group D, by Lemma 5.1.)

PROOF. The part for E_R a formula will follow as before, after we prove the part for E_R a term or functor. We do this (taking both E_R and the specified occurrence of R to be variable; cf. IM p. 90) by course-of-values induction on the *depth* of (the specified occurrence of) R in E_R, defined to be 0 if E_R is (CASE 1) R itself, and to be $r+1$ if E_R is as follows: (CASE 2) $u_R(t)$ where R is at *depth* r in u_R, (CASE 3) $f_i(t_1, \ldots, t_{k_i}, u_1, \ldots, u_{l_i})$ where R is in one of $t_1, \ldots, t_{k_i}, u_1, \ldots, u_{l_i}$ at *depth* r, (CASE 4) $\alpha(t_R)$ where R is at *depth* r in t_R, (CASE 5) $\{\lambda x s(x)\}(t_R)$ where r is the maximum of the *depths* of R in t_R and of each of the free occurrences of x in s(x), (CASE 6) $\lambda x s_R$ where R is at *depth* r in s_R. Case 1 is trivial. In Case 2, by the hypothesis of the induction, $u_R{=}u_S$, which unabbreviated is $\forall x(u_R(x){=}u_S(x))$ with x not free in u_R or u_S; thence by $\forall$-elim., $u_R(t){=}u_S(t)$, i.e. $E_R{=}E_S$. The equality axioms for $f_0, \ldots, f_p$ (cf. the proviso) with the hyp. ind. (and substitution IM *66 p. 147, and $\supset$-elim.) take care of Case 3; Axiom ^{x}1.1 similarly of Case 4; and *0.2 of Case 6. In Case 5, say s(x) contains n free occurrences of x, and let a, b be distinct number variables not occurring in s(x). By n successive applications of the hyp. ind., $a{=}b \vdash s(a){=}s(b)$; thence using Axiom Schema ^{x}0.1 twice, $a{=}b \vdash \{\lambda x s(x)\}(a){=}\{\lambda x s(x)\}(b)$; and thence by $\supset$-introd., substitution and

$\supset$-elim., $t_R = t_S \vdash \{\lambda x s(x)\}(t_R) = \{\lambda x s(x)\}(t_S)$. Also, by hyp. ind. $R = S \vdash^{x_1 \dots x_n} t_R = t_S$. —

The definition of congruence IM p. 153 extends obviously to the present system in which variables may be bound by λ-prefixes as well as by quantifiers. Since by Lemma 4.2 we (will after we obtain Lemma 5.1) have the replacement property of equality as well as of equivalence (also the symmetric, reflexive and transitive properties), and *0.3 as well as *73, *74, congruent terms or functors are equal and congruent formulas are equivalent. (This adapts IM Lemma 15b.)

If x is any number variable, A(x) is any formula, and α is any function variable free for x in A(x) and not occurring free in A(x):

*0.5. $\vdash \forall x A(x) \sim \forall \alpha A(\alpha(0))$. *0.6. $\vdash \exists x A(x) \sim \exists \alpha A(\alpha(0))$.

PROOFS. *0.5. Assume (preparatory to $\supset$-introd.) (a) $\forall \alpha A(\alpha(0))$. Thence by $\forall$-elim., $A((\lambda y x)(0))$ (y a variable distinct from x); by ˣ0.1 and replacement (Lemma 4.2), A(x); and by $\forall$-introd., $\forall x A(x)$. (Since $\forall \alpha A(\alpha(0))$ does not contain x free, x has not been varied.) By the $\supset$-introd. (discharging the assumption (a)), $\forall \alpha A(\alpha(0)) \supset \forall x A(x)$. Using simply the predicate calculus, $\forall x A(x) \supset \forall \alpha A(\alpha(0))$. By &-introd. (*16), $\forall x A(x) \sim \forall \alpha A(\alpha(0))$.

4.6. For all our essential purposes, we could have postulated in place of ˣ2.1 the following consequence of it *2.2 (cf. 7.15 below).

If x and y are any distinct number variables, A(x, y) is any formula in which x is free for y, and α is any function variable free for y in A(x, y) and not occurring free in A(x, y):

*2.2. $\vdash \forall x \exists y A(x, y) \supset \exists \alpha \forall x A(x, \alpha(x))$.

PROOF. Assume (preparatory to $\supset$-introd.) $\forall x \exists y A(x, y)$. Thence by *0.6 and replacement $\forall x \exists \alpha A(x, \alpha(0))$, whence by ˣ2.1 $\exists \alpha \forall x A(x, (\lambda y \alpha(2^x \cdot 3^y))(0))$. Assume (preparatory to $\exists$-elim.) $\forall x A(x, (\lambda y \alpha(2^x \cdot 3^y))(0))$. By ˣ0.1 and replacement $\forall x A(x, \{\lambda x (\lambda y \alpha(2^x \cdot 3^y))(0)\}(x))$, whence by $\exists$-introd. $\exists \alpha \forall x A(x, \alpha(x))$ (which does not contain α free). Completing the $\exists$-elim. and $\supset$-introd., $\forall x \exists y A(x, y) \supset \exists \alpha \forall x A(x, \alpha(x))$.

Simply the predicate calculus gives the converses of ˣ2.1 and *2.2; so they can be strengthened to:

*2.1a. $\vdash \forall x \exists \alpha A(x, \alpha) \sim \exists \alpha \forall x A(x, \lambda y \alpha(2^x \cdot 3^y))$.
*2.2a. $\vdash \forall x \exists y A(x, y) \sim \exists \alpha \forall x A(x, \alpha(x))$.

4.7. We say a term, functor or formula E is *convertible into* one F, and write "E conv F", if E can be transformed into F by a series of zero or more replacements (I) of a part λxr(x) by λbr(b) under the conditions of *0.3, or (II) of a part {λxr(x)}(t) by r(t) under the conditions of ×0.1, or (III) inversely. The relation 'E conv F' is reflexive, symmetric and transitive. Since we (will after proving Lemma 5.1) have the replacement theorem, etc., if E conv F, then $\vdash$ E=F if E, F are terms or functors ($\vdash$ E $\sim$ F if E, F are formulas).

A term, functor or formula is (*lambda*) *normal* if it contains no part of the form {λxs}(t) where x is a variable, s is a term, and t is a term (not necessarily free for x in s). If E conv F, and F is normal, F is a *normal form of* E.

LEMMA 4.3. *Each term, functor or formula* E *has a normal form.*

PROOF. It clearly suffices to prove this for any term or functor E, by (course-of-values) induction on its *rank* defined thus: 0, if E consists of a single symbol; the maximum of the *ranks* of $t_1, \ldots, t_{k_i}$, $u_1, \ldots, u_{l_i}$, if E is $f_i(t_1, \ldots, t_{k_i}, u_1, \ldots, u_{l_i})$; the *rank* of t, if E is u(t), unless u is λxs, in which case it is $1+\max(s, t)$ where s, t are the *ranks* of s, t; the *rank* of s, if E is λxs. E is normal exactly if its rank is 0. If E is not normal, its rank $r > 0$ is the maximum of the ranks of its parts {λxs}(t), and is possessed by $n > 0$ such parts, which are outermost, i.e. not contained in the s or t of another such part. Within the induction on r we use one on n. Select a part $\{\lambda x_1 s_1\}(t_1)$ of rank r. By hyp. ind. on r, s_1 has a normal form q_1; pick a term $r_1(x_1)$ congruent to q_1 in which t_1 is free for x_1, and replace the part $\{\lambda x_1 s_1\}(t_1)$ of E successively by $\{\lambda x_1 q_1\}(t_1)$, $\{\lambda x_1 r_1(x_1)\}(t_1)$ and $r_1(t_1)$ to obtain E_1. Since $r_1(x_1)$ is normal, the parts {λxs}(t) in $r_1(t_1)$ will be exactly those in the occurrences of t_1 that replace the free occurrences of x_1 in $r_1(x_1)$ (no part u(t) of $r_1(x_1)$ with u a one-symbol functor becomes a part {λxs}(t) by the substitution of t_1 for x_1, since x_1 is a number variable), so their ranks will be $< r$. Outside the part $\{\lambda x_1 s_1\}(t_1)$ replaced by $r_1(t_1)$, nothing will have been changed (no part u(t) of the whole with u a one-symbol functor becomes a part {λxs}(t) by the replacement). So the number n of the parts {λxs}(t) of the maximum rank r will have been decreased by 1, unless n was 1, in which case the maximum rank r of such parts will have been decreased. So by hyp. ind. on n or on r, E_1 has a normal form; and hence E does.

REMARK 4.4. In the general theory of λ-conversion (Church 1932,

Kleene 1934; exposition in Church 1941, Curry-Feys **1958**), where no type distinction is enforced between terms and functors, an expression E may have no normal form; e.g. $\{\lambda xx(x)\}(\lambda xx(x))$ does not.

REMARK 4.5. By the Church-Rosser theorem **1936** Theorem 1 Corollary 2 p. 479 for their third kind of conversion p. 482, any two normal forms of a given expression are congruent.

REMARK 4.6. An application of (II), with zero or more preliminary applications of (I), we call a *reduction*. The proof of Lemma 4.3 shows that a certain sequence of reductions starting from E will lead to a normal expression. By Kleene **1962** Theorem 1 (b) (adapting Church-Rosser **1936** Theorem 2 p. 479), there is a number m such that any sequence of reductions starting from E will lead to a normal expression in at most m reductions.

§ 5. Postulates for certain primitive recursive functions, and consequences (Postulate Group D). **5.1.** In this section we shall specify more of the list $f_0, \ldots, f_p$ of symbols expressing primitive recursive functions, and introduce as axioms the recursion equations for, or equations defining explicitly, the functions expressed. (The symbols f_0–f_4 were already specified in 3.2, 4.2, 4.3, and the recursion equations for f_2, f_3 are already included in Postulate Group B.) The further axioms thus introduced will constitute Postulate Group D, but we leave it open whether the list $f_0, \ldots, f_p$ and Postulate Group D are concluded in this section or are to be augmented as may be convenient later.

Properties and applications of suitable primitive recursive functions beyond 0, $'$, $+$, $\cdot$ are an essential part of our subject matter in developing intuitionistic mathematics in the formal system. It is simplest now to add the symbols as primitive symbols, and the recursion equations as axioms, which are entirely acceptable from the intuitionistic standpoint (as well as classically). We could avoid these additions and still in effect develop the same theory in the system (adapting IM § 74, especially pp. 415–416), but that can better be left till later.

We begin with some generalities, independent of the particular list of symbols and axioms to be added. The recursion equations need not be in the standard form of IM (Va) or (Vb) p. 219, but may have steps of explicit definition lumped with them (as in IM bottom p. 221). Now that we have function variables, the steps of explicit definition may include substitutions of functions formed using the λ-operator (cf. Lemma 3.2).

Specifically, the axiom(s) for a function symbol f_i for $i \geq 4$ will be (and for $i = 2, 3$ already are) of one of the following two forms, which we illustrate for the case of three variables y, a, α:

(a) $f_i(y, a, \alpha) = p(y, a, \alpha)$,

(b) $\begin{cases} f_i(0, a, \alpha) = q(a, \alpha), \\ f_i(y', a, \alpha) = r(y, f_i(y, a, \alpha), a, \alpha), \end{cases}$

where $p(y, a, \alpha)$, $q(a, \alpha)$ and $r(y, z, a, \alpha)$ are terms containing only the distinct variables shown and only function symbols from among $f_0, \ldots, f_{i-1}$, and y, a, α are free for z in $r(y, z, a, \alpha)$.

5.2. LEMMA 5.1. *The equality axioms* IM *p.* 403 *are provable for each of the function symbols* $f_0, \ldots, f_p$ *introduced successively with axioms as described* (*Postulate Group* B, *and* 5.1).

PROOF, by (informal) induction on i. For $i = 0$ there are no equality axioms, and for $i = 1$ there is one, which is postulated as Axiom 17 of Group B. For $i > 1$, assume the equality axioms for $f_0, \ldots, f_{i-1}$ provable. Those for $q(a, \alpha)$, i.e. (with b, β free for a, α, resp.)

$$a = b \supset q(a, \alpha) = q(b, \alpha), \qquad \alpha = \beta \supset q(a, \alpha) = q(a, \beta),$$

and for $r(y, z, a, \alpha)$, or for $p(y, a, \alpha)$, are provable by applications of Lemma 4.2 and $\supset$-introd., since by the hyp. ind. on i the equality axioms for the function symbols in them are provable. The equality axioms for f_i introduced by (a) follow immediately. Those for f_i introduced by (b) follow by the method used for a+b (our f_2) in IM *104, *105 pp. 183–184, namely: The conclusions of the formulas

$$a = b \supset f_i(y, a, \alpha) = f_i(y, b, \alpha), \qquad \alpha = \beta \supset f_i(y, a, \alpha) = f_i(y, a, \beta)$$

expressing the replaceability in the position of either parameter are deducible from the antecedents by (formal) induction on the recursion variable y. To prove the replaceability in the position of the recursion variable, write "A(y, x)" for the formula

$$y = x \supset f_i(y, a, \alpha) = f_i(x, a, \alpha)$$

to be proved. We prove $\forall x A(y, x)$ by induction on y. BASIS: to prove $\forall x A(0, x)$. We prove $A(0, x)$ by ind. cases on x (IM bottom p. 186), and $\forall x A(0, x)$ follows by $\forall$-introd. IND. STEP.: assuming $\forall x A(y, x)$, to deduce $\forall$xA(y′, x). We get A(y, x) by $\forall$-elim. from the hyp. ind. Thence we deduce A(y′, x) by ind. cases on x, and $\forall$xA(y′, x) follows by $\forall$-introd.

5.3. Next we generalize IM *(175)–*(177) p. 201 (cf. pp. 195, 199, 200), and extend the theory to function variables. As in IM p. 298 but with the function variables $\alpha_1, \ldots, \alpha_l$ as the function letters $g_1, \ldots, g_l$ and with uniformity, we say a formula $A(a_1, \ldots, a_k, \alpha_1, \ldots, \alpha_l, w)$ containing free only the (distinct) variables shown *numeralwise represents* a function $s(a_1, \ldots, a_k, \alpha_1, \ldots, \alpha_l)$, if, for each $a_1, \ldots, a_k, \alpha_1, \ldots, \alpha_l$,

(v) if $s(a_1, \ldots, a_k, \alpha_1, \ldots, \alpha_l)=w$,
then $E^{\alpha_1\ldots\alpha_l}_{\alpha_1\ldots\alpha_l} \vdash A(\boldsymbol{a}_1, \ldots, \boldsymbol{a}_k, \alpha_1, \ldots, \alpha_l, \boldsymbol{w})$,

(vi) $E^{\alpha_1\ldots\alpha_l}_{\alpha_1\ldots\alpha_l} \vdash \exists!wA(\boldsymbol{a}_1, \ldots, \boldsymbol{a}_k, \alpha_1, \ldots, \alpha_l, w)$,

with $\alpha_1, \ldots, \alpha_l$ held constant in each deduction.

LEMMA 5.2. *Consider any term* $s(a_1, \ldots, a_k, \alpha_1, \ldots, \alpha_l)$ (*briefly* s) *containing free only the* (*distinct*) *variables shown, and let* $s(a_1, \ldots, a_k, \alpha_1, \ldots, \alpha_l)$ *be the function expressed by it* (*Lemma* 3.3). *The formula* $s(a_1, \ldots, a_k, \alpha_1, \ldots, \alpha_l)=w$ (*where* w *is another number variable*) *numeralwise represents* $s(a_1, \ldots, a_k, \alpha_1, \ldots, \alpha_l)$.

PROOF. I. (vi) is immediate by IM *171 p. 199.

II. *If* (v) *holds for each of the function symbols* f_i *in* s (i.e., if, for each $a_1, \ldots, a_{k_i}, \alpha_1, \ldots, \alpha_{l_i}$, whenever $f_i(a_1, \ldots, a_{k_i}, \alpha_1, \ldots, \alpha_{l_i})=w$ then $E^{\alpha_1\ldots\alpha_{l_i}}_{\alpha_1\ldots\alpha_{l_i}} \vdash f_i(\boldsymbol{a}_1, \ldots, \boldsymbol{a}_{k_i}, \alpha_1, \ldots, \alpha_{l_i})=\boldsymbol{w}$ with $\alpha_1, \ldots, \alpha_{l_i}$ held constant), *then* (v) *holds for* s *itself*. By Lemma 4.3, we can without loss of generality take s to be normal. Now we use induction on the number of (occurrences of) function symbols and function variables in s. If s is a single symbol, (v) is trivial. Next take the case s is $f_i(t_1, \ldots, t_{k_i}, u_1, \ldots, u_{l_i})$. Consider fixed values $a_1, \ldots, a_k, \alpha_1, \ldots, \alpha_l$ of $a_1, \ldots, a_k, \alpha_1, \ldots, \alpha_l$, and say that for these values $t_1, \ldots, t_{k_i}, u_1, \ldots, u_{l_i}$, s express $t_1, \ldots, t_{k_i}, \beta_1, \ldots, \beta_{l_i}, w$. By hyp. ind.,

(1) $E^{\alpha_1\ldots\alpha_l}_{\alpha_1\ldots\alpha_l} \vdash t_j^*=\boldsymbol{t}_j \qquad (j = 1, \ldots, k_i)$

with $\alpha_1, \ldots, \alpha_l$ held constant, where * indicates the substitution of $\boldsymbol{a}_1, \ldots, \boldsymbol{a}_k$ for $a_1, \ldots, a_k$. By the hypothesis that (v) holds for the f_i in s, $E^{\beta_1\ldots\beta_{l_i}}_{\beta_1\ldots\beta_{l_i}} \vdash f_i(\boldsymbol{t}_1, \ldots, \boldsymbol{t}_{k_i}, \beta_1, \ldots, \beta_{l_i})=\boldsymbol{w}$ with $\beta_1, \ldots, \beta_{l_i}$ held constant, whence by &-elims. and $\supset$-introd.,

(2) $\vdash F(\beta_1, \ldots, \beta_{l_i}) \supset f_i(\boldsymbol{t}_1, \ldots, \boldsymbol{t}_{k_i}, \beta_1, \ldots, \beta_{l_i})=\boldsymbol{w}$,

where $F(\beta_1, \ldots, \beta_{l_i})$ is a conjunction of the finitely many equations of $E^{\beta_1 \ldots \beta_{l_i}}_{\beta_1 \ldots \beta_{l_i}}$ used in a given such deduction. But $F(\beta_1, \ldots, \beta_{l_i})$ is a conjunction of equations of the form $\beta_m(\boldsymbol{a})=\boldsymbol{b}$ where $\beta_m(a)=b$ $(1 \leq m \leq l_i)$. If u_m is $\lambda a v_m(a)$, $E^{\alpha_1 \ldots \alpha_l}_{\alpha_1 \ldots \alpha_l} \vdash v^*_m(\boldsymbol{a})=\boldsymbol{b} \vdash (u^*_m)(\boldsymbol{a})=\boldsymbol{b}$ with $\alpha_1, \ldots, \alpha_l$ held constant, by hyp. ind. and ${}^{\times}0.1$. If u_m is a one-symbol functor, similarly $E^{\alpha_1 \ldots \alpha_l}_{\alpha_1 \ldots \alpha_l} \vdash (u^*_m)(\boldsymbol{a})=\boldsymbol{b}$ (u^*_m being u_m simply), by hyp. ind. So

$$(3) \quad E^{\alpha_1 \ldots \alpha_l}_{\alpha_1 \ldots \alpha_l} \vdash F(u^*_1, \ldots, u^*_{l_i})$$

with $\alpha_1, \ldots, \alpha_l$ held constant. By substitution in (2),

$$(4) \quad \vdash F(u^*_1, \ldots, u^*_{l_i}) \supset f_i(\boldsymbol{t}_1, \ldots, \boldsymbol{t}_{k_i}, u^*_1, \ldots, u^*_{l_i})=\boldsymbol{w},$$

whence by replacements using (1),

$$(5) \quad E^{\alpha_1 \ldots \alpha_l}_{\alpha_1 \ldots \alpha_l} \vdash F(u^*_1, \ldots, u^*_{l_i}) \supset f_i(t^*_1, \ldots, t^*_{k_i}, u^*_1, \ldots, u^*_{l_i})=\boldsymbol{w}$$

with $\alpha_1, \ldots, \alpha_l$ held constant. Now (v) comes from (3) and (5) by $\supset$-elim. The remaining case that s is $\alpha(t)$ for α one of $\alpha_1, \ldots, \alpha_l$ is similar but simpler.

III. (v) *holds for each of* $f_0, \ldots, f_p$ *as the* s. By induction on i, using II and substitution into the axioms (a) or (b) 5.1 introducing f_i, and in the case of (b) induction on y.

5.4. Similarly (using IM p. 298 with function variables and uniformity), the notion of when a formula *numeralwise expresses* a predicate (IM p. 195) extends to the present situation.

We elected (end 3.3) to use symbols of abbreviation for primitive recursive predicates other than $=$. Using Lemma 5.2 and Ax. 15 IM p. 82, each predicate $P_j(a_1, \ldots, a_{m_j}, \alpha_1, \ldots, \alpha_{n_j})$ for which we introduce a symbol P_j of abbreviation will be numeralwise expressed by $P_j(a_1, \ldots, a_{m_j}, \alpha_1, \ldots, \alpha_{n_j})$.

5.5. There are in the literature a number of formal treatments of number-theoretic functions beyond 0, ′, +, ·, particularly in Skolem 1923, Hilbert-Bernays 1934 and Nelson 1947. We give another, in order that the present monograph combined with IM be (nearly) self-contained, to confirm that the proofs can be given in an intuitionistic system (as in Nelson 1947), and to keep close agreement between the formal notations and certain informal ones.

The informal notations are those which we have been using in the theory of recursive functions. In particular, we have been following Gödel 1931 in correlating single natural numbers to finite sequences of natural numbers via the unique factorization of positive integers (the "fundamental theorem of arithmetic"). It is possible, and advantageous at least for purposes of machine computation (which isn't an objective here), to use a correlation based instead on n-adic representation of numbers, which does not require as rapidly increasing functions as the exponential function, as in Smullyan **1959**. By using this alternative, we could probably spare the proof of "Euclid's first theorem" and the fundamental theorem of arithmetic; but we would have to develop some other things instead. However the work of proving these theorems is not so very great, and puts us in a position where we can use formally the notations used in the informal literature to which the present investigations are most closely related, which has its advantages too. (Our choice to do this is anticipated in Axiom Schema ˣ2.1; thus *2.1a formalizes Kleene **1955** (5).)

Accordingly we now introduce formal notations for functions (and symbols of abbreviation for predicates), the same as in IM except with formal variables, taking over IM §§ 44, 45 #3–#21, #A–#F and also as #22 Seq (Kleene **1955a** p. 416), #23 $\bar{\alpha}(x)$ (Kleene 1950a p. 680), #24 $\tilde{\alpha}(x)$ (IM § 46). We know a priori that we can have function symbols and predicate symbols-of-abbreviation expressing, and numeralwise representing or expressing, all these functions and predicates. What we are to add now is the verification that, with the recursion equations selected (usually) as in IM but now formalized, the formulas expressing various useful properties of the functions and predicates are provable.

We already have #1, #2 (addition and multiplication) with Axioms 18–21, and properties in IM § 39. At least until #15 is treated, we use "a<b", "a≤b" etc. as in IM p. 187. We abbreviate a<c & b<c as "a, b<c", etc. Note that:

*137a. $\vdash a \leq b \sim \exists c(c+a=b)$. *145c. $\vdash a \leq b \supset ac \leq bc$.

For #3 (exponentiation), we already have the symbol (as f_4), we now take the recursion equations as Axioms ˣ3.1, ˣ3.2 (with a, b particular distinct number variables), and we give further properties as the provable formulas of *3.3–*3.13. Then for #4 (the factorial function), we introduce the symbol a! as f_5 (with $k_5 = 1$, $l_5 = 0$), the recursion equations as Axioms ×4.1, ×4.2, and a property in *4.3; and similarly through #13 [a/b] as f_{14}. In #14 no new symbol is involved, and in #15, #16 predicate symbols of abbreviation.

×3.1. $a^0=1$. ×3.2. $a^{b'}=a^b a$.
*3.3. $\vdash a^b a^c=a^{b+c}$. *3.4. $\vdash (a^b)^c=a^{bc}$. *3.5. $\vdash (ab)^c=a^c b^c$.
*3.6. $\vdash a^1=a$. *3.7. $\vdash b>0 \sim 0^b=0$. *3.8. $\vdash 1^b=1$.
*3.9. $\vdash a>0 \supset a^b>0$. *3.10. $\vdash a>1 \supset a^b>b$.
*3.11. $\vdash a, b>1 \supset a^b>a$. *3.12. $\vdash a>1 \supset (b<c \sim a^b<a^c)$.
*3.13. $\vdash a, c>0 \supset (a<b \sim a^c<b^c)$.

Proofs. *3.3–*3.5. Induction on c (the necessary replacement properties are available, by Lemmas 4.2, 5.1).

*3.7. I. Prove $0^{c'}=0$ (cf. the proof of IM *143b p. 189). II. By ×3.1.

*3.9. Ind. on b (using *143a).

*3.10. Ind. on b (using *143b, *138b).

*3.11. We prove $a>1 \supset a^{c''}>a$, using *3.10 (with *145a), *143a.

*3.12. Assume $a>1$. I. Assume $d'+b=c$ (for ∃-elim. from $b<c$), and use *3.3, *143b (with *3.9, *3.10). II. The converse follows as in the proof of *145a.

*3.13. I. Prove $a>0$ & $a<b \supset a^{c'}<b^{c'}$ by ind. on c, using *145a twice (and *3.9).

×4.1. $0!=1$. ×4.2. $a'!=a!a'$. *4.3. $\vdash a!>0$.

×5.1. $pd(0)=0$. ×5.2. $pd(a')=a$. *5.3. $\vdash pd(a)\leq a$.

×6.1. $a \dot{-} 0=a$. ×6.2. $a \dot{-} b'=pd(a \dot{-} b)$.
*6.3. $\vdash (a+b) \dot{-} b=a$. *6.3a. $\vdash a \dot{-} a=0$. *6.4. $\vdash 0 \dot{-} a=0$.
*6.5. $\vdash a \dot{-} (b+c)=(a \dot{-} b) \dot{-} c$. *6.6. $\vdash b\geq c \supset (a+b) \dot{-} c=a+(b \dot{-} c)$.
*6.7. $\vdash a\geq b \sim (a \dot{-} b)+b=a$. *6.8. $\vdash (a+c) \dot{-} (b+c)=a \dot{-} b$.
*6.9. $\vdash b\geq c \sim a \dot{-} (b \dot{-} c)=(a+c) \dot{-} b$. *6.10. $\vdash a\geq b \sim a \dot{-} (a \dot{-} b)=b$.
*6.11. $\vdash a\leq b \sim a \dot{-} b=0$. *6.12. $\vdash a>b \sim a \dot{-} b>0$.
*6.13. $\vdash c>0 \supset (a \dot{-} b\geq c \sim a\geq b+c)$. *6.13a. $\vdash a \dot{-} b>c \sim a>b+c$.
*6.13b. $\vdash c>0 \supset (a \dot{-} b=c \sim a=b+c)$. *6.14. $\vdash c(a \dot{-} b)=ca \dot{-} cb$.
*6.15. $\vdash a \dot{-} b\leq a$. *6.16. $\vdash a, b>0 \sim a \dot{-} b<a$.
*6.17. $\vdash a\geq b \supset a \dot{-} c\geq b \dot{-} c$. *6.18. $\vdash a\leq b \supset c \dot{-} a\geq c \dot{-} b$.

*6.19. $\vdash a>b, c \sim a \dot{-} c>b \dot{-} c$. *6.20. $\vdash a<b, c \sim c \dot{-} a>c \dot{-} b$.

*6.21. $\vdash a \dot{-} c \leq (a \dot{-} b)+(b \dot{-} c)$.

PROOFS. *6.3. We prove $\forall a((a+b) \dot{-} b=a)$ by ind. on b. IND. STEP. Assume $\forall a((a+b) \dot{-} b=a)$, whence $(a'+b) \dot{-} b=a'$. Now $(a+b') \dot{-} b' = pd((a+b') \dot{-} b) = pd((a'+b) \dot{-} b)$ [Ax. 19, *118] $=$ $pd(a') = a$, whence $\forall a((a+b') \dot{-} b'=a)$.

*6.5. Ind. on c.

*6.6. $b \geq c$ gives $\exists d(d+c=b)$; assume $d+c=b$. Now $(a+b) \dot{-} c =$ $(a+d+c) \dot{-} c = a+d$ [*6.3] $= a+((d+c) \dot{-} c)$ [*6.3] $= a+(b \dot{-} c)$.

*6.7. I. By *6.6, *6.3. II. By *137a (above).

*6.9. I. $a \dot{-} (b \dot{-} c) = (a+c) \dot{-} ((b \dot{-} c)+c)$ [*6.8] $= (a+c) \dot{-} b$ [*6.7]. II. $b<c$ gives $a \dot{-} (b \dot{-} c)<(a+c) \dot{-} b$ (put $d'+b=c$, and use *6.5 etc.).

*6.11, *6.12. First prove $a \leq b \supset a \dot{-} b=0$, $a>b \supset a \dot{-} b>0$. The converses then follow by *139, *140.

*6.14. By *139, $a>b \vee a \leq b$. CASE 1: $a>b$. Use *6.3. CASE 2: $a \leq b$. Use *145c (above), *6.11.

*6.17. Use cases ($b \leq c$, $b>c$).

×7.1. $\min(a, b)=b \dot{-} (b \dot{-} a)$.	×8.1. $\max(a, b)=(a \dot{-} b)+b$.
*7.2. $\vdash a \leq b \sim \min(a, b)=a=\min(b, a)$.	*8.2. $\vdash a \geq b \sim \max(a, b)=a=\max(b, a)$.
*7.3. $\vdash \min(a, b)=\min(b, a)$.	*8.3. $\vdash \max(a, b)=\max(b, a)$.
*7.4. $\vdash \min(a, b) \leq a, b$.	*8.4. $\vdash \max(a, b) \geq a, b$.
*7.5. $\vdash a, b \geq c \sim \min(a, b) \geq c$.	*8.5. $\vdash a, b \leq c \sim \max(a, b) \leq c$.
*7.6. $\vdash a, b>c \sim \min(a, b)>c$.	*8.6. $\vdash a, b<c \sim \max(a, b)<c$.
*7.7. $\vdash \min(c+a, c+b) =c+\min(a, b)$.	*8.7. $\vdash \max(c+a, c+b) =c+\max(a, b)$.

*8.8. $\vdash \min(a, b) \leq \max(a, b)$.

*8.9. $\vdash \min(a, b)+\max(a, b)=a+b$.

PROOFS. *7.2. I. $\min(a, b) = b \dot{-} (b \dot{-} a) = a$ [*6.10] $= a \dot{-} 0$ [×6.1] $=$ $a \dot{-} (a \dot{-} b)$ [*6.11] $= \min(b, a)$.

*7.3–*7.7. By cases ($a \leq b$, $a \geq b$) from *7.2.

×9.1. $\overline{sg}(0)=1$.	×10.1. $sg(0)=0$.
×9.2. $\overline{sg}(a')=0$.	×10.2. $sg(a')=1$.
*9.3. $\vdash \overline{sg}(a)=0 \sim a>0$.	*10.3. $\vdash sg(a)=0 \sim a=0$.
*9.4. $\vdash \overline{sg}(a)=1 \sim a=0$.	*10.4. $\vdash sg(a)=1 \sim a>0$.
*9.5. $\vdash \overline{sg}(a)=0 \vee \overline{sg}(a)=1$.	*10.5. $\vdash sg(a)=0 \vee sg(a)=1$.

PROOFS. By induction cases (IM p. 186).

×11.1. $|a-b|=(a\dot{-}b)+(b\dot{-}a)$.

*11.2. $\vdash |a-b|=0 \sim a=b$.

*11.3. $\vdash |a-b|>0 \sim a\neq b$.

*11.4. $\vdash |a-b|=|b-a|$.

*11.5. $\vdash |a-c|\leq|a-b|+|b-c|$.

*11.6. $\vdash |a-0|=a$.

*11.7. $\vdash |(a+c)-(b+c)|=|a-b|$.

*11.8. $\vdash |(a+b)-b|=a$.

*11.9. $\vdash c|a-b|=|ca-cb|$.

*11.10. $c\leq a, b \supset |(a\dot{-}c)-(b\dot{-}c)|=|a-b|$.

*11.11. $\vdash c\geq a, b \supset |(c\dot{-}a)-(c\dot{-}b)|=|a-b|$.

*11.12. $\vdash a\dot{-}b, b\dot{-}a\leq|a-b|\leq \max(a, b)\leq a+b$.

*11.13. $\vdash |(a\dot{-}c)-(b\dot{-}c)|, |(c\dot{-}a)-(c\dot{-}b)|, ||a-c|-|b-c||\leq|a-b|$.

*11.14. $\vdash |a-b|\geq c \sim a\geq b+c \vee b\geq a+c$.

*11.14a, b. Similarly with $>$, $=$.

*11.15. $\vdash |a-b|<c \sim a<b+c \;\&\; b<a+c$.

*11.15a, b. Similarly with $\leq$, $\neq$.

PROOFS. 11.2, 11.3. Using *6.3a, $a=b \supset |a-b|=0$. Using cases $(a<b, a>b)$, *6.12 and *142a, $a\neq b \supset |a-b|>0$. The converses follow by *158, *140.

*11.5. Use *6.21.

*11.10. Use *11.7, *6.7.

*11.11. Use *6.5, *6.10.

*11.14. I. By cases $(a\geq b, a\leq b)$. II. By cases.

*11.15. From *11.14 by *30, *63, *139.

×12.1. $\mathrm{rm}(0, b)=0$.

×12.2. $\mathrm{rm}(a', b)=(\mathrm{rm}(a, b))'\cdot \mathrm{sg}|b-(\mathrm{rm}(a, b))'|$.

×13.1. $[0/b]=0$.

×13.2. $[a'/b]=[a/b]+\overline{\mathrm{sg}}|b-(\mathrm{rm}(a, b))'|$.

*12.3. $\vdash b>0 \sim \mathrm{rm}(a, b)<b$.

*13.3. $\vdash [a/0]=0 \;\&\; \mathrm{rm}(a, 0)=a$.

*13.4. $\vdash a=b[a/b]+\mathrm{rm}(a, b)$.

*13.5. $\vdash a=bq+r \;\&\; r<b \supset q=[a/b] \;\&\; r=\mathrm{rm}(a, b)$.

*13.6. $\vdash [a/1]=a \;\&\; \mathrm{rm}(a, 1)=0$.

*13.7. $\vdash a<b \supset [a/b]=0 \;\&\; \mathrm{rm}(a, b)=a$.

*13.8. $\vdash b>0 \supset [ab+c/b]=a+[c/b] \;\&\; \mathrm{rm}(ab+c, b)=\mathrm{rm}(c, b)$.

*13.9. $\vdash b>0 \supset [ab\dot{-}c/b]=a\dot{-}([c/b]+\mathrm{sg}\ \mathrm{rm}(c, b))$.

*13.10. $\vdash a\leq b \supset [a/c]\leq[b/c]$.

*13.11. $\vdash a\geq b>0 \supset [c/a]\leq[c/b]$.

PROOFS. *12.3. I. Ind. on a. In the ind. step, use cases by *10.5: $\mathrm{sg}|b-(\mathrm{rm}(a, b))'|=0$, $\mathrm{sg}|b-(\mathrm{rm}(a, b))'|=1$ (then by *10.4 and *11.3, $b\neq(\mathrm{rm}(a, b))'$).

*13.4. Similarly.

*13.5. By *146b with *13.4, *12.3.

*13.6, *13.7. By *13.5.

*13.8. By *12.3 (assuming $b>0$), $\mathrm{rm}(c, b)<b$. Also $b(a+[c/b])+\mathrm{rm}(c, b) = ab+(b[c/b]+\mathrm{rm}(c, b)) = ab+c$ [*13.4]. So *13.5 applies.

*13.9. Assume $b>0$. CASE 1: $\mathrm{rm}(c, b)=0$. Then $[ab \dot{-} c/b] = [ab \dot{-} b[c/b]/b]$ [*13.4] $= [b(a \dot{-} [c/b])/b]$ [*6.14] $= a \dot{-} [c/b]$ [*13.8, ${}^{\times}$13.1] $= a \dot{-} ([c/b]+\mathrm{sg}\,\mathrm{rm}(c, b))$. CASE 2: $\mathrm{rm}(c, b)>0$. SUBCASE 1: $[c/b]<a$. Then $a \dot{-} [c/b] \geq 1$ [*6.12, *138b], whence $ab \dot{-} b[c/b] \geq b$ [*145c, *6.14]. So $[ab \dot{-} c/b] = [ab \dot{-} (b[c/b]+\mathrm{rm}(c, b))/b]$ [*13.4] $= [(ab \dot{-} b[c/b]) \dot{-} \mathrm{rm}(c, b)/b]$ [*6.5] $= [(((ab \dot{-} b[c/b]) \dot{-} b)+b) \dot{-} \mathrm{rm}(c, b)/b]$ [*6.7] $= [(ab \dot{-} (b[c/b]+b))+(b \dot{-} \mathrm{rm}(c, b))/b]$ [*6.5; *6.6 with *12.3] $= [b(a \dot{-} ([c/b]+1))+(b \dot{-} \mathrm{rm}(c, b))/b] = a \dot{-} ([c/b]+1)$ [*13.8; *13.7 with *6.16] $= a \dot{-} ([c/b]+\mathrm{sg}\,\mathrm{rm}(c, b))$. SUBCASE 2: $[c/b] \geq a$. Then $b[c/b] \geq ab$. So $[ab \dot{-} c/b] = [ab \dot{-} (b[c/b]+\mathrm{rm}(c, b))/b] = [0/b]$ [*6.11] $= 0$ [${}^{\times}$13.1] $= a \dot{-} ([c/b]+\mathrm{sg}\,\mathrm{rm}(c, b))$ [*6.11].

*13.10. By ind. on p, $[a/c] \leq [a+p/c]$.

*13.11. Assume $a \geq b>0$ and $[c/a]>[c/b]$, and use *13.4 to deduce $\mathrm{rm}(c, b) \geq b$, contradicting *12.3.

By a *standard* formula, we mean a (prime) formula of the form $t=0$ where $\vdash t=0 \vee t=1$ (equivalently, $\vdash t \leq 1$). The following result *14.1 (by *11.2, *10.3) with *10.5 shows that any prime formula $t_1=t_2$ is equivalent to a standard formula. This will be useful in building other standard formulas to express primitive recursive predicates under the constructions of #D–#F.

*14.1. $\vdash a=b \sim \mathrm{sg}|a-b|=0$.

Hitherto "$a<b$" has been an abbreviation for $\exists c(c'+a=b)$. Now *15.1 shows that hereafter we can take it alternatively as an abbreviation for the standard (prime) formula $\mathrm{sg}(a' \dot{-} b)=0$, and all formulas proved in IM (intuitionistically) and above will still hold. (Either the first or the third member in *15.1 is merely the unabbreviation of the middle one, according to which way we interpret the abbreviation "$a<b$".)

*15.1. $\vdash \exists c(c'+a=b) \sim a<b \sim \mathrm{sg}(a' \dot{-} b)=0$.

PROOF. $\mathrm{sg}(a' \dot{-} b)=0 \sim a' \dot{-} b=0$ [*10.3] $\sim a' \leq b$ [*6.11] $\sim a<b$ [*138b].

In IM p. 191, "a|b" abbreviates $\exists c(ac=b)$. Now *16.8 gives an alternative. (The proofs of *16.1–*16.5 are good in the system of IM.)

*16.1. $\vdash a|0$. *16.2. $\vdash 0|b \sim b=0$. *16.3. $\vdash 1|a$.
*16.4. $\vdash a|c \& b|d \supset ab|cd$. *16.5. $\vdash a|b \supset (a|c \sim a|b+c)$.
*16.6. $\vdash b>0 \supset a|a^b$. *16.7. $\vdash 0<b\leq a \supset b|a!$.
*16.8. $\vdash \exists c(ac=b) \sim a|b \sim sg(rm(b, a))=0$.

PROOFS. *16.2. By *16.1, *156.

*16.5. To show $a|b$, $a|b+c \vdash a|c$, assume $ad=b$ and $ae=b+c$, whence $c = ae \dot{-} b$ [*6.3] $= ae \dot{-} ad = a(e \dot{-} d)$ [*6.14]; or (avoiding $\dot{-}$) we may proceed as follows. CASE 1: $a=0$. Use *16.2. CASE 2: $a\neq 0$. Assume for $\exists$-elim. $ap=b$ and $aq=b+c$. Then $ap = b \leq b+c = aq$, so by *145b, $p\leq q$. Assume $p+r=q$. Now $ap+ar = aq = b+c = ap+c$, and by *132, $ar=c$.

*16.6. Prove $a|a^{c'}$.

*16.7. Ind. on a (using *138a).

*16.8. I. Assume $a|b$, and for $\exists$-elim. $aq=b$. CASE 1: $a=0$. Then $b=0$, so $rm(b, a)=0$ by ˣ12.1; etc. CASE 2: $a>0$. Then $b=aq+0$ & $0<a$, so $rm(b, a)=0$ by *13.5. II. Assume $sg(rm(b, a))=0$. Then $rm(b, a)=0$ by *10.3, hence $b=a[b/a]$ by *13.4, hence $a|b$.

#A, #C need no comment. To take over #B, we let the next function symbol f_{15} express finite sum ($k_{15} = l_{15} = 1$), with the axioms:

ˣB1a. $f_{15}(0, \alpha)=0$. ˣB2a. $f_{15}(z', \alpha)=f_{15}(z, \alpha)+\alpha(z)$.

This meets the requirements of Lemma 5.1. So, introducing "$\Sigma_{y<s}t(y)$" for any variable y and terms t(y), s as abbreviation for $f_{15}(s, \lambda y t(y))$, and using *0.4, the two axioms are equivalent to the formulas of *B1, *B2 below. This puts the development in familiar notation. Similarly we introduce the finite product with two axioms:

ˣB3a. $f_{16}(0, \alpha)=1$. ˣB4a. $f_{16}(z', \alpha)=f_{16}(z, \alpha)\cdot\alpha(z)$.

For any variable y and formulas A(y), R(y), we may use "$\forall y_{A(y)}R(y)$" as abbreviation for $\forall y(A(y) \supset R(y))$, and "$\exists y_{A(y)}R(y)$" for $\exists y(A(y) \& R(y))$. We also want versions of *B5–*B13, *B16–*B21 with the inequality $y<z$ replaced by $v\leq y<z$ throughout. (Alternative versions of *B14, *B15 are given explicitly.) We interpret "$\Sigma_{v\leq y<z}t(y)$" as an abbreviation for $\Sigma_{y<z\dot{-}v}t(v+y)$ if v is free for y in t(y) (otherwise

for $\Sigma_{y<z\dot{-}v}t_1(v+y)$ where $t_1(y)$ is congruent to $t(y)$ and v is free for y in $t_1(y)$). These versions follow from those given using $\vdash \forall y_{v\leq y<z}R(y) \sim \forall y_{y<z\dot{-}v}R(v+y)$, $\vdash \exists y_{v\leq y<z}R(y) \sim \exists y_{y<z\dot{-}v}R(v+y)$ (by *144a, *6.7 with *6.12; *6.19, *6.7). "$\Sigma_{v<y<z}$" shall be $\Sigma_{v'\leq y<z}$, etc. (cf. *138a, *138b). In citing *B5–*B23 we shall mean the modified versions when appropriate. We shall abbreviate $(\Sigma_{y<s}t(y))+r$ as "$\Sigma_{y<s}t(y)+r$".

*B1. $\vdash \Sigma_{y<0}\alpha(y)=0$. *B2. $\vdash \Sigma_{y<z'}\alpha(y)=\Sigma_{y<z}\alpha(y)+\alpha(z)$.
*B3. $\vdash \Pi_{y<0}\alpha(y)=1$. *B4. $\vdash \Pi_{y<z'}\alpha(y)=(\Pi_{y<z}\alpha(y))\cdot\alpha(z)$.
*B5. $\vdash \forall y_{y<z}\alpha(y)=0 \sim \Sigma_{y<z}\alpha(y)=0$.
*B6. $\vdash \forall y_{y<z}\alpha(y)>0 \sim \Pi_{y<z}\alpha(y)>0$.
*B7. $\vdash \exists y_{y<z}\alpha(y)>0 \sim \Sigma_{y<z}\alpha(y)>0$.
*B8. $\vdash \exists y_{y<z}\alpha(y)=0 \sim \Pi_{y<z}\alpha(y)=0$.
*B9. $\vdash \forall y_{y<z}\alpha(y)\leq 1 \supset \Sigma_{y<z}\alpha(y)\leq z$.
*B10. $\vdash \forall y_{y<z}\alpha(y)\leq 1 \supset \Pi_{y<z}\alpha(y)\leq 1$.
*B11. $\vdash \forall y_{y<z}\alpha(y)=1 \supset \Sigma_{y<z}\alpha(y)=z$.
*B12. $\vdash \forall y_{y<z}\alpha(y)\leq 1 \supset [\Sigma_{y<z}\alpha(y)=z \supset \forall y_{y<z}\alpha(y)=1]$.
*B13. $\vdash \forall y_{y<z}\alpha(y)=1 \sim \Pi_{y<z}\alpha(y)=1$.
*B14. $\vdash w\leq z \supset \Sigma_{y<z}\alpha(y)=\Sigma_{y<w}\alpha(y)+\Sigma_{w\leq y<z}\alpha(y)$,
$\vdash v\leq w\leq z \supset \Sigma_{v\leq y<z}\alpha(y)=\Sigma_{v\leq y<w}\alpha(y)+\Sigma_{w\leq y<z}\alpha(y)$.
*B15. $\vdash w\leq z \supset \Pi_{y<z}\alpha(y)=(\Pi_{y<w}\alpha(y))\cdot\Pi_{w\leq y<z}\alpha(y)$,
$\vdash v\leq w\leq z \supset \Pi_{v\leq y<z}\alpha(y)=(\Pi_{v\leq y<w}\alpha(y))\cdot\Pi_{w\leq y<z}\alpha(y)$.
*B16. $\vdash \Sigma_{y<z}(\alpha(y)+\beta(y))=\Sigma_{y<z}\alpha(y)+\Sigma_{y<z}\beta(y)$.
*B17. $\vdash \Pi_{y<z}\alpha(y)\beta(y)=(\Pi_{y<z}\alpha(y))\cdot\Pi_{y<z}\beta(y)$.
*B18. $\vdash \forall y_{y<z}\alpha(y)=\beta(y) \supset \Sigma_{y<z}\alpha(y)=\Sigma_{y<z}\beta(y)$.
*B19. $\vdash \forall y_{y<z}\alpha(y)=\beta(y) \supset \Pi_{y<z}\alpha(y)=\Pi_{y<z}\beta(y)$.
*B20. $\vdash \forall y_{y<z}\alpha(y)\leq\beta(y) \supset \Sigma_{y<z}\alpha(y)\leq\Sigma_{y<z}\beta(y)$.
*B21. $\vdash \forall y_{y<z}\alpha(y)\leq\beta(y) \supset \Pi_{y<z}\alpha(y)\leq\Pi_{y<z}\beta(y)$.
*B22. $\vdash \forall y_{y\leq z}\alpha(y)<\beta(y) \supset \Sigma_{y\leq z}\alpha(y)<\Sigma_{y\leq z}\beta(y)$.
*B23. $\vdash \forall y_{y\leq z}\alpha(y)<\beta(y) \supset \Pi_{y\leq z}\alpha(y)<\Pi_{y\leq z}\beta(y)$.

PROOFS. *B5. Ind. on z. BASIS. Use $\neg y<0$, *10a, *11. IND. STEP. Use *138a, *128.

*B7. I. Assume $\exists y_{y<z}\alpha(y)>0$. By *136, $\Sigma_{y<z}\alpha(y)\geq 0$. But if $\Sigma_{y<z}\alpha(y)=0$, then by *B5 $\forall y_{y<z}\alpha(y)=0$, whence (using *84a with Remark 4.1) $\neg\exists y_{y<z}\alpha(y)>0$, contradicting the assumption. II. Assume $\Sigma_{y<z}\alpha(y)>0$. By Remark 4.1, $\exists y_{y<z}\alpha(y)>0 \vee \neg\exists y_{y<z}\alpha(y)>0$. But if $\neg\exists y_{y<z}\alpha(y)>0$, then (using *86, *58b) $\forall y_{y<z}\alpha(y)=0$, and by *B5 $\Sigma_{y<z}\alpha(y)=0$.

*B12. In the ind. step $\alpha(z)=1$ or *B9 would be contradicted.

*B13. Note *131.

*B14. Assume $w \leq z$, and for $\exists$-elim. $z=w+c$. Then *6.3 gives $c=z \dot{-} w$. So we have to deduce $\Sigma_{y<w+c}\alpha(y)=\Sigma_{y<w}\alpha(y)+\Sigma_{y<c}\alpha(w+y)$, which is straightforward by ind. on c.

*B21, *B23. Using $a \leq b \supset ac \leq bc$ and (for *B23) *145a.

We treat #D, #E as schemata concerning certain forms of definitions. The results will enable us, for any formulas $P_1, \ldots, P_m$ to each of which we have chosen an equivalent standard formula and any formula E composed out of $P_1, \ldots, P_m$ by propositional connectives and bounded quantifiers, to build a standard formula equivalent to E. By Remark 4.1 now, it suffices under #D to treat $\vee$ and $\neg$ (as in IM). Via the alternative versions of *B5 etc., we have alternative versions of *E1–*E4 with $v \leq y < z$ as the bound. For any variable y and formula R(y) such that $\vdash R(y) \sim r(y)=0$ for some term r(y) with $\vdash r(y) \leq 1$, we take "$\mu y_{y<z}R(y)$" to be an abbreviation for $\Sigma_{x<z}\Pi_{y<x'}r(y)$ for some such term r(y) (by *B18, *B19 it will be immaterial which) and a variable x not occurring free in r(y).

Let Q, R *be formulas, and* q, r *terms, such that* $\vdash Q \sim q=0$ *with* $\vdash q \leq 1$, *and* $\vdash R \sim r=0$ *with* $\vdash r \leq 1$. *Then*:

*D1. $\vdash Q \vee R \sim qr=0$.

*D2. $\vdash qr \leq 1$.

*D3. $\vdash \neg Q \sim \overline{sg}(q)=0$.

*D4. $\vdash \overline{sg}(q) \leq 1$.

Let y *be a variable,* R(y) *a formula, and* r(y) *a term, such that* $\vdash R(y) \sim r(y)=0$ *with* $\vdash r(y) \leq 1$. *In* *E5–*E7, *let* M_z *be* $\mu y_{y<z}R(y)$. *Then*:

*E1. $\vdash \exists y_{y<z}R(y) \sim \Pi_{y<z}r(y)=0$.

*E2. $\vdash \Pi_{y<z}r(y) \leq 1$.

*E3. $\vdash \forall y_{y<z}R(y) \sim sg(\Sigma_{y<z}r(y))=0$.

*E4. $\vdash sg(\Sigma_{y<z}r(y)) \leq 1$.

*E5. $\vdash \exists y_{y<z}R(y) \supset M_z<z \;\&\; R(M_z) \;\&\; \forall w_{w<M_z}\neg R(w)$.

*E6. $\vdash M_z<z \sim \exists y_{y<z}R(y)$.

*E7. $\vdash M_z=z \sim \forall y_{y<z}\neg R(y)$.

PROOFS. *D1. By *129 etc.

*D4. By *9.5.

*E1. This is essentially *B8 (merely $\lambda y r(y)$ has been substituted for α.)

*E2. By *B10.

*E3. By *B5 with *10.3.

Before proving *E5–*E7, we obtain by *B13 with *B11

(a) $\vdash \forall y_{y<z} \neg R(y) \supset M_z = z$.

Also by Remark 4.1 (or *150), *86 and *58b,

(b) $\vdash \exists y_{y<z} R(y) \vee \forall y_{y<z} \neg R(y)$.

*E5. Assume $\exists y_{y<z} R(y)$. By *149a (with Remark 4.1, *159, or #D, #15), $\exists v[v<z \ \& \ R(v) \ \& \ \forall w_{w<v} \neg (w<z \ \& \ R(w))]$. Omitting the $\exists v$ (preparatory to $\exists$-elim.), and $w<z$ (by *45), $v<z \ \& \ R(v) \ \& \ \forall w_{w<v} \neg R(w)$. It will suffice now to show that $M_z = v$. Using $v<z$ with *B14, $M_z = M_v + \Sigma_{v \leq x < z} \Pi_{y<x'} r(y)$. By $R(v)$, whence $r(v)=0$, with *B8 and *B5, $\Sigma_{v \leq x < z} \Pi_{y<x'} r(y) = 0$. So $M_z = M_v = v$ (using (a)).

*E6. I. Assume $M_z < z$. Then by (b) $\exists y_{y<z} R(y)$, since $\forall y_{y<z} \neg R(y)$ with (a) would give $M_z = z$. II. By *E5.

In #F (for any $m \geq 1$), we simply take $\varphi_{m+1} = 0$ (IM p. 229).

Let $Q_1, \ldots, Q_m$ *be formulas such that* $\vdash \neg(Q_i \ \& \ Q_j)$ $(i \neq j)$, $q_1, \ldots, q_m$ *terms such that* $\vdash Q_i \sim q_i = 0$, *and* $p_1, \ldots, p_m$ *any terms. Let* p *be the term* $\overline{sg}(q_1) \cdot p_1 + \ldots + \overline{sg}(q_m) \cdot p_m$. *Then*:

*F1. $\vdash Q_i \supset p = p_i$. *F2. $\vdash \neg Q_1 \ \& \ \ldots \ \& \ \neg Q_m \supset p = 0$.

#G (IM § 46) will be taken over by Lemma 5.3 (c) in 5.6.

*H1–*H4 are equivalent to axioms ˣH1a–ˣH4a (with function symbols f_{17}, f_{18}), as *B1–*B4 to ˣB1a–ˣB4a.

*H1. $\vdash \min_{y \leq 0} \alpha(y) = \alpha(0)$.

*H2. $\vdash \min_{y \leq z'} \alpha(y) = \min(\min_{y \leq z} \alpha(y), \alpha(z'))$.

*H3. $\vdash \max_{y \leq 0} \alpha(y) = \alpha(0)$.

*H4. $\vdash \max_{y \leq z'} \alpha(y) = \max(\max_{y \leq z} \alpha(y), \alpha(z'))$.

*H5. $\vdash \forall y_{y \leq z} \alpha(y) \geq \min_{y \leq z} \alpha(y)$. *H6. $\vdash \exists y_{y \leq z} \alpha(y) = \min_{y \leq z} \alpha(y)$.

*H7. $\vdash \forall y_{y \leq z} \alpha(y) \leq \max_{y \leq z} \alpha(y)$. *H8. $\vdash \exists y_{y \leq z} \alpha(y) = \max_{y \leq z} \alpha(y)$.

By #15, #16, #E and #D ($\exists c(1<c<a \ \& \ c|a)$ being equivalent to $\exists c_{2 \leq c < a} c|a$), we find a standard (prime) formula $Pr(a)$ such that:

*17.1. $\vdash Pr(a) \sim a>1 \ \& \ \neg \exists c(1<c<a \ \& \ c|a)$.

Thus the prime formula $Pr(a)$ is equivalent to the composite formula taken as $Pr(a)$ in IM p. 191. (The proofs of *17.3, *17.4, *17.5 are good in the system of IM.) By 5.4,

$\vdash \neg Pr(0)$, $\vdash \neg Pr(1)$, $\vdash Pr(2)$, $\vdash Pr(3)$, $\vdash \neg Pr(4)$, $\vdash Pr(5)$, ...

As we now have the function $a!$ formally, we can improve *161 to:

*17.2. $\vdash \exists b(a<b\leq a!+1 \ \&\ Pr(b))$. (Euclid's theorem.)

PROOF. By *4.3 and *16.7, $\vdash a!>0 \ \&\ \forall b(0<b\leq a \supset b|a!)$; so a! can play the role of d in the arguments on IM p. 192 from (1) on. In particular, we had there (as part of (2)) $b|d'$, so now $b|a!+1$, and by *156 $b\leq a!+1$, which is all that remained to be shown.

*17.3. $\vdash Pr(a) \ \&\ Pr(b) \ \&\ a\neq b \supset \neg a|b$.

PROOF. By cases ($a<b$, $a>b$), using *17.1, *156.

*17.4. $\vdash a>1 \supset \exists p[Pr(p) \ \&\ p|a]$.

PROOF. Assume $a>1$. Using *153, $a>1 \ \&\ a|a$, whence $\exists b[b>1 \ \&\ b|a]$. Hence by *149a (with Remark 4.1, or in the system of IM Remark 1 (b) p. 134 with *158–*160, *150), $\exists p[p>1 \ \&\ p|a \ \&\ \forall c_{c<p}\neg(c>1 \ \&\ c|a)]$. Assume for $\exists$-elim. $p>1 \ \&\ p|a \ \&\ \forall c_{c<p}\neg(c>1 \ \&\ c|a)$. Now assume $\exists c(1<c<p \ \&\ c|p)$, and for $\exists$-elim. $1<c<p \ \&\ c|p$. Thence with $p|a$ and *154, $c>1 \ \&\ c|a$; but also using $c<p$ with $\forall c_{c<p}\neg(c>1 \ \&\ c|a)$, $\neg(c>1 \ \&\ c|a)$. By reductio ad absurdum (cf. IM Remark bottom p. 188), $\neg\exists c(1<c<p \ \&\ c|p)$.

*17.5. $\vdash Pr(p) \ \&\ p|ab \supset p|a \vee p|b$. (Euclid's first theorem.)

PROOF. We use the method of Hardy-Wright **1954** p. 21 2.11 (but only to get Euclid's first theorem, not the fundamental theorem of arithmetic). Thus informally $p|a \vee p|b$ holds trivially for $ab = 0$. So we show the absurdity that there exist n, p, a, b such that $n=ab$ & $n\neq 0$ & $Pr(p)$ & $p|n$ & $\overline{p|a}$ & $\overline{p|b}$. But if there exist any four such numbers n, p, a, b, we can in particular pick the least n for which the other three p, a, b exist; then pick the least p for which with this n the other two a, b exist; and finally pick the least a for which with this n and p the b exists; then the b is determined by $n=ab$ & $n\neq 0$. This start of the proof is formalized, with the precautions necessary for working in an intuitionistic system, in I–III below. The deductions about the n, p, a, b by which we then reach a contradiction, in IV–VII, are hardly different formally than informally, except for our having to verify meticulously that various familiar propositions we need to apply are expressed by formulas previously shown to be provable in the present system.

I. For $ab=0$, use *129, *16.1. It remains for us to prove $ab\neq 0 \ \&\ Pr(p) \ \&\ p|ab \supset p|a \vee p|b$. By the classical propositional calculus (applicable by Remark 4.1, or in the system of IM by Remark 1 (a) p. 134 with *158–*160, *150), this is equivalent to $\neg(ab \neq 0$ & $Pr(p)$ & $p|ab$ & $\neg p|a$ & $\neg p|b)$. So it will suffice, using $\vdash \exists n\, n=ab$ (by *100) and ∃-elim., to deduce a contradiction from

$$(1) \qquad n=ab \;\&\; n\neq 0 \;\&\; Pr(p) \;\&\; p|n \;\&\; \neg p|a \;\&\; \neg p|b,$$

call this $R(n, p, a, b)$, or by ∃-introds., from

$$(2) \qquad \exists n \exists p \exists a \exists b R(n, p, a, b).$$

II. We show that $R(n, p, a, b) \vdash p<n$ & $a<n$ & $b<n$. By $n=ab$ & $n\neq 0$, $a\neq 0$ & $b\neq 0$. Also $a\neq 1$ (as $a=1$ with $n=ab$ and $p|n$ contradicts $\neg p|b$); similarly $b\neq 1$. So $a>1$ & $b>1$, and also by *143b, $a<ab=n$ & $b<ab=n$. By $Pr(p)$, $p>1$. Using $n=ab$, $n\neq 0$ and $p|n$ in *156, $p\leq ab$. But if $p=ab$ (so $a|p$), then by *156 (with $p>1$, $a>1$), $1<a\leq p$; but $a=p$ contradicts $\neg p|a$ by *153, so $1<a<p$; but this (with $a|p$) contradicts $Pr(p)$. So $p<ab=n$.

III. Now we take (2) as the $\exists x A(x)$ for *149a. To get the $A(x) \vee \neg A(x)$, observe that by II with $A \supset B \vdash A \sim B \;\&\; A$ and *91, $\exists b R(n, p, a, b) \sim \exists b_{b<n} R(n, p, a, b)$, $\exists a \exists b R(n, p, a, b) \sim \exists a_{a<n} \exists b_{b<n} R(n, p, a, b)$ and $\exists p \exists a \exists b R(n, p, a, b) \sim \exists p_{p<n} \exists a_{a<n} \exists b_{b<n} R(n, p, a, b)$. So three uses of *150 (with IM Remark 1 (b) p. 134 etc.) give the $A(x) \vee \neg A(x)$, and we are led by *149a to $\exists y[A(y)$ & $\forall z_{z<y} \neg A(z)]$. Omitting the ∃y preparatory to ∃-elim., we assume

$$(3) \qquad \exists p \exists a \exists b R(n, p, a, b) \;\&\; \forall m_{m<n} \neg \exists p \exists a \exists b R(m, p, a, b).$$

Repeating the procedure twice, we further assume

$$(4) \qquad \exists a \exists b R(n, p, a, b) \;\&\; \forall q_{q<p} \neg \exists a \exists b R(n, q, a, b),$$

$$(5) \qquad \exists b R(n, p, a, b) \;\&\; \forall e_{e<a} \neg \exists b R(n, p, e, b).$$

Preparatory to ∃-elim. from the first member of (5), assume

$$(6)\ (=(1)) \qquad R(n, p, a, b).$$

Finally, preparatory to ∃-elim. from $p|n$ (in (6)), assume

$$(7) \qquad pd=n.$$

By the anticipated ∃-elims., a contradiction deduced from (3)–(7) will result in one deduced from (2) and thence from (1) (cf. IM Remark bottom p. 188).

IV. We deduce $p \leq d$. By (7) $d \neq 0$ (since $n \neq 0$ is in (6)), and $d \neq 1$ (since $p<n$, by (6) with II). Thus $d>1$. To show $p \leq d$, assume $p>d$. By $d>1$ with *17.4, $\exists q[\Pr(q) \& q|d]$. Assume for $\exists$-elim., $\Pr(q) \& q|d$. By *156, $q \leq d$, so $q<p$. By *154 and (7), $q|n$. If we can deduce also $\neg q|a \& \neg q|b$, we will have $R(n, q, a, b)$ whence $\exists a \exists b R(n, q, a, b)$, besides $q<p$, contradicting the second member of (4). To show $\neg q|a$, suppose $q|a$. For $\exists$-elims. from this and $q|d$, assume $qa_1=a$, $qd_1=d$. Then $a_1 \neq 0$, $d_1 \neq 0$ (by $a \neq 0$ in II, and $d \neq 0$). Now $pqd_1 = pd = n = ab = qa_1b$, whence $pd_1=a_1b$ (by *133, with $q>1$ from $\Pr(q)$); so $p|a_1b$, but $\neg p|a_1$ (or we would have $p|a$). Also by *143b with $q>1$ and $d_1 \neq 0$, $d_1 < qd_1 = d$; so by *145a, $a_1b = pd_1 < pd = n$. Thus $a_1b<n$, and $R(a_1b, p, a_1, b)$ whence $\exists p \exists a \exists b R(a_1b, p, a, b)$, contradicting the second member of (3). Symmetrically, $\neg q|b$.

V. We deduce $a \leq b$. For $b<a$, with $R(n, p, b, a)$ whence $\exists a R(n, p, b, a)$, would contradict (5).

VI. We deduce $pa<n$. From $p \leq d$ (in IV) by *145c (5.5 ¶ 4), $pp \leq pd = n$; and from $a \leq b$ (in V), $aa \leq ab = n$. Now if $pp=n \& aa=n$, we would have $pp=aa$, which by *145a (with $p>1$, $a \neq 0$) would make $p<a$ and $p>a$ absurd, and thus would imply $p=a$; but $p \neq a$ by $\neg p|a$ with *153. So $pp<n \& aa \leq n$ or $pp \leq n \& aa<n$, each of which leads by *145a, *145c to $papa<nn$, whence (by contradicting $pa \geq n$) $pa<n$.

VII. We finally deduce a contradiction. We have $pa \neq 0$, also $pa<n$ (in VI). Assume (for $\exists$-elim. from $pa<n$) $q'+pa=n$. Now $0 \neq q' < n$. Also $p|q'$ and $a|q'$ by *16.5; so assume $ar=q'$. Now $ab = n = q'+pa = ar+pa = a(r+p)$, so by *133 $b=r+p$. But $\neg p|r$, or we would contradict $\neg p|b$ by *16.5, *153. Thus $q'<n$, and $R(q', p, a, r)$ whence $\exists p \exists a \exists b R(q', p, a, b)$, contradicting (3).

*17.6. $\vdash \Pr(p) \& p|a^n \supset p|a$.

PROOF. Ind. on n. BASIS. Assume $\Pr(p) \& p|a^0$. By ˣ3.1 and *156, $p=1$. But $\neg\Pr(1)$. IND. STEP. Use *17.5 with ˣ3.2.

We use #4, #15, #17, #D, #E, #A to select the term $\mu b_{b<a!+2}[a<b \& \Pr(b)]$ used in the second recursion equation ˣ18.2.

ˣ18.1. $p_0=2$. ˣ18.2. $p_{i'}=\mu b_{b<p_i!+2}[p_i<b \& \Pr(b)]$.
*18.3. $\vdash p_i<p_{i'} \leq p_i!+1 \& \Pr(p_{i'}) \& \forall a(p_i<a<p_{i'} \supset \neg\Pr(a))$.
*18.4. $\vdash \Pr(p_i)$. *18.5. $\vdash p_i>i'$. *18.6. $\vdash i<j \sim p_i<p_j$.
*18.7. $\vdash \Pr(a) \sim \exists i(a=p_i)$.

PROOFS. *18.3. By *17.2 (and *138a), $\exists b_{b<p_i!+2}[p_i<b \,\&\, Pr(b)]$, whence *18.3 follows by *E5 and ˣ18.2.

*18.6. I. Prove $p_i<p_{i+c'}$ by ind. on c.

*18.7. I. Assume Pr(a). By *18.5, $p_a > a' > a$, so $\exists j p_j>a$. By *149a, $\exists j[p_j>a \,\&\, \forall i_{i<j}\neg p_i>a]$. Assume $p_j>a \,\&\, \forall i_{i<j}\neg p_i>a$, whence $\forall i_{i<j}p_i\leq a$. CASE 1: j=0. Then $p_j = 2 > a$, which with $\neg Pr(0)$, $\neg Pr(1)$ contradicts Pr(a). CASE 2: j>0. Write j=i′. Then i<j, so $p_i\leq a$. But if $p_i<a$, we would have $p_i < a < p_j = p_{i'}$, and by the last member of *18.3, $\neg Pr(a)$.

ˣ19.1. $(a)_i=\mu x_{x<a}[p_i^x|a \,\&\, \neg p_i^{x'}|a]$.

*19.2. $\vdash a>0 \sim (a)_i<a \,\&\, p_i^{(a)_i}|a \,\&\, \neg p_i^{(a)_i+1}|a$.

*19.3. $\vdash h\leq(a)_i \supset p_i^h|a$. *19.4. $\vdash a>0 \,\&\, h>(a)_i \sim \neg p_i^h|a$.

*19.5. $\vdash a>0 \,\&\, p_i^h|a \,\&\, \neg p_i^{h'}|a \supset (a)_i=h$. *19.6. $\vdash (0)_i=0$.

*19.7. $\vdash (a)_i>0 \sim a>0 \,\&\, p_i|a$. *19.8. $\vdash i\geq a \supset (a)_i=0$.

*19.9. $\vdash (p_i^h)_i=h$. *19.9a. $\vdash (1)_i=0$. *19.10. $\vdash i\neq j \supset (p_i^h)_j=0$.

*19.11. $\vdash ab>0 \supset (ab)_i=(a)_i+(b)_i$.

*19.12. $\vdash \forall y_{y<z}\alpha(y)>0 \supset (\Pi_{y<z}\alpha(y))_i=\Sigma_{y<z}(\alpha(y))_i$.

*19.13. $\vdash j\geq k \supset (\Pi_{i<k}p_i^{\alpha(i)})_j=0$, $\vdash j<h \vee j\geq k \supset (\Pi_{h\leq i<k}p_i^{\alpha(i)})_j=0$.

*19.14. $\vdash j<k \supset (\Pi_{i<k}p_i^{\alpha(i)})_j=\alpha(j)$, $\vdash h\leq j<k \supset (\Pi_{h\leq i<k}p_i^{\alpha(i)})_j=\alpha(j)$.

*19.15. $\vdash 0<a\leq b \,\&\, \forall i_{i<b}(a)_i=(b)_i \supset a=b$.

*19.16. $\vdash a>0 \sim a=\Pi_{i<a}p_i^{(a)_i}$.

PROOFS. *19.2. I. Assume a>0. The conclusion will follow from ˣ19.1 by *E5 if we get $\exists x_{x<a}[p_i^x|a \,\&\, \neg p_i^{x'}|a]$. By *18.5, $p_i>1$. By *3.10 $p_i^a>a$, so by *156 $\neg p_i^a|a$, whence $\exists y \neg p_i^y|a$. By *149a, $\exists y(\neg p_i^y|a \,\&\, \forall x_{x<y}\neg\neg p_i^x|a)$. Assume $\neg p_i^y|a \,\&\, \forall x_{x<y}\neg\neg p_i^x|a$, whence $\forall x_{x<y}p_i^x|a$. CASE 1: y=0. Then $\neg p_i^y|a$ contradicts $p_i^0|a$ (obtained by *16.3, ˣ3.1). CASE 2: y>0. Put y=x′. Then $p_i^x|a$ by $\forall x_{x<y}p_i^x|a$. Also x<a; for if x≥a, then by *3.12 $p_i^x\geq p_i^a>a$, so by *156 $\neg p_i^x|a$. Thus $x<a \,\&\, p_i^x|a \,\&\, \neg p_i^{x'}|a$. Use ∃-introd.

*19.3, *19.4. By *16.1, *19.2, *154 and $h\leq m \supset c^h|c^m$ (from *3.3).

*19.6. From ˣ19.1 by *E7.

*19.8. CASE 1: a=0. By *19.6. CASE 2: a>0. By *19.7 (for i≥a with *18.5, *156 gives $\neg p_i|a$).

*19.9. By *18.5, $p_i>1$; so $p_i^h>0$ by *3.9, and $p_i^h<p_i^{h'}$ by *3.12. So by *156, $\neg p_i^{h'}|p_i^h$. By *153, $p_i^h|p_i^h$. Use *19.5.

*19.10. Assume i≠j and $(p_i^h)_j\neq 0$. Then by *19.7 $p_j|p_i^h$, whence by *17.6 (with *18.4) $p_j|p_i$. But by *17.3 (with *18.6) $\neg p_j|p_i$.

*19.11. Assume ab>0. By *19.2 (with *129), $p_i^{(a)_i}|a$, $\neg p_i^{(a)_i+1}|a$,

$p_i^{(b)_i}|b$, $\neg p_i^{(b)_i+1}|b$. Hence (by *16.4, *3.3), $p_i^{(a)_i+(b)_i}|ab$. The conclusion will follow by *19.5, if we deduce $\neg p_i^{(a)_i+(b)_i+1}|ab$. So assume $p_i^{(a)_i+(b)_i+1}|ab$, and for ∃-elims., $p_i^{(a)_i}c=a$, $p_i^{(b)_i}d=b$, $p_i^{(a)_i+(b)_i+1}e=ab$. Then $\neg p_i|c$ (or we would contradict $\neg p_i^{(a)_i+1}|a$, by *16.4, *153, ×3.2); similarly, $\neg p_i|d$. Now $p_i^{(a)_i}cp_i^{(b)_i}d = ab = p_i^{(a)_i+(b)_i+1}e$, whence by *133 (with ×3.2, *18.5, *3.9) $cd=p_ie$, so $p_i|cd$. By *17.5 (with *18.4), this contradicts $\neg p_i|c$ & $\neg p_i|d$.

*19.12. Ind. on z, with *19.9a, *19.11, *B6, etc. (There is an alternative version with $v \leq y < z$ as the bound.)

*19.13. Use *19.12 (with *18.5, *3.9), *19.10, *B5.

*19.14. Use *B15 (with j′ as the w), *B4, *19.11 (with *B6), *19.9, *19.13.

*19.15. Using *86, it will suffice to prove ¬A(b) where A(b) is $\exists a[0<a<b \ \& \ \forall i_{i<b}(a)_i=(b)_i]$. This we do by infinite descent *163. Assume A(b), and for ∃-elim., $0<a<b \ \& \ \forall i_{i<b}(a)_i=(b)_i$. Then $b>1$, and by *17.4 $\exists p[\mathrm{Pr}(p) \ \& \ p|b]$; assume $\mathrm{Pr}(p) \ \& \ p|b$. By *18.7, $\exists j(p=p_j)$; assume $p=p_j$, so $p_j|b$. By *19.7, $(b)_j>0$. By *18.5 and *156, $j < j' < p_j \leq b$; so by $\forall i_{i<b}(a)_i=(b)_i$, $(a)_j=(b)_j$. Thence by $(b)_j>0$ and *19.7, $p_j|a$. Assume $p_jc=a$ and $p_jd=b$. By *145a, $0<c<d$. By *143b, $d<b$. So assuming $i<d$, *19.11 and $\forall i_{i<b}(a)_i=(b)_i$ give $(p_j)_i+(c)_i = (p_jc)_i = (a)_i = (b)_i = (p_jd)_i = (p_j)_i+(d)_i$; and by *132, $(c)_i=(d)_i$. Thus $\forall i_{i<d}(c)_i=(d)_i$, which with $0<c<d$ and $d<b$ gives $d<b$ & A(d).

*19.16. I. By cases ($i<a, i\geq a$) using *19.14, *19.13 and *19.8, $\forall i(a)_i=(\Pi_{i<a}p_i^{(a)_i})_i$, whence the implication follows by *19.15 (with *B6 etc.).

We obtain the function lh more simply than in IM p. 230.

×20.1. $\mathrm{lh}(a)=\Sigma_{i<a}\mathrm{sg}((a)_i)$.

*20.2. $\vdash a>1 \sim \mathrm{lh}(a)>0$.

*20.3. $\vdash \mathrm{lh}(p_i^{s+1})=1$.

*20.4. $\vdash a>0 \sim a\geq 2^{\mathrm{lh}(a)}$.

*20.5. $\vdash a>0 \sim \mathrm{lh}(a)<a$.

Proofs. *20.2. I. Were $\mathrm{lh}(a)=0$, by *B5, *10.3, ×3.1 and *B13, we would have $\Pi_{i<a}p_i^{(a)_i}=1$, and so by *19.16 $a\leq 1$.

*20.3. By *18.5, *3.6 and *3.12, $i'<p_i^{s+1}$. Now use *B14 (with i′ as the w), *B2, *19.10, ×10.1, *B5, *19.9, ×10.2.

*20.4. I. Assume $a>0$. To infer the conclusion from *19.16, deduce $\Pi_{i<k}p_i^{(a)_i} \geq 2 \exp \Sigma_{i<k}\mathrm{sg}((a)_i)$ by ind. on k, using in the ind. step cases ($(a)_k=0$, $(a)_k>0$) and *18.5, *3.12, *3.13 etc.

*20.5. I. Immediate for $a=1$; and for $a>1$, $lh(a) \geq a$ would contradict *20.4 via *3.12, *3.10.

×21.1. $a*b = a \cdot \Pi_{i<lh(b)} p_{lh(a)+i}^{(b)_i}$.

Let Seq(a) be a standard formula such that *22.1 holds.

*22.1. $\vdash \mathrm{Seq}(a) \sim a>0 \ \&\ \forall i_{i<lh(a)}(a)_i>0$.
*22.2. $\vdash \mathrm{Seq}(a) \sim a>0 \ \&\ \forall i_{i \geq lh(a)}(a)_i=0$.
*22.3. $\vdash \mathrm{Seq}(a) \sim a=\Pi_{i<lh(a)} p_i^{(a)_i}$.
*22.4. $\vdash a>0 \ \&\ \forall i_{i<k}(a)_i>0 \ \&\ \forall i_{i \geq k}(a)_i=0 \sim lh(a)=k \ \&\ \mathrm{Seq}(a)$.
*22.5. $\vdash \mathrm{Seq}(2^{s+1})$. *22.6. $\vdash a*1=a$. *22.7. $\vdash \mathrm{Seq}(b) \sim 1*b=b$.
*22.8. $\vdash \mathrm{Seq}(a) \ \&\ \mathrm{Seq}(b) \supset lh(a*b)=lh(a)+lh(b) \ \&\ \mathrm{Seq}(a*b)$.
*22.9. $\vdash \mathrm{Seq}(a) \ \&\ \mathrm{Seq}(b) \ \&\ \mathrm{Seq}(c) \supset (a*b)*c=a*(b*c)$.

Proofs. *22.2. When $a>0$, then by ×20.1 and *B14 with *20.5 $lh(a) = \Sigma_{i<a} sg((a)_i) = \Sigma_{i<lh(a)} sg((a)_i) + \Sigma_{lh(a) \leq i<a} sg((a)_i)$. I. Assume Seq(a), so $a>0 \ \&\ \forall i_{i<lh(a)}(a)_i>0$. Then by *B11 (with ×10.2), $\Sigma_{i<lh(a)} sg((a)_i)=lh(a)$, so by *132 the second term $\Sigma_{lh(a) \leq i<a} sg((a)_i)$ is 0. Thence by *B5 with *10.3 $\forall i_{lh(a) \leq i<a}(a)_i=0$, and by *19.8 $\forall i_{i \geq lh(a)}(a)_i=0$. II. Assume $a>0 \ \&\ \forall i_{i \geq lh(a)}(a)_i=0$. Then by *B5 (with ×10.1) the second term is 0, hence the first term is lh(a), hence by *B12 (with *10.5, *10.4) $\forall i_{i<lh(a)}(a)_i>0$.

*22.3. When $a>0$, then by *19.16 and *B15 with *20.5, $a = \Pi_{i<a} p_i^{(a)_i} = (\Pi_{i<lh(a)} p_i^{(a)_i}) \cdot \Pi_{lh(a) \leq i<a} p_i^{(a)_i}$. I. Assume Seq(a). Then by *22.2 $\forall i_{lh(a) \leq i<a}(a)_i=0$, so by *B13 the second factor $\Pi_{lh(a) \leq i<a} p_i^{(a)_i}$ is 1. II. Assume $a=\Pi_{i<lh(a)} p_i^{(a)_i}$. By *B6 (with *18.5, *3.9) $a =$ (the first factor) > 0. So by *133 the second factor is 1. Hence by *B13 with *3.10 and *19.8 $\forall i_{lh(a) \leq i}(a)_i=0$, and so by *22.2 Seq(a).

*22.4. I. Assume the hyp. By *19.8, $k \leq a$; so by ×20.1, *B14, *B11, *B5 etc., $lh(a)=k$; so by *22.2, Seq(a).

*22.7. By *22.3 (with $lh(1)=0$).

*22.8. Assume Seq(a) & Seq(b). Then $a*b \neq 0$ (*B6 etc.). Using *6.3 (with *B19) $\Pi_{i<lh(b)} p_{lh(a)+i}^{(b)_i} = \Pi_{lh(a) \leq i<lh(a)+lh(b)} p_i^{(b)_{i \dot{-} lh(a)}}$. So using *22.3, *19.11, *19.13, *19.14, *6.19 and *6.3, we obtain the evaluations

$i<lh(a) \supset (a*b)_i=(a)_i$,
$lh(a) \leq i<lh(a)+lh(b) \supset (a*b)_i=(b)_{i \dot{-} lh(a)} \ \&\ i \dot{-} lh(a)<lh(b)$,
$i \geq lh(a)+lh(b) \supset (a*b)_i=0$.

Hence by *22.4, $lh(a*b)=lh(a)+lh(b) \ \&\ \mathrm{Seq}(a*b)$.

*22.9. Assume the hyp. By *22.8 and the evaluations in its proof, *6.17, *6.3 and *6.19,

$$i<lh(a) \supset ((a*b)*c)_i=(a)_i=(a*(b*c))_i,$$
$$lh(a)\leq i<lh(a)+lh(b) \supset ((a*b)*c)_i=(b)_{i\dot{-}lh(a)}=(a*(b*c))_i,$$
$$lh(a)+lh(b)\leq i<lh(a)+lh(b)+lh(c) \supset$$
$$((a*b)*c)_i=(c)_{i\dot{-}(lh(a)+lh(b))}=(a*(b*c))_i,$$
$$i\geq lh(a)+lh(b)+lh(c) \supset ((a*b)*c)_i=0=(a*(b*c))_i.$$

Hence by *19.15, $(a*b)*c=a*(b*c)$.

$^{\times}$23.1. $\bar{\alpha}(x)=\Pi_{i<x}p_i^{\alpha(i)+1}$.

*23.2. $\vdash i<x \sim (\bar{\alpha}(x))_i=\alpha(i)+1$.

*23.3. $\vdash i\geq x \sim (\bar{\alpha}(x))_i=0$.

*23.4. $\vdash y\leq x \sim \bar{\alpha}(y)=\Pi_{i<y}p_i^{(\bar{\alpha}(x))_i}$.

*23.5. $\vdash lh(\bar{\alpha}(x))=x \;\&\; Seq(\bar{\alpha}(x))$.

*23.6. $\vdash Seq(a) \sim \exists\alpha\exists x\, a=\bar{\alpha}(x)$.

*23.7. $\vdash \bar{\alpha}(t+u)=\bar{\alpha}(t)*\overline{(\lambda x\alpha(t+x))}(u)$.

*23.8. $\vdash \bar{\alpha}(x')=\bar{\alpha}(x)\cdot p_x^{\alpha(x)+1}=\bar{\alpha}(x)*2^{\alpha(x)+1}$.

$^{\times}$24.1. $\tilde{\alpha}(x)=\Pi_{i<x}p_i^{\alpha(i)}$.

*24.2. $\vdash i<x \supset (\tilde{\alpha}(x))_i=\alpha(i)$.

*24.3. $\vdash i\geq x \supset (\tilde{\alpha}(x))_i=0$.

*24.4. $\vdash y\leq x \supset \tilde{\alpha}(y)=\Pi_{i<y}p_i^{(\tilde{\alpha}(x))_i}$.

PROOFS. *23.2, *23.3, *23.4. I. By *19.14, *19.13, *B19 (with *23.2), respectively.

*23.5. By *22.4.

*23.6. I. Assume Seq(a). Then $a=\Pi_{i<lh(a)}p_i^{(a)_i}$ [*22.3] $=\Pi_{i<lh(a)}p_i^{((a)_i\dot{-}1)+1}$ [*6.7 with $(a)_i>0$ from Seq(a), *B19] $=\Pi_{i<lh(a)}p_i^{(\lambda x(a)_x\dot{-}1)(i)+1}$ [$^{\times}$0.1, *B19] $=\overline{(\lambda x(a)_x\dot{-}1)}(lh(a))$, whence $\exists\alpha\exists x a=\bar{\alpha}(x)$.

*23.7. $\bar{\alpha}(t+u)=(\Pi_{i<t}p_i^{\alpha(i)+1})\cdot\Pi_{i<u}p_{t+i}^{\alpha(t+i)+1}$ [*B15, *6.3] $=(\Pi_{i<t}p_i^{\alpha(i)+1})\cdot\Pi_{i<u}p_{t+i}^{(\overline{(\lambda x\alpha(t+x))}(u))_i}$ [*23.2, $^{\times}$0.1, *B19] $=\bar{\alpha}(t)*\overline{(\lambda x\alpha(t+x))}(u)$ [$^{\times}$21.1 with *23.5].

*23.8. First use *B4; then $^{\times}$21.1 with *20.3, *B4, *B3, *23.5, *19.9.

5.6. We may want a function defined by recursion for temporary use only (so that we prefer not to add a new formal symbol and axioms), or even for use only in constructing a deduction under assumption formulas preparatory to using a subsidiary deduction rule (cf. IM Chapter V). We can in effect adjoin such a function temporarily by assuming the formula $A(\alpha)$ expressing the recursion equations, preparatory to $\exists$-elim. from the proved formula $\exists\alpha A(\alpha)$ of Lemma 5.3 (b) (α and any other variables free in $A(\alpha)$ to be held constant during this temporary use of α). An illustration occurs in the proof of Lemma 5.3 (c). This serves the purpose of IM Example 9 p. 415; but this comes at a later stage of development, and the greater resources of the present system make it easier to obtain.

Using instead Lemma 5.3 (c), the recursion can be of the course-of-values type (IM § 46 #G).

We may even find it convenient notationally to introduce similarly a variable to stand for an explicitly defined function, preparatory to $\exists$-elim. from Lemma 5.3 (a). An illustration occurs in the proof of Lemma 5.3 (c). (The same effect could be gained by introducing a symbol of abbreviation for the functor $\lambda y p(y)$, for which however we prefer not to use a Greek letter.)

LEMMA 5.3. *Let* y, z *be distinct number variables, and* α *a function variable. Let* p(y), q, r(y, z), r(z) *be terms not containing* α *free, with* α *and* y *free for* z *in* r(y, z) *and in* r(z). *Then*:

(a) $\vdash \exists\alpha\forall y\alpha(y)=p(y)$.
(b) $\vdash \exists\alpha[\alpha(0)=q \;\&\; \forall y\alpha(y')=r(y, \alpha(y))]$.
(c) $\vdash \exists\alpha\forall y\alpha(y)=r(\bar{\alpha}(y))$ *and* $\vdash \exists\alpha\forall y\alpha(y)=r(y, \tilde{\alpha}(y))$.

(If q contains y free, some occurrences of y in the proof and applications of (b) will have to be changed to other variables. By *23.5, $r(y, \bar{\alpha}(y))=r(\bar{\alpha}(y))$ when r(z) is r(lh(z), z).)

PROOFS. (a) By *100 and $^{\times}$0.1 $\{\lambda y p(y)\}(y)=p(y)$, whence by $\forall$- and $\exists$-introd. $\exists\alpha\forall y\alpha(y)=p(y)$.

(b) Let B(c, i, w) be the formula $(c)_i=w$. By IM *171 p. 199,

$$\vdash \exists! w B(c, i, w).$$

By *19.9 $\vdash (p_0^w)_0=w$, whence

$$(\alpha) \qquad \vdash \exists c B(c, 0, w).$$

Using *19.11 etc., *19.14, *19.10, *19.13, *19.9,

$$\vdash \forall i_{i<y'}(c_1)_i=((\Pi_{i<y'}p_i^{(c_1)_i})\cdot p_{y'}^w)_i \;\&\; ((\Pi_{i<y'}p_i^{(c_1)_i})\cdot p_{y'}^w)_{y'}=w,$$

whence

$$(\beta) \qquad \vdash \exists c_2\{\forall i_{i\leq y}\exists u[B(c_1, i, u) \;\&\; B(c_2, i, u)] \;\&\; B(c_2, y', w)\}.$$

Let Q(w), R(y, z, w) be q=w, r(y, z)=w. By *171,

$$\vdash \exists! w Q(w), \qquad\qquad \vdash \exists! w R(y, z, w).$$

Form $P(y, w)$ as $P(y, x_2, \ldots, x_n, w)$ was formed on IM p. 243, but with the present B, Q, R. By Remark 1 p. 244,

(1) $\vdash P(0, w) \sim Q(w)$,
(2) $\vdash P(y', w) \sim \exists z[P(y, z) \;\&\; R(y, z, w)]$,
(3) $\vdash \exists! w P(y, w)$,

whence $\vdash \forall y \exists w P(y, w)$, and by *2.2 $\vdash \exists\alpha\forall y P(y, \alpha(y))$. Assume for $\exists$-elim., $\forall y P(y, \alpha(y))$. Now as in IM top p. 416, $\alpha(0)=q \;\&\; \forall y \alpha(y')=r(y, \alpha(y))$, whence (b) follows by $\exists$-introd. (and the $\exists$-elim.).

(c) WITH $\tilde{\alpha}(y)$. Assume for $\exists$-elim. from (a case of) (b),

(i) $\beta(0)=1 \;\&\; \forall y \beta(y')=\beta(y)\cdot p_y^{r(y,\beta(y))}$.

Assume for $\exists$-elim. from (a),

(ii) $\forall y \alpha(y)=(\beta(y'))_y$.

Using *19.11 (with ˣ24.1, *B6 etc.), *24.3, *19.9,

(iii) $(\tilde{\alpha}(y)\cdot p_y^{r(y,\tilde{\alpha}(y))})_y=r(y, \tilde{\alpha}(y))$.

Now we deduce by induction

(iv) $\beta(y)=\tilde{\alpha}(y)$.

Basis. $\beta(0) = 1$ [(i)] $= \tilde{\alpha}(0)$ [ˣ24.1, *B3]. Ind. step. $\beta(y') = \beta(y)\cdot p_y^{r(y,\beta(y))}$ [(i)] $= \tilde{\alpha}(y)\cdot p_y^{r(y,\tilde{\alpha}(y))}$ [hyp. ind.] $= \tilde{\alpha}(y)\cdot p_y^{(\beta(y'))_y}$ [(iii)] $= \tilde{\alpha}(y)\cdot p_y^{\alpha(y)}$ [(ii)] $= \tilde{\alpha}(y')$ [ˣ24.1, *B4]. — So $\alpha(y) = (\beta(y)\cdot p_y^{r(y,\beta(y))})_y$ [(ii), (i)] $= (\tilde{\alpha}(y)\cdot p_y^{r(y,\tilde{\alpha}(y))})_y$ [(iv)] $= r(y, \tilde{\alpha}(y))$ [(iii)]. By $\forall$- and $\exists$-introd. (and the two $\exists$-elims.), $\exists\alpha\forall y \alpha(y)=r(y, \tilde{\alpha}(y))$.

(c) WITH $\bar{\alpha}(y)$. Apply (c) with $\tilde{\alpha}(y)$, for $r(\Pi_{i<y} p_i^{(z)_i+1})$ as the $r(y, z)$.

5.7. We bring together results permitting alterations of quantifiers, based on #19 in 5.5. Cf. *0.5, *0.6, *2.1a, *2.2a in 4.5–4.6; IM p. 285; Kleene **1955** p. 315, **1959** 2.1. For each $m \geq 0$, let "$\langle a_0, \ldots, a_m\rangle$" abbreviate $\boldsymbol{p}_0^{a_0}\cdot\ldots\cdot\boldsymbol{p}_m^{a_m}$, where $\boldsymbol{p}_i$ is the numeral for the prime number p_i (and "$[a_0, \ldots, a_m]$" abbreviate $\langle a_0+1, \ldots, a_m+1\rangle$, after **1955** 9.2); let "$\langle\alpha_0, \ldots, \alpha_m\rangle$" abbreviate $\lambda x\langle\alpha_0(x), \ldots, \alpha_m(x)\rangle$; and let "$(\alpha)_i$" abbreviate $\lambda x(\alpha(x))_i$ (and "$(\alpha)_{i,j}$" abbreviate $((\alpha)_i)_j$, etc.). The letters here are subject to the obvious stipulations, under the blanket rule of 3.2.

*25.1. $\vdash (\langle a_0, \ldots, a_m\rangle)_{\boldsymbol{i}}=a_i$ $(i = 0, \ldots, m)$.
*25.2. $\vdash (\langle \alpha_0, \ldots, \alpha_m\rangle)_{\boldsymbol{i}}=\alpha_i$ $(i = 0, \ldots, m)$.
*25.3. $\vdash \forall a_0 \ldots \forall a_m A(a_0, \ldots, a_m) \sim \forall a A((a)_0, \ldots, (a)_{\boldsymbol{m}})$.

*25.4. $\vdash \exists a_0 \ldots \exists a_m A(a_0, \ldots, a_m) \sim \exists a A((a)_0, \ldots, (a)_m)$.
*25.5. $\vdash \forall \alpha_0 \ldots \forall \alpha_m A(\alpha_0, \ldots, \alpha_m) \sim \forall \alpha A((\alpha)_0, \ldots, (\alpha)_m)$.
*25.6. $\vdash \exists \alpha_0 \ldots \exists \alpha_m A(\alpha_0, \ldots, \alpha_m) \sim \exists \alpha A((\alpha)_0, \ldots, (\alpha)_m)$.
*25.7. $\vdash \forall a_0 \ldots \forall a_m \exists b A(a_0, \ldots, a_m, b)$
$\sim \exists \alpha \forall a_0 \ldots \forall a_m A(a_0, \ldots, a_m, \alpha(\langle a_0, \ldots, a_m \rangle))$.
*25.8. $\vdash \forall a_0 \ldots \forall a_m \exists \alpha A(a_0, \ldots, a_m, \alpha)$
$\sim \exists \alpha \forall a_0 \ldots \forall a_m A(a_0, \ldots, a_m, \lambda y \alpha(\langle a_0, \ldots, a_m, y \rangle))$.
*25.9. $\vdash \forall i_{i<a} \exists x A(i, x) \sim \exists x \forall i_{i<a} A(i, (x)_i)$.
*25.10. $\vdash \forall i_{i<a} \exists \alpha A(i, \alpha) \sim \exists \alpha \forall i_{i<a} A(i, (\alpha)_i)$.

Proofs. *25.1. By *19.11 (with *18.5, *3.9, *129), *19.9, *19.10 (and Lemma 5.2, by which $p_i = \boldsymbol{p}_i$).

*25.2. By ˣ0.1, *25.1, *0.4.

*25.3. I. Simply by the predicate calculus. II. Assume $\forall a A((a)_0, \ldots, (a)_m)$, and apply $\forall$-elim. with $\langle a_0, \ldots, a_m \rangle$ as the t (IM p. 99), *25.1 and $\forall$-introd. (IM *64).

*25.7. I. Use *25.3 and *2.2; then, prior to $\exists$-elim., assume $\forall a A((a)_0, \ldots, (a)_m, \alpha(a))$ and continue as for *25.3.

*25.8. I. Similarly, using ˣ2.1, and transforming $A(a_0, \ldots, a_m, \lambda y \alpha(\langle\langle a_0, \ldots, a_m \rangle, y \rangle))$ into $A(a_0, \ldots, a_m, \lambda y \{\lambda z \alpha(\langle\langle (z)_0, \ldots, (z)_m \rangle, (z)_{m'} \rangle)\}(\langle a_0, \ldots, a_m, y \rangle))$.

*25.9. I. Assume $\forall i_{i<a} \exists x A(i, x)$. By cases ($i<a$, $\neg i<a$) $\exists x[i<a \supset A(i, x)]$, whence by $\forall$-introd. $\forall i \exists x [i<a \supset A(i, x)]$, whence by *2.2 $\exists \alpha \forall i_{i<a} A(i, \alpha(i))$. Assume $\forall i_{i<a} A(i, \alpha(i))$. Using *24.2 $i<a \supset A(i, (\tilde{\alpha}(a))_i)$, whence by $\forall$- and $\exists$-introd. $\exists x \forall i_{i<a} A(i, (x)_i)$.

Remark 5.4. Alternatively *25.9 can be proved by induction on a, whereupon ˣ2.1 will be used since 4.6 only as follows: via *2.2 in proving Lemma 5.3 (b) (and thence (c)) and *25.7; directly in proving *25.8 (which for $m = 0$ is *2.1a). Cf. 7.15 below.

Now we combine #F (e.g. for $m = 2$) with Lemma 5.3.

Lemma 5.5. (stated for $m = 2$). *Under stipulations like those for Lemma* 5.3:

(a) $Q_1(y) \vee Q_2(y), \neg(Q_1(y) \;\&\; Q_2(y)) \vdash^{y} \exists \alpha \forall y \alpha(y) = \begin{cases} p_1(y) \text{ if } Q_1(y), \\ p_2(y) \text{ if } Q_2(y). \end{cases}$

The first assumption formula expresses that the cases are exhaustive. The other $\binom{m}{2}$ ($= 1$ for $m = 2$) assumption formulas express that the cases are mutually exclusive. The conclusion formula shall be an abbreviation for

$$\exists \alpha \forall y [(Q_1(y) \vee Q_2(y)) \;\&\; (Q_1(y) \supset \alpha(y) = p_1(y)) \;\&\; (Q_2(y) \supset \alpha(y) = p_2(y))].$$

(b) $Q_1(y, \alpha(y)) \vee Q_2(y, \alpha(y)), \neg(Q_1(y, \alpha(y)) \,\&\, Q_2(y, \alpha(y))) \vdash^{y\alpha}$

$$\exists\alpha[\alpha(0)=q \,\&\, \forall y \alpha(y')= \begin{cases} r_1(y, \alpha(y)) \text{ if } Q_1(y, \alpha(y)), \\ r_2(y, \alpha(y)) \text{ if } Q_2(y, \alpha(y))]. \end{cases}$$

(c) $Q_1(\bar{\alpha}(y)) \vee Q_2(\bar{\alpha}(y)), \neg(Q_1(\bar{\alpha}(y)) \,\&\, Q_2(\bar{\alpha}(y))) \vdash^{y\alpha}$

$$\exists\alpha\forall y \alpha(y)= \begin{cases} r_1(\bar{\alpha}(y)) \text{ if } Q_1(\bar{\alpha}(y)), \\ r_2(\bar{\alpha}(y)) \text{ if } Q_2(\bar{\alpha}(y)) \end{cases}$$

and

$$Q_1(y, \tilde{\alpha}(y)) \vee Q_2(y, \tilde{\alpha}(y)), \neg(Q_1(y, \tilde{\alpha}(y)) \,\&\, Q_2(y, \tilde{\alpha}(y))) \vdash^{y\alpha}$$

$$\exists\alpha\forall y \alpha(y)= \begin{cases} r_1(y, \tilde{\alpha}(y)) \text{ if } Q_1(y, \tilde{\alpha}(y)), \\ r_2(y, \tilde{\alpha}(y)) \text{ if } Q_2(y, \tilde{\alpha}(y)). \end{cases}$$

PROOFS. (a) In the SPECIAL CASE that $Q_1(y)$, $Q_2(y)$ are prime formulas, or equivalent to prime formulas by applications of #D and #E, we need only apply Lemma 5.3 (a) for $p(y)$ the term p obtained by using #F with $\vdash Q_i(y) \sim q_i=0$ and with $p_i(y)$ as the p_i ($i = 1, 2$).

However, the GENERAL CASE can be treated directly, thus. The first assumption formula gives two cases. CASE 1: $Q_1(y)$. Then $\neg Q_2(y)$. So

$$(Q_1(y) \vee Q_2(y)) \,\&\, (Q_1(y) \supset p_1(y)=p_1(y)) \,\&\, (Q_2(y) \supset p_1(y)=p_2(y)).$$

By $\exists$-introd.,

$$\exists a[(Q_1(y) \vee Q_2(y)) \,\&\, (Q_1(y) \supset a=p_1(y)) \,\&\, (Q_2(y) \supset a=p_2(y))].$$

CASE 2: $Q_2(y)$. Similarly. — By $\forall$-introd. and *2.2,

$$\exists\alpha\forall y[(Q_1(y) \vee Q_2(y)) \,\&\, (Q_1(y) \supset \alpha(y)=p_1(y)) \,\&\, (Q_2(y) \supset \alpha(y)=p_2(y))].$$

(b) Substituting $(y)_0$, $\lambda t(y)_1$ for y, α, and using ˣ0.1: $Q_1((y)_0, (y)_1) \vee Q_2((y)_0, (y)_1)$ and $\neg(Q_1((y)_0, (y)_1) \,\&\, Q_2((y)_0, (y)_1))$. So, using (a), we can assume for $\exists$-elim.

(i) $$\forall y \rho(y)= \begin{cases} r_1((y)_0, (y)_1) \text{ if } Q_1((y)_0, (y)_1), \\ r_2((y)_0, (y)_1) \text{ if } Q_2((y)_0, (y)_1). \end{cases}$$

Applying Lemma 5.3 (b) with $\rho(\langle y, z\rangle)$ as the $r(y, z)$, assume for $\exists$-elim.

(ii) $$\alpha(0)=q \,\&\, \forall y \alpha(y')=\rho(\langle y, \alpha(y)\rangle).$$

Taking $\langle y, \alpha(y)\rangle$ for y in (i) (by $\forall$-elim.) and using *25.1, the result with (ii) gives $Q_i(y, \alpha(y)) \supset \alpha(y')=r_i(y, \alpha(y))$. By &-, $\forall$- and $\exists$-introds., we obtain the required formula. It does not contain ρ or α free, so the $\exists$-elims. can be completed.

(c) WITH $\bar{\alpha}(y)$. By *158 since Seq(z) is prime, Seq(z) $\vee$ $\neg$Seq(z). Using cases thence, and in the first case *23.6 with $Q_1(\bar{\alpha}(y)) \vee Q_2(\bar{\alpha}(y))$: (i) $(\mathrm{Seq}(z) \;\&\; Q_1(z)) \vee (\mathrm{Seq}(z) \;\&\; Q_2(z)) \vee \neg\mathrm{Seq}(z)$. Using *23.6 with $\neg(Q_1(\bar{\alpha}(y)) \;\&\; Q_2(\bar{\alpha}(y)))$: $\mathrm{Seq}(z) \supset \neg(Q_1(z) \;\&\; Q_2(z))$. Using this and *50, the three cases in (i) are mutually exclusive. Assume for $\exists$-elim. from the result of an application of (a) with $m = 3$,

$$\forall z \rho(z) = \begin{cases} r_1(z) & \text{if } \mathrm{Seq}(z) \;\&\; Q_1(z), \\ r_2(z) & \text{if } \mathrm{Seq}(z) \;\&\; Q_2(z), \\ 0 & \text{if } \neg\mathrm{Seq}(z). \end{cases}$$

Now use Lemma 5.3 (c) with $\rho(z)$ as the $r(z)$ (and later *23.5).

(c) WITH $\tilde{\alpha}(y)$. Apply (c) with $\bar{\alpha}(y)$, for $Q_i(\mathrm{lh}(z), \Pi_{i<\mathrm{lh}(z)} p_i^{(z)_i \dot{-} 1})$, $r_i(\mathrm{lh}(z), \Pi_{i<\mathrm{lh}(z)} p_i^{(z)_i \dot{-} 1})$ as the $Q_i(z)$, $r_i(z)$.

LEMMA 5.6. *Let* x *be a variable, and* A(x) *a formula. Then*

$$\exists! x A(x) \vdash A(x) \vee \neg A(x).$$

PROOF. Assume preparatory to $\exists$-elim. from $\exists! w A(w)$, $A(w) \;\&\; \forall x(A(x) \supset w=x)$. By *158, $w=x \vee w \neq x$. CASE 1: $w=x$. Then A(x), whence $A(x) \vee \neg A(x)$. CASE 2: $w \neq x$. Then $\neg A(x)$, whence again $A(x) \vee \neg A(x)$.

§ 6. Postulate on spreads (the bar theorem). 6.1. In the intuitionistic set theory or analysis of Brouwer, a fundamental role is played by what he called a "set (Menge)" in his early papers on the subject (**1918–9** I p. 3, **1919** pp. 204–205 or 950–951, **1924–7** I pp. 244–245) and more recently a "spread" (**1954** p. 8). There are several versions of the notion of 'spread', differing in details. We begin with a version differing from that of Brouwer's early papers, reproduced in Kleene 1950a § 1 (p. 680 end line 8, add "> 0"), by the omission of what Brouwer called "sterilized (gehemmt)" sequences. In 6.9, we shall consider other versions.

A given *spread* is generated by (i) *choosing* natural numbers in sequence, (either freely or) under an effective restriction which says, given the (numbers chosen in the respective) previous choices if any and any number, whether that number may be chosen next, and (ii) after each choice *correlating* effectively an object (depending on the previous choices if any and that choice) from a fixed countable set. Furthermore, under (i) it is effectively determined after each choice whether (depending on the previous choices if any and that choice) the sequence of choices is to *terminate* therewith or shall continue; in the latter case, the restriction governing the choices must allow at least one natural number to be chosen next.

When a sequence of choices terminates, the *element* of the set or spread correlated to the sequence is the finite sequence of the objects correlated to the choices up to its termination. When a sequence of choices continues unterminated ad infinitum, the *element* correlated to the sequence is the infinite sequence of the objects correlated to the choices; intuitionistically this element is not considered as completed, but only as in process of growth as the choices proceed.

The word "effectively" in the foregoing is intended to convey what Brouwer expressed (in his early papers) by speaking of a "law (Gesetz)"; and indeed in **1924** § 1, **1927** § 2 he used "algorithm (Algorithmus)" in a related connection. What choices are permitted, and whether termination takes place, is determined by a law, which we call the *choice law*. What object is correlated is determined by another law, which we call the *correlation law*. (Cf. Kleene 1950a p. 680, and Heyting **1956** p. 34, where the terminology is a little different.) These two laws each operate upon the finite sequence of the choices (natural numbers) up to and including the one which is under consideration (i.e. the natural number about to be chosen, when the question is whether the choice of it after the choices already made if any is permissible; the one just chosen, when the question is whether the sequence of choices thereupon terminates, or what object is thereupon correlated).

A set or spread is not thought of intuitionistically as the "totality" of its elements, not even in the case all (permitted) choice sequences terminate so that the elements themselves become intuitionistically completed objects. To do so would (in general) involve the completed infinite (IM p. 48); e.g. the spread in which all choice sequences terminate after one choice which is completely free, with the number chosen correlated, is simply the set of all (unit sequences of) natural numbers. A spread from the intuitionistic standpoint is the pair of laws governing the generation process under which its elements grow. Through his notion of 'spread', Brouwer found a way, while maintaining the standpoint of the potential infinite, to deal with collections some of which are even uncountably infinite (of classical cardinal number $2^{\aleph_0}$).

The objects correlated to the choices in Brouwer's applications may be, e.g., natural numbers, rational numbers, intervals with rational endpoints. Since for a given spread they must be chosen from a given countable class of objects, abstractly we can always take them to be natural numbers. When we do so, the notations available in the formal system suffice for the theory of spreads.

Indeed, these notions include the fundamental constituents for dealing with spreads. These constituents can be combined under the formation rules of the system in a flexible manner, so that the particular way of combining them that gives a spread loses some of its preeminence in this formalism. Cf. however 7.8 below.

6.2. In this section, we shall concentrate on the choice sequences, which may underlie a spread, and which can be regarded as themselves constituting a spread by taking for the correlation law the trivial one which correlates the last natural number chosen. If then there is no restriction on the choices, the spread consists simply of all the infinite sequences of natural numbers in process of growth. This Brouwer called the *universal spread*. We study it now.

When exactly t (≥ 0) natural numbers $a_0, a_1, \ldots, a_{t-1}$ have been chosen successively, we have in other words chosen the first t values $\alpha(0), \alpha(1), \ldots, \alpha(t-1)$ of a number-theoretic function $\alpha(x)$, the remaining values of which are still undetermined. Now we may associate with any finite sequence $a_0, \ldots, a_{t-1}$ of natural numbers the natural number $a = p_0^{a_0+1} \cdot \ldots \cdot p_{t-1}^{a_{t-1}+1} = \langle a_0+1, \ldots, a_{t-1}+1 \rangle = [a_0, \ldots, a_{t-1}] = \bar{\alpha}(t)$, whereupon $t = \mathrm{lh}(\bar{\alpha}(t))$, and $a_i = \alpha(i) = (\bar{\alpha}(t))_i \dot{-} 1$ for $i < t$ (cf. #18–#23 in 5.5, and 5.7). This maps the finite sequences of choices 1–1 onto the natural numbers a such that $\mathrm{Seq}(a)$, which we call *sequence numbers*. The theory of choice sequences can now be dealt with in terms of the sequence numbers. The fundamental relation between sequence numbers is that of a sequence number a to the sequence numbers $a*2^{s+1}$ ($s = 0, 1, 2, \ldots$) which represent the sequences $a_0, \ldots, a_{t-1}, s$ coming from $a_0, \ldots, a_{t-1}$ by choosing one more number s; the numbers $a*2^{s+1}$ are thus exactly the numbers $\bar{\alpha}(t+1)$ for the various functions α such that $a = \bar{\alpha}(t)$.

6.3. Because the sequences of choices ("Wahlfolgen" in Brouwer **1918–9** and **1924–7**, "infinitely proceeding sequences" in Brouwer **1952** and Heyting **1956**) are considered intuitionistically as in process of growing by new choices, especial prominence is given in intuitionism to those properties of choice sequences which if possessed can be recognized effectively as possessed at some (finite) stage in the growth of the choice sequence. Such a property of a choice sequence α is of the form $(Ex)R(\bar{\alpha}(x))$ where $R(a)$ is a number-theoretic predicate, effective at least when applied to sequence numbers a.

With respect to such a predicate $R(a)$, we say that, as a choice sequence $\alpha(0), \alpha(1), \alpha(2), \ldots$ is generated, the finite sequence of the choices $\alpha(0), \ldots, \alpha(t-1)$, or the sequence number $\bar{\alpha}(t)$ representing these first t choices, is *secured*, if it is known already from these t choices by the test of the predicate R that α possesses the property $(Ex)R(\bar{\alpha}(x))$, i.e. if $(Ex)_{x \leq t}R(\bar{\alpha}(x))$; *past secured*, if this was known already without the last choice, i.e. if $(Ex)_{x<t}R(\bar{\alpha}(x))$; *immediately secured*, if this is known only after the last choice, i.e. if $(x)_{x<t}\bar{R}(\bar{\alpha}(x))$ & $R(\bar{\alpha}(t))$ (the first conjunctive member can be omitted if R is taken so that, for any α, $R(\bar{\alpha}(x))$ is true of at most one x). We say $\alpha(0), \ldots, \alpha(t-1)$, or $\bar{\alpha}(t)$, is *securable*, if, no matter how the future choices (the $t+1$-st, $t+2$-nd, $t+3$-rd, ...) are made, α will possess the property $(Ex)R(\bar{\alpha}(x))$, i.e. if $(\beta)[\bar{\beta}(t)=\bar{\alpha}(t) \rightarrow (Ex)R(\bar{\beta}(x))]$ or equivalently $(Ex)_{x<t}R(\bar{\alpha}(x)) \vee (\beta)(Ex)R(\bar{\alpha}(t)*\bar{\beta}(x))$. In particular (changing the bound variable β to α), 1 is securable exactly if $(\alpha)(Ex)R(\bar{\alpha}(x))$. We have stated these notions with respect to a fixed predicate $R(a)$. A sequence number w not past secured is securable with respect to $R(a)$ exactly if 1 is securable with respect to $\lambda a\, R(w*a)$. A sequence number w is *barred*, if 1 is securable with respect to $\lambda a\, R(w*a)$, i.e. if $(\alpha)(Ex)R(w*\bar{\alpha}(x))$. (Numbers other than sequence numbers are to be *unsecured, unsecurable, unbarred.*)

6.4. We have used function variables in expressing these notions. But there is a basic difference between the classical and the intuitionistic concepts; for the intuitionists, the functions are not completed. The universal function quantifier (β) or (α), with its scope, in the expression for securability cannot be considered intuitionistically as a conjunction extended over all completed one-place number-theoretic functions, as it is classically. The intuitionistic meaning of $(\alpha)(Ex)R(\bar{\alpha}(x))$ is that, whenever one chooses successively natural numbers $\alpha(0), \alpha(1), \alpha(2), \ldots$ in any way, one must eventually encounter an x such that $R(\bar{\alpha}(x))$.

How then can the intuitionists utilize the notion of securability? To begin with, they can particularize, compatibly with their interpretation of (α), from $(\alpha)(Ex)R(\bar{\alpha}(x))$ to $(Ex)R(\bar{\alpha}_1(x))$ for such particular choice sequences α_1 as they can specify; these, in connection with which Brouwer (**1952** p. 143, **1954** p. 7) uses the term "sharp arrows", are ones whose growth can be completely governed in advance by a law (after any $t \geq 0$ choices, the law allows exactly one next choice). We have the formal counterpart of this in Axiom Schema 10F, where the functors u express primitive recursive functions in the case they contain no function variables (by Lemma 3.3).

But it would seem that this makes rather weak use of $(\alpha)(Ex)R(\bar{\alpha}(x))$. In fact, under the interpretation that an α_1 giving a sharp arrow is a general recursive function, $(\alpha_1)(Ex)R(\bar{\alpha}_1(x))$ is in general weaker than $(\alpha)(Ex)R(\bar{\alpha}(x))$; and the important "fan theorem" (in 6.10 below) fails when its hypothesis is weakened in the corresponding manner (Kleene 1950a § 3, or Lemma 9.8 below). The intuitionists may refrain from adopting this interpretation, but they are in no position to refute it, since their actual constructions or laws conform to it (Chapter II below). Pursuing the matter further from the classical standpoint, while the fan theorem becomes true upon enlarging the class of α's to the arithmetical functions (those such that $\alpha(x)=w$ is an arithmetical predicate IM p. 239; cf. Lemma 9.12 below), in order to exhaust the full force of $(\alpha)(Ex)R(\bar{\alpha}(x))$ not even all the hyperarithmetical functions suffice (Kleene **1955b** pp. 210, 208 with **1959** p. 48; or **1959b**).

REMARK 6.1. In the intuitionistic system, using *158, we can prove $\forall\alpha\forall x(\alpha(x)=0 \vee \alpha(x)\neq 0)$, which seems to imply that any function α taking only 0 and 1 as values is recursive. (More generally, we can prove $\forall\alpha\forall x\forall w(\alpha(x)=w \vee \alpha(x)\neq w)$, which seems to say that, for each α, the predicate $\alpha(x)=w$ is decidable, so presumable recursive, so by IM Theorem III p. 279 the function α, $= \lambda x\, \mu w \alpha(x)=w$, is recursive.) On the other hand, as noted, we cannot interpret the universal quantifier (α) to mean "for all recursive functions α" without making the fan theorem of intuitionism false. This apparent contradiction is explained thus. As we choose the numbers $\alpha(0), \alpha(1), \alpha(2), \ldots$ making up any choice sequence α, it will be known after each choice what number has been chosen; it is in this sense that $\forall\alpha\forall x(\alpha(x)=0 \vee$

$\alpha(x) \neq 0)$ is true. But as a choice sequence $\alpha(0), \alpha(1), \alpha(2), \ldots$ grows, in advance of each choice any number in the case of the universal spread (any number ≤ 1 in the case of the spread of choice sequences governed by $(x)\alpha(x) \leq 1$) is eligible to be chosen; so the α is not restricted to be a recursive function.

REMARK 6.2. In the present classical system with the same formation rules as the intuitionistic, the functors u available for Axiom Schema 10F are the same. (This is not so in classical systems like the ones in Hilbert-Bernays 1939 Supplement IV having a choice operator ε or descriptive operator ι.) Fuller use of assumptions $\forall\alpha A(\alpha)$ is obtained in the classical system via indirect proofs.

6.5. Brouwer found a solution to the problem of how to utilize an hypothesis of securability more fully than by Axiom Schema 10F. This consists in looking at the situation from the opposite direction, proceeding backwards from those sequence numbers $\bar{\alpha}(x)$ for which $R(\bar{\alpha}(x))$ to the other sequence numbers having such numbers in all their (sufficiently continued) extensions.

To fix our ideas, let us confine our attention for the moment to sequence numbers not past secured (so that, in any sequence α of choices, we don't overrun the first x at which we find $R(\bar{\alpha}(x))$ true). Then, slightly paraphrasing Brouwer **1927** FOOTNOTE 7 to make it read in our notation and terminology: Thought through intuitionistically, this securability is nothing else than the property which is defined thus. It holds for every sequence number a such that $R(a)$. It holds for any sequence number a, if for every s $(s = 0, 1, 2, \ldots)$ it holds for $a*2^{s+1}$. This remark draws after it immediately the well-orderedness property ....

In other words, Brouwer's Footnote 7 says that securability is that property (of sequence numbers not past secured) which originates at the immediately secured sequence numbers, and propagates back to the unsecured but securable numbers across the junctions between a sequence number a and its immediate extensions $a*2^{s+1}$ $(s = 0, 1, 2, \ldots)$.

Let us review the situation using a geometrical picture (Figure 1). We can represent the universal spread 6.2 by a "tree", with the sequence numbers $a = p_0^{a_0+1} \cdot \ldots \cdot p_{t-1}^{a_{t-1}+1} = [a_0, \ldots, a_{t-1}]$ at the vertices. The initial (leftmost) vertex is occupied by the sequence number $1 = [\,] = \bar{\alpha}(0)$. From any vertex, occupied by the sequence number a,

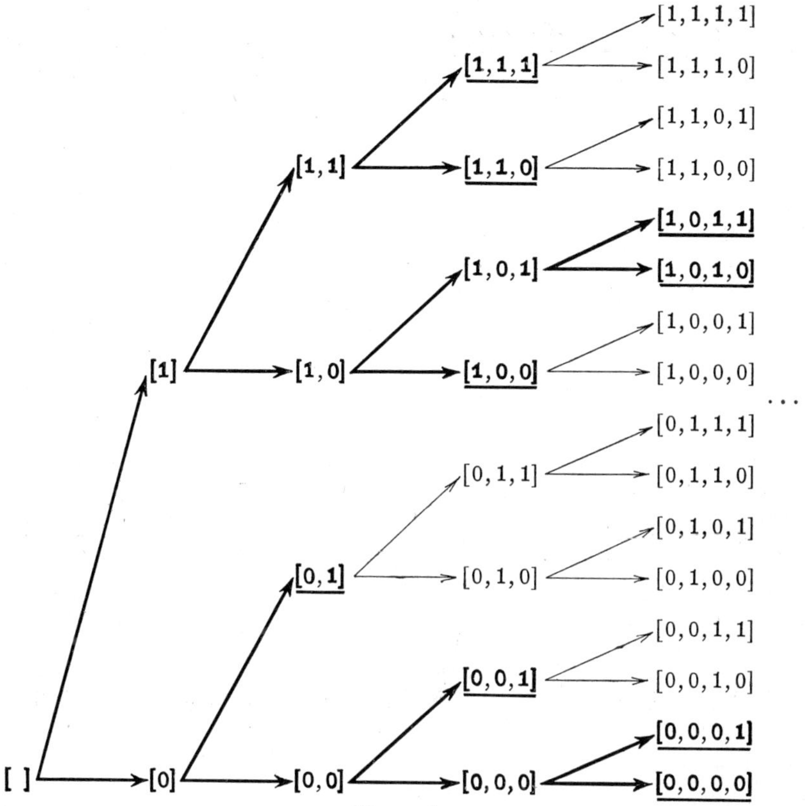

Figure 1.

infinitely many arrows lead to the next vertices, occupied by the sequence numbers $a*2^{s+1}$ ($s = 0, 1, 2, \ldots$). A part of this tree is shown in Figure 1; but the arrows for $s > 1$ are left to our imagination, as well as the vertices for $t = \text{lh}(a) > 4$ suggested by the dots. (The figure actually shows the "binary spread" or "binary fan" 6.10 as far as its vertices with $\text{lh}(a) \leq 4$.)

An infinite choice sequence α or $\alpha(0), \alpha(1), \alpha(2), \ldots$ is represented by an infinite path in the tree, starting at the leftmost vertex (occupied by) [ ] and following arrows; a finite sequence of choices by an initial segment of such a path, or by the vertex $\bar{\alpha}(t)$ at the (right) end of that segment. Thus, before $\alpha(0)$ is chosen, we are at the vertex [ ]; then if we choose $\alpha(0) = 1$, we move to the vertex [1]; choosing next $\alpha(1) = 0$, we continue to [1, 0]; choosing $\alpha(2) = 1$, to [1, 0, 1]; choosing $\alpha(3) = 1$, to [1, 0, 1, 1]; etc.

Consider a predicate $R(a)$, effective at least when applied to sequence numbers a. For each α, let us follow the corresponding path in the tree (starting from $[\,]$) until we first encounter a vertex $\bar{\alpha}(x)$ for which $R(\bar{\alpha}(x))$, if we ever do, whereupon we underline that vertex. In the language of 6.3, we underline (the vertices occupied by) the immediately secured sequence numbers.

Now $(\alpha)(Ex)R(\bar{\alpha}(x))$, as we considered it in 6.3 and (intuitionistically) in 6.4, means geometrically that, along each infinite path starting from the leftmost vertex $[\,]$ and following arrows, we will encounter an underlined vertex. This is illustrated in Figure 1, so far as it can be shown with only the arrows for $s = 0, 1$. More generally, $(\beta)(Ex)R(a*\bar{\beta}(x))$ or in words *a is securable* (*but not past secured*) means geometrically that, along each infinite path starting from the vertex occupied by a and following the arrows, we will encounter an underlined vertex.

Brouwer's reversal of the direction consists in replacing this meaning of *a is securable* (*but not past secured*) by that of belonging to the class of sequence numbers which is defined to include the ones underlined, and to include a whenever it includes all $a*2^{s+1}$ for $s = 0, 1, 2, \ldots$, but to include no other sequence numbers. (This definition is an example of an inductive definition, in the terminology of IM § 53.)

In Figure 1, the securable but not past secured sequence numbers are those which are in bold face (heavy type), if we suppose appropriate behavior along paths containing arrows with $s \geq 2$. But under the first meaning of *securable* (*but not past secured*), which we now call the *explicit sense*, a vertex's being in bold face means that proceeding rightward from it in the direction of arrows along all possible divergent paths an underlined vertex will be encountered. Under the second meaning (the *inductive sense*), a vertex's being in bold face signifies its membership in the class of vertices generated by putting into the class the underlined vertices, and proceeding in the leftward or convergent direction (reverse to arrows) to include a in the class whenever all $a*2^{s+1}$ ($s = 0, 1, 2, \ldots$) are included in the class.

Since the explicit sense, which our symbols in 6.3 directly express, had already been used before Brouwer's **1927** Footnote 7 was introduced, that Footnote 7 must come to this: The two meanings of *securable* (*but not past secured*) are equivalent; and this equivalence is given by intuition (by thinking the matter through intuitionistically). We agree with him.

In the figure, whether one puts vertices in bold face by the criterion of finding an underlined vertex at or to the right of them along all paths, or moves leftward across the figure putting vertices in bold face by the two principles generating a class of vertices, the result is the same.

One of the implications in this equivalence is actually unproblematical, i.e. easily proved (cf. end 6.7). The other implication, that by *securable* (*but not past secured*) in the explicit sense of *securable* (*but not past secured*) in the inductive sense (or of the "well-orderedness property", which the latter entails immediately) is essentially what Brouwer subsequently called the "bar theorem" (**1954** p. 14, cf. Remark 6.3 below).

In **1924** § 1 (cf. **1924a** §§ 1, 2), in the text of **1927** § 2, and in **1954**, he used a more complicated analysis to prove the bar theorem. Footnote 7 of **1927** concluded, "The proof carried through in the text for the latter property [well-orderedness] seems to me nevertheless of interest on account of the propositions included in its line of thought."

We shall simply introduce what is needed here by an axiom schema ×26.3 which gives the effect of the bar theorem for the case of the universal spread. This schema takes the form of a principle of induction which attributes to the securable (but not past secured) sequence numbers any property expressible in the symbolism of the system which originates and propagates in the same way as the securability property itself (under the inductive sense). Our procedure amounts to adopting Brouwer's **1927** Footnote 7 in place of the more elaborate treatment in the text of **1927**.

We thus quickly get over a moot point in Brouwer's deduction of his analysis by postulating an axiom schema. This may strike some as an evasion. But this axiom schema is independent of the other intuitionistic postulates, as we shall see in Corollary 9.9 (and 9.2, by which its negation is unprovable). So there can be a question of deriving the axiom schema (the bar theorem), only if we first substitute another postulate to derive it from. We are unconvinced that any known substitute is more fundamental and intuitive. However, in view of the attention which the proof in Brouwer's text of **1927** has continued to receive, we shall also examine that, in 6.12.

6.6. We consider now just how to state the bar theorem in the formal symbolism.

The definition of a property in Brouwer's Footnote 7 reads, under the restriction there to sequence numbers not past secured, as an inductive definition of the securable (sequence) numbers. If we substitute "*a* which is secured" for "*a* such that $R(a)$", then without the restriction it reads as an inductive definition of all the securable numbers (Kleene **1955a** p. 416). If we omit the restriction, but require R to be a predicate such that, for any α, $R(\bar{\alpha}(x))$ for at most one x, it reads as an inductive definition of the numbers securable but not past secured. If we simply omit the restriction, it reads as an inductive definition of the barred numbers. It makes little difference to us here which reading we use, and the last is the simplest.

We also obtain some simplification by stating the induction principle corresponding to the inductive definition only for inferring properties of 1 (for which 'securable', 'securable but not past secured' and 'barred' are equivalent). We do not lose thereby, as we shall verify in 6.11.

For securability in the explicit sense of 6.3 we now write "securable$_{\mathrm{E}}$", in the inductive sense of Footnote 7 "securable$_{\mathrm{I}}$". The bar theorem is then the implication

$$(*) \qquad \text{securable}_{\mathrm{E}} \rightarrow \text{securable}_{\mathrm{I}},$$

when the right side is rendered by the principle of induction corresponding to the inductive definition (cf. IM § 53). We want to formalize this, applied to 1, with respect to R.

The left side of (*) is then simply $(\alpha)(Ex)R(\bar{\alpha}(x))$.

Let $\mathfrak{J}(R, A)$ be $(a)[\mathrm{Seq}(a) \,\&\, R(a) \rightarrow A(a)] \,\&\, (a)[\mathrm{Seq}(a) \,\&\, (s)A(a{*}2^{s+1}) \rightarrow A(a)] \rightarrow A(1)$; and for any formulas $\mathrm{A(a)}$ and $\mathrm{R(a)}$, let $\mathfrak{J}(\mathrm{R, A})$ be the correspondingly constructed formula. The principle of induction rendering the right side of (*) is $(A)\mathfrak{J}(R, A)$.

Thus we render (*) in informal symbolism as $(\alpha)(Ex)R(\bar{\alpha}(x)) \rightarrow (A)\mathfrak{J}(R, A)$. Expressing this in the formal symbolism as nearly as we can in the absence of predicate variables (cf. IM p. 432), we are led to $\forall\alpha\exists \mathrm{x}\mathrm{R}(\bar{\alpha}(\mathrm{x})) \supset \mathfrak{J}(\mathrm{R, A})$, which (trivially rearranged) is *26.1.

6.7. Before postulating a slight restriction of this for the basic system or the intuitionistic system, we verify that it is provable in the classical system. The proof is a formalization of the classical proof of (*) in Kleene **1955a** (E) p. 417.

If a, s, x are any number variables (a and s distinct), α is any function variable, A(a) is any formula not containing s free in which s is free for a, and R(a) is any formula not containing α or x free in which α and x are free for a:

*26.1°. $\vdash \forall\alpha\exists x R(\bar{\alpha}(x))\ \&\ \forall a[\mathrm{Seq}(a)\ \&\ R(a) \supset A(a)]\ \&$
$\forall a[\mathrm{Seq}(a)\ \&\ \forall s A(a*2^{s+1}) \supset A(a)] \supset A(1)$.

Proof. By the classical propositional calculus, it will suffice to assume

(a) $\forall a[\mathrm{Seq}(a)\ \&\ R(a) \supset A(a)]$,
(b) $\forall a[\mathrm{Seq}(a)\ \&\ \forall s A(a*2^{s+1}) \supset A(a)]$,
(c) $\neg A(1)$,

and deduce $\neg\forall\alpha\exists x R(\bar{\alpha}(x))$, which by the classical predicate calculus (*85, *86) is equivalent to $\exists\alpha\forall x\neg R(\bar{\alpha}(x))$. Likewise (b) is equivalent to $\forall a[\mathrm{Seq}(a)\ \&\ \neg A(a) \supset \exists s\neg A(a*2^{s+1})]$, whence by *97 $\forall a\exists s[\mathrm{Seq}(a)\ \&\ \neg A(a) \supset \neg A(a*2^{s+1})]$, whence by *2.2 $\exists\sigma\forall a[\mathrm{Seq}(a)\ \&\ \neg A(a) \supset \neg A(a*2^{\sigma(a)+1})]$. Assume for $\exists$-elim. from this

(d) $\forall a[\mathrm{Seq}(a)\ \&\ \neg A(a) \supset \neg A(a*2^{\sigma(a)+1})]$.

By Lemma 5.3 (c), $\exists\alpha\forall x\alpha(x)=\sigma(\bar{\alpha}(x))$; so assume

(e) $\forall x\alpha(x)=\sigma(\bar{\alpha}(x))$.

Now we deduce by induction

(f) $\neg A(\bar{\alpha}(x))$.

Basis. By *23.1 and *B3, $\bar{\alpha}(0)=1$. So by (c), $\neg A(\bar{\alpha}(0))$. Ind. step. $\bar{\alpha}(x') = \bar{\alpha}(x)*2^{\alpha(x)+1}$ [*23.8] $= \bar{\alpha}(x)*2^{\sigma(\bar{\alpha}(x))+1}$ [(e)]. So by (d) with the hyp. ind. and *23.5, $\neg A(\bar{\alpha}(x'))$. – By (f) and (a) with *23.5, $\neg R(\bar{\alpha}(x))$. By $\forall$- and $\exists$-introd., $\exists\alpha\forall x\neg R(\bar{\alpha}(x))$.

The converse implication

(**) $$\text{securable}_{\mathrm{I}} \rightarrow \text{securable}_{\mathrm{E}}$$

(Kleene **1955a** (D) p. 416) is $(A)\mathfrak{J}(R, A) \rightarrow (\alpha)(Ex)R(\bar{\alpha}(x))$. This holds intuitionistically, a fortiori from $\mathfrak{J}(R, A_1) \rightarrow (\alpha)(Ex)R(\bar{\alpha}(x))$ where $A_1 = \lambda a\ (\alpha)(Ex)R(a*\bar{\alpha}(x))$. So *26.2 can be considered as giving (**) in the basic system.

If a, s, x are any distinct number variables, α is any function variable, and R(a) is any formula not containing x, s, α free in which x, s, α are free for a:

*26.2. $\vdash \{\forall a[\mathrm{Seq}(a) \;\&\; R(a) \supset \forall\alpha\exists x R(a*\bar{\alpha}(x))] \;\&$
$\forall a[\mathrm{Seq}(a) \;\&\; \forall s\forall\alpha\exists x R((a*2^{s+1})*\bar{\alpha}(x)) \supset \forall\alpha\exists x R(a*\bar{\alpha}(x))]$
$\supset \forall\alpha\exists x R(1*\bar{\alpha}(x))\} \supset \forall\alpha\exists x R(\bar{\alpha}(x))$.

PROOF. Using $a=a*\bar{\alpha}(0)$ (by *22.6 with ^{x}23.1, *B3),

(a) $\forall a[\mathrm{Seq}(a) \;\&\; R(a) \supset \forall\alpha\exists x R(a*\bar{\alpha}(x))]$.

Toward (b) below, assume **(i)** $\mathrm{Seq}(a)$ and **(ii)** $\forall s\forall\alpha\exists x R((a*2^{s+1})*\bar{\alpha}(x))$. Using (ii), $\exists x R((a*2^{\alpha(0)+1})*\overline{\{\lambda x\alpha(1+x)\}}(x))$. Assume preparatory to $\exists$-elim. **(iii)** $R((a*2^{\alpha(0)+1})*\overline{\{\lambda x\alpha(1+x)\}}(x))$. But $(a*2^{\alpha(0)+1})*\overline{\{\lambda x\alpha(1+x)\}}(x)$ $= a*(2^{\alpha(0)+1}*\overline{\{\lambda x\alpha(1+x)\}}(x))$ [*22.9 with (i), *22.5, *23.5] $=$ $a*\bar{\alpha}(1+x)$ [23.7 with ^{x}23.1, *B4, *B3, *127]. So by $\exists$-introd., (completing) the $\exists$-elim., and $\forall$-introd., $\forall\alpha\exists x R(a*\bar{\alpha}(x))$. By &-elim. and $\supset$- and $\forall$-introd.,

(b) $\forall a[\mathrm{Seq}(a) \;\&\; \forall s\forall\alpha\exists x R((a*2^{s+1})*\bar{\alpha}(x)) \supset \forall\alpha\exists x R(a*\bar{\alpha}(x))]$.

Assuming the antecedent of the main implication of *26.2, and using (a) and (b), we obtain $\forall\alpha\exists x R(1*\bar{\alpha}(x))$, whence the consequent $\forall\alpha\exists x R(\bar{\alpha}(x))$ follows by *22.7 with *23.5.

6.8. The restriction that *R* be an effective predicate, introduced beginning 6.3 (but immaterial from the classical standpoint), must be made explicit in postulating the bar theorem (*) for the basic system or the intuitionistic system. As expressed by *26.1 simply, (*) is inconsistent with the further intuitionistic postulate ^{x}27.1 to be introduced in § 7, by *27.23. We give four forms ^{x}26.3a–^{x}26.3d of the new axiom schema. Whichever one is introduced now as the postulate, all axioms by each of the others become provable. When it is immaterial which one we cite, we call it simply ^{x}26.3. The stipulations for ^{x}26.3a and ^{x}26.3c are the same as for *26.1. For ^{x}26.3b, α and ρ are any distinct function variables, etc.

^{x}26.3a. $\forall a[\mathrm{Seq}(a) \supset R(a) \vee \neg R(a)] \;\&\; \forall\alpha\exists x R(\bar{\alpha}(x)) \;\&$
$\forall a[\mathrm{Seq}(a) \;\&\; R(a) \supset A(a)] \;\&\; \forall a[\mathrm{Seq}(a) \;\&\; \forall s A(a*2^{s+1}) \supset A(a)]$
$\supset A(1)$.

^{x}26.3b. $\forall\alpha\exists x \rho(\bar{\alpha}(x))=0 \;\&$
$\forall a[\mathrm{Seq}(a) \;\&\; \rho(a)=0 \supset A(a)] \;\&\; \forall a[\mathrm{Seq}(a) \;\&\; \forall s A(a*2^{s+1}) \supset A(a)]$
$\supset A(1)$.

^{x}26.3c. $\forall\alpha\exists!x R(\bar{\alpha}(x))$ &
$\forall a[\mathrm{Seq}(a) \;\&\; R(a) \supset A(a)] \;\&\; \forall a[\mathrm{Seq}(a) \;\&\; \forall s A(a*2^{s+1}) \supset A(a)]$
$\supset A(1)$.

^{x}26.3d. $\forall\alpha\exists x[R(\bar{\alpha}(x)) \;\&\; \forall y_{y<x}\neg R(\bar{\alpha}(y))]$ &
$\forall\alpha\forall x[R(\bar{\alpha}(x)) \;\&\; \forall y_{y<x}\neg R(\bar{\alpha}(y)) \supset A(\bar{\alpha}(x))]$ &
$\forall a[\mathrm{Seq}(a) \;\&\; \forall s A(a*2^{s+1}) \supset A(a)] \supset A(1)$.

DERIVATION OF ^{x}26.3b FROM ^{x}26.3a. Taking $R(a)$ in ^{x}26.3a as $\rho(a)=0$, we have $R(a) \vee \neg R(a)$ by *158, a fortiori $\forall a[\mathrm{Seq}(a) \supset R(a) \vee \neg R(a)]$.

^{x}26.3a FROM ^{x}26.3b. Assume the four hypotheses **(a)–(d)** of ^{x}26.3a. By *158, because $\mathrm{Seq}(a)$ is prime, $\mathrm{Seq}(a) \vee \neg\mathrm{Seq}(a)$. Using cases thence, and in the first case subcases from (a), $(\mathrm{Seq}(a) \;\&\; R(a)) \vee \neg(\mathrm{Seq}(a) \;\&\; R(a))$. Using this with *50 to apply Lemma 5.5 (a), assume preparatory to $\exists$-elim. from the result

$$\forall a \rho(a) = \begin{cases} 0 \text{ if } \mathrm{Seq}(a) \;\&\; R(a), \\ 1 \text{ if } \neg(\mathrm{Seq}(a) \;\&\; R(a)). \end{cases}$$

Now $\mathrm{Seq}(a) \supset (R(a) \sim \rho(a)=0)$, using which and *23.5 the three hypotheses of ^{x}26.3b follow from (b)–(d).

^{x}26.3c FROM ^{x}26.3a. Assume $\forall\alpha\exists!x R(\bar{\alpha}(x))$. Assume $\mathrm{Seq}(a)$, so via *23.6 we can put $a=\bar{\alpha}(x)$ (i.e. we assume this preparatory to $\exists$-elims.). Using Lemma 5.6, $R(\bar{\alpha}(x)) \vee \neg R(\bar{\alpha}(x))$, whence $R(a) \vee \neg R(a)$. By (completing) the $\exists$-elims., $\supset$- and $\forall$-introd., $\forall a[\mathrm{Seq}(a) \supset R(a) \vee \neg R(a)]$. Also, $\forall\alpha\exists x R(\bar{\alpha}(x))$.

^{x}26.3a FROM ^{x}26.3c. Assume the four hyps. **(a)–(d)** of ^{x}26.3a. Let $R'(a)$ be $R(a) \;\&\; \forall y_{y<lh(a)}\neg R(\Pi_{i<y} p_i^{(a)_i})$, so using *23.5 and *23.4 $R'(\bar{\alpha}(x)) \sim R(\bar{\alpha}(x)) \;\&\; \forall y_{y<x}\neg R(\bar{\alpha}(y))$. By *23.5 $\mathrm{Seq}(\bar{\alpha}(x))$, so (a) gives $R(\bar{\alpha}(x)) \vee \neg R(\bar{\alpha}(x))$. Thence by *149a and *174b $\forall\alpha[\exists x R(\bar{\alpha}(x)) \supset \exists!x R'(\bar{\alpha}(x))]$, and by *69 $\forall\alpha\exists x R(\bar{\alpha}(x)) \supset \forall\alpha\exists!x R'(\bar{\alpha}(x))$. So we have $\forall\alpha\exists!x R'(\bar{\alpha}(x))$. Since $R'(a) \supset R(a)$, we also have $\forall a[\mathrm{Seq}(a) \;\&\; R'(a) \supset A(a)]$. Now we can apply ^{x}26.3c with R' as the R.

^{x}26.3d FROM ^{x}26.3c. Use *174b.

6.9. The foregoing induction principle ^{x}26.3 takes care of the bar theorem for the universal spread. We should like it also for other spreads of choice sequences.

So instead of dealing with the class of all the sequence numbers *a*, characterized by Seq(*a*), we shall now deal with any suitable subclass of them, which we shall characterize by $\sigma(a)=0$ for some function σ.

For simplicity, we may omit from consideration terminated sequences of choices (cf. 6.1), so this σ will serve as the choice law (the other function of the choice law in 6.1, to say when a sequence of choices terminates, is suppressed). We may do this here without loss, since we are interested only in what happens up to an x such that $R(\bar{\alpha}(\mathrm{x}))$. Indeed in general, with a simplified choice law σ that doesn't provide for termination, we can still obtain the effect of termination, either (a) by using a predicate R and considering $\alpha(0), \alpha(1), \alpha(2), \ldots$ to terminate at $\alpha(x-1)$ for the least x if any such that $R(\bar{\alpha}(x))$, or (b) for spreads with a non-trivial correlation law, by using positive integers as (or to represent) the objects which we are interested in correlating, and correlating 0 otherwise (essentially Brouwer **1924–7** I Footnote 1).

In our theory of choice sequences we have been using to advantage the empty sequence, represented by the sequence number $\bar{\alpha}(0) = 1$. (Brouwer employed neither the empty sequence, nor sequence numbers.) For spreads all of whose elements are to be sequences with the same first member, we find it convenient to correlate that first member to the empty choice sequence. Then the correlation law ρ operates simply on all sequence numbers a with $\sigma(a) = 0$. When we don't want the elements all to begin with the same first member (correlated to 1), we may simply ignore what $\rho(1)$ is. But whether we do or do not wish to consider $\rho(1)$ as first member of the elements, it seems to us natural to take advantage of our empty sequence by letting the spread be non-empty exactly when the empty sequence is permitted, i.e. when $\sigma(1) = 0$. Thus the choice law suffices itself for deciding whether a spread is empty or not.

When we thus both omit terminated sequences and use the empty sequence to test for a spread's not being empty, we are led to the following formula Spr(σ) expressing in the formal symbolism the restrictions on σ that it characterize the choice sequences for a spread.

$$\mathrm{Spr}(\sigma): \quad \forall \mathrm{a}[\sigma(\mathrm{a})=0 \supset \mathrm{Seq}(\mathrm{a})] \;\&\; \forall \mathrm{a}[\sigma(\mathrm{a})=0 \supset \exists \mathrm{s}\sigma(\mathrm{a}*2^{\mathrm{s}+1})=0]$$
$$\& \; \forall \mathrm{a}[\mathrm{Seq}(\mathrm{a}) \;\&\; \sigma(\mathrm{a})>0 \supset \forall \mathrm{s}\sigma(\mathrm{a}*2^{\mathrm{s}+1})>0].$$

In *26.4 we state the bar theorem for spreads generally, using this version of the notion of 'spread'. The second hypothesis $\sigma(1)=0$ expresses that the spread is not empty.

If we were simply to omit terminated sequences (which would give the version of Heyting **1956** pp. 34–35, = essentially Brouwer **1924–7**

I Footnote 2), we would use instead of Spr(σ) the formula Spd(σ) obtained from it by prefixing $\sigma(1)=0$ & and replacing the second $\forall a$ by $\forall a_{a>1}$. *26.4 would become *26.4′ with $\exists s \sigma(2^{s+1})=0$ replacing $\sigma(1)=0$ to express the non-emptiness of the spread.

Under the version of 'spread' in Brouwer **1954**, all spreads are non-empty.

That a choice sequence α is permitted by the choice law σ of a spread is expressed formally by $\forall x \sigma(\bar{\alpha}(x))=0$, which we abbreviate as "$\alpha \in \sigma$".

The form *26.4a of *26.4 corresponds to ×26.3a and is proved from it; using instead ×26.3b–×26.3d, corresponding forms *26.4b–*26.4d are obtained (not written out when clear). Also, from any one of *26.4a–*26.4d the others can be derived (using only Postulate Groups A–D), as with *26.3. Similarly with *26.6, *26.7 and *26.8 below.

*26.4a. $\vdash \mathrm{Spr}(\sigma) \ \& \ \sigma(1)=0 \ \& \ \forall a[\sigma(a)=0 \supset R(a) \vee \neg R(a)] \ \&$
$\forall \alpha_{\alpha \in \sigma} \exists x R(\bar{\alpha}(x)) \ \& \ \forall a[\sigma(a)=0 \ \& \ R(a) \supset A(a)] \ \&$
$\forall a[\sigma(a)=0 \ \& \ \forall s\{\sigma(a*2^{s+1})=0 \supset A(a*2^{s+1})\} \supset A(a)] \supset A(1).$

*26.4d. $\vdash \mathrm{Spr}(\sigma) \ \& \ \sigma(1)=0 \ \& \ \forall \alpha_{\alpha \in \sigma} \exists x[R(\bar{\alpha}(x)) \ \& \ \forall y_{y<x} \neg R(\bar{\alpha}(y))] \ \&$
$\forall \alpha \forall x[\sigma(\bar{\alpha}(x))=0 \ \& \ R(\bar{\alpha}(x)) \ \& \ \forall y_{y<x} \neg R(\bar{\alpha}(y)) \supset A(\bar{\alpha}(x))] \ \&$
$\forall a[\sigma(a)=0 \ \& \ \forall s\{\sigma(a*2^{s+1})=0 \supset A(a*2^{s+1})\} \supset A(a)] \supset A(1).$

Proof of *26.4a. In I, we shall set up a mapping of the universal spread onto the spread characterized by σ. Thus, to each element α of the universal spread, the function α_γ $(= \lambda t\, (\gamma(\bar{\alpha}(t')))_t \dot{-} 1)$ will belong to the spread σ, as shown by (ε). If α already belongs to σ, $\alpha_\gamma = \alpha$, as shown by (η). (We give (ζ) and (η) for use in proving *26.7a, *27.4 etc.) In II, this mapping carries the bar theorem for the universal spread into the bar theorem for σ.

I. Assume the first two hypotheses of *26.4a, call them **(1)** and **(2)**. By cases from $\sigma(a)=0 \vee \sigma(a)\neq 0$ (by *158), using (1), $\exists s[\sigma(a)=0 \supset \sigma(a*2^{s+1})=0]$, whence by $\forall$-introd. and *2.2 $\exists \pi \forall a[\sigma(a)=0 \supset \sigma(a*2^{\pi(a)+1})=0]$. Assume

(α) $\quad \forall a[\sigma(a)=0 \supset \sigma(a*2^{\pi(a)+1})=0].$

In the following formula (β), the case hypotheses are exhaustive (by cases from applications of *158, since the components are prime) and mutually exclusive (using *50). So Lemma 5.5 (c) applies (indeed, the special case), and we assume (preparatory to $\exists$-elim. from the result)

$$(\beta)\quad \forall a \gamma(a)=\begin{cases} 0 \text{ if } \neg\mathrm{Seq}(a), \\ 1 \text{ if } \mathrm{Seq}(a) \,\&\, \mathrm{lh}(a)=0, \\ (\tilde{\gamma}(a))_B * 2^{S+1} \text{ if } \mathrm{Seq}(a) \,\&\, \mathrm{lh}(a)\neq 0 \,\&\, \sigma((\tilde{\gamma}(a))_B * 2^{S+1})=0, \\ (\tilde{\gamma}(a))_B * 2^{\pi((\tilde{\gamma}(a))_B)+1} \\ \qquad\qquad \text{if } \mathrm{Seq}(a) \,\&\, \mathrm{lh}(a)\neq 0 \,\&\, \sigma((\tilde{\gamma}(a))_B * 2^{S+1})\neq 0 \end{cases}$$

where B is $\Pi_{i<\mathrm{lh}(a)\dot{-}1} p_i^{(a)_i}$ and S is $(a)_{\mathrm{lh}(a)\dot{-}1}\dot{-}1$. If in (β) we use $\bar{\alpha}(0)$ for a (via ∀-elim.), the second case applies and gives $\gamma(\bar{\alpha}(0)) = \gamma(1) = 1$ (using *23.5, ˣ23.1, *B3). If in (β) we use $\bar{\alpha}(x')$ for a, then the third or fourth case applies; furthermore using *23.4, *23.2, *23.8 etc., $B=\bar{\alpha}(x)$, $S=\alpha(x)$, $\bar{\alpha}(x') = a = B\cdot p_x^{S+1} = B*2^{S+1}=\bar{\alpha}(x)*2^{\alpha(x)+1}$, so $B < a$ (using *143b, *3.10 etc.) and $(\tilde{\gamma}(a))_B = \gamma(B) = \gamma(\bar{\alpha}(x))$ (by *24.2). Now by ind., using (2) in the basis, and (α) to deal with the fourth case of (β) in the ind. step,

$$(\gamma)\quad \sigma(\gamma(\bar{\alpha}(x)))=0.$$

Let "α_γ" abbreviate $\lambda t(\gamma(\bar{\alpha}(t')))_t\dot{-}1$. Now we deduce by induction

$$(\delta)\quad \overline{\alpha_\gamma}(x)=\gamma(\bar{\alpha}(x)).$$

BASIS: trivial. IND. STEP. $\overline{\alpha_\gamma}(x') = \overline{\alpha_\gamma}(x)*2^{((\gamma(\bar{\alpha}(x')))_x\dot{-}1)+1}$ [*23.8, ˣ0.1] $= \gamma(\bar{\alpha}(x))*2^{((\gamma(\bar{\alpha}(x')))_x\dot{-}1)+1}$ [hyp. ind.], which (using $(\gamma(\bar{\alpha}(x))*2^{A+1})_x = (\overline{\alpha_\gamma}(x)*2^{A+1})_x$ [hyp. ind.] $= A+1$), if the third case of (β) applies to $a=\bar{\alpha}(x')$, $= \gamma(\bar{\alpha}(x))*2^{\alpha(x)+1} = \gamma(\bar{\alpha}(x'))$ {if the fourth case applies, $= \gamma(\bar{\alpha}(x))*2^{\pi(\gamma(\bar{\alpha}(x)))+1} = \gamma(\bar{\alpha}(x'))$}. — By (γ) and (δ),

$$(\varepsilon)\quad \alpha_\gamma \in \sigma.$$

We also deduce by induction

$$(\zeta)\quad \sigma(\bar{\alpha}(x))=0 \supset \gamma(\bar{\alpha}(x))=\bar{\alpha}(x).$$

IND. STEP. Assuming $\sigma(\bar{\alpha}(x'))=0$, the third member of (1) gives $\sigma(\bar{\alpha}(x))=0$, so by hyp. ind. $\gamma(\bar{\alpha}(x))=\bar{\alpha}(x)$, and the third case of (β) applies. — By (δ), (ζ), *23.2 and *6.3, $\sigma(\bar{\alpha}(x'))=0 \supset \alpha_\gamma(x)=\alpha(x)$, whence

$$(\eta)\quad \alpha\in\sigma \supset \alpha_\gamma=\alpha.$$

II. Assume also the remaining hyps. **(3)–(6)** of *26.4a. We shall apply ˣ26.3a with $R(\gamma(a))$, $A(\gamma(a))$ as the $R(a)$, $A(a)$. If we can then verify the four hyps. of ˣ26.3a, the concl. of *26.4a will follow using $\gamma(1)=1$ (in I). We get the first hyp. by (γ) with (3) (using *23.6 to put $a=\bar{\alpha}(x)$ preparatory to ∃-elims.). For the second, by (ε) and (4)

$\exists x R(\overline{\alpha_\gamma}(x))$, whence by (δ) $\exists x R(\gamma(\bar{\alpha}(x)))$. We get the third (putting $a=\bar{\alpha}(x)$) by (γ) with (5). For the fourth, assume $\mathrm{Seq}(a)\ \&\ \forall s A(\gamma(a*2^{s+1}))$. By (γ) with *23.6, $\sigma(\gamma(a))=0$. Put $x=\mathrm{lh}(a)$. Assuming $\sigma(\gamma(a)*2^{s+1})=0$, and using *22.8, *22.5, *23.6 to put $a*2^{s+1}=\bar{\alpha}(y)$ (then $y=x'$ [*22.8, *20.3, *23.5], $a \cdot p_x^{s+1} = a*2^{s+1}$ [$^{\times}$21.1 etc.] $= \bar{\alpha}(x') = \bar{\alpha}(x) \cdot p_x^{\alpha(x)+1}$ [*23.8], so $s=\alpha(x)$ [*19.11, *22.2, *19.9, *6.3] and $a=\bar{\alpha}(x)$ [*133]), the third case of (β) applies to $\bar{\alpha}(x')$ and gives $\gamma(\bar{\alpha}(x'))=\gamma(a)*2^{s+1}$, so $\forall s A(\gamma(a*2^{s+1}))$ gives $A(\gamma(a)*2^{s+1})$; thus $\forall s\{\sigma(\gamma(a)*2^{s+1})=0 \supset A(\gamma(a)*2^{s+1})\}$. By (6), $A(\gamma(a))$.

6.10. From his bar theorem Brouwer inferred his "fan theorem" (implicit in **1923a** p. 4 (II); **1924** Theorem 2; **1927** Theorem 2; **1954** § 5). A "finite set" or "finitary spread", most recently called a *fan*, is a spread in which each choice must be from a finite collection of numbers. Say e.g. that, for $t = 0, 1, 2, \ldots$, the number $\alpha(t)$ must be chosen from among $0, 1, \ldots, \beta(\bar{\alpha}(t))$; i.e. $(t)\alpha(t) \leq \beta(\bar{\alpha}(t))$. We shall here be considering only the choice sequences underlying a fan, which constitute a fan by taking for the correlation law ρ the trivial correlation $\rho(\bar{\alpha}(x')) = \alpha(x)$. According to one version of the fan theorem (classically true), if, for all choice sequences α restricted to this fan (determined by β), $(Ex)R(\bar{\alpha}(x))$, then there is a finite upper bound z to the least x's for which $R(\bar{\alpha}(x))$. In this "pure" version, symbolized by *26.6a (or *26.6b–*26.6d), we can prove the fan theorem from the bar theorem with no further postulate. Another version *27.7 (classically false), favored by Brouwer, will follow from this by the new intuitionistic postulate $^{\times}$27.1 of § 7. A classical contrapositive of the present version is König's lemma **1926**, which we shall give in Remark 9.11.

First, we give a proof of the present version of the fan theorem informally. Consider any sequence number a belonging to the given fan, i.e. representing a finite choice sequence belonging to that fan; by the *subfan issuing from a* we mean the fan of those choice sequences α by which a can be extended in the given fan, i.e. such that, for each x, the sequence number $a*\bar{\alpha}(x)$ represents a finite choice sequence belonging to that fan. We apply Brouwer's **1927** Footnote 7 in 6.5 above, but considering only sequence numbers not past secured belonging to the given fan: "for every s ($s = 0, 1, 2, \ldots$)" becomes "for every $s \leq \beta(a)$". We use the corresponding form of induction to prove as follows that, under the hyp. of the fan theorem for the given fan and the given predicate R, the conclusion of the fan theorem holds for the subfan issuing from any sequence number $a = \bar{\delta}(y)$ securable but not past secured (in the given fan with respect to the given R) and the predicate $\lambda w\, R(a*w)$. The subfan issuing from a sequence number a such that R(a) has 0 as a z for the fan theorem. Consider a sequence number a whose securability follows from that of all $a*2^{s+1}$ for $s \leq \beta(a)$; by the hyp. ind., for each $s \leq \beta(a)$ the subfan issuing from $a*2^{s+1}$ has a z, call it z_s, for the fan theorem. So the subfan issuing from a has $1+\max(z_0, \ldots, z_{\beta(a)})$ as a z for the fan theorem. This completes the induction. But under the hyp. of the fan theorem, 1 is securable but not past secured. So the conclusion of the fan theorem holds for the subfan issuing from 1 and the predicate $\lambda w\, R(1*w)$, i.e. for the given fan and R.

This is easily pictured geometrically. Our fan is represented by a tree in which from each vertex, occupied by the sequence number a, finitely many arrows (namely $\beta(a)+1$ of them) lead to vertices, occupied by $a*2^{0+1}, \ldots, a*2^{\beta(a)+1}$. This is illustrated by Figure 1 in 6.5 for the case $(a)[\beta(a)=2]$ (the *binary fan*), where now we are not to imagine arrows for $s > 1$. Again consider a predicate $R(a)$; and suppose that, for each α, we underline the first $\bar{\alpha}(x)$ (if any) for which $R(\bar{\alpha}(x))$. Figure 1 illustrates a case in which $(\alpha)(Ex)R(\bar{\alpha}(x))$. To simplify terminology, let us suppress in each branch all vertices to the right of an underlined $\bar{\alpha}(x)$; so in Figure 1 only the part of the tree printed in bold face remains. The hypothesis of the fan theorem then says that all paths are finite. The conclusion says that there is a finite upper bound to their lengths. The proof is by induction, corresponding to the inductive definition of the class of the securable (but not past secured) sequence numbers a (6.5, but now in the fan rather than in the universal spread). The induction proposition is that there is a finite upper bound to the lengths of paths in the subtree issuing from a. As basis of the induction, this upper bound is 1 (the z is 0) for a at the end of any branch. As induction step, in proceeding leftward from all $a*2^{s+1}$ ($s = 0, 1$ in Figure 1) to a, we graft finitely many subtrees (2 in our Figure 1) with respective finite upper bounds onto a to obtain a subtree with upper bound the maximum of the respective upper bounds increased by one.

In formalizing this proof, we first prove a lemma *26.5, in which b, s, z, w are any distinct number variables, and B(s, z) is any formula not containing b, w free in which w is free for z.

*26.5. $\forall s \forall z \forall w[B(s, z) \,\&\, w \geq z \supset B(s, w)]$
$\vdash \forall s_{s \leq b} \exists z B(s, z) \supset \exists z \forall s_{s \leq b} B(s, z)$.

PROOF. We assume **(a)** $\forall s \forall z \forall w[B(s, z) \,\&\, w \geq z \supset B(s, w)]$, and deduce the rest by ind. on b. IND. STEP. Assume $\forall s_{s \leq b'} \exists z B(s, z)$, whence $\exists z B(b', z)$ and $\forall s_{s \leq b} \exists z B(s, z)$. By hyp. ind., $\exists z \forall s_{s \leq b} B(s, z)$. Assume for $\exists$-elim., $B(b', z_1)$ and $\forall s_{s \leq b} B(s, z_2)$. Using (a) and *8.4, $B(b', \max(z_1, z_2))$ and $\forall s_{s \leq b} B(s, \max(z_1, z_2))$, whence $\forall s_{s \leq b'} B(s, \max(z_1, z_2))$, whence $\exists z \forall s_{s \leq b'} B(s, z)$.

*26.6a. $\vdash \forall a[\mathrm{Seq}(a) \supset R(a) \vee \neg R(a)] \,\&\, \forall \alpha_{B(\alpha)} \exists x R(\bar{\alpha}(x)) \supset$
$\exists z \forall \alpha_{B(\alpha)} \exists x_{x \leq z} R(\bar{\alpha}(x))$

where $B(\alpha)$ is $\forall t \alpha(t) \leq \beta(\bar{\alpha}(t))$.

*26.6d. $\vdash \forall \alpha_{B(\alpha)} \exists x[R(\bar{\alpha}(x)) \,\&\, \forall y_{y<x} \neg R(\bar{\alpha}(y))] \supset$
$\exists z \forall \alpha_{B(\alpha)} \exists x_{x \leq z}[R(\bar{\alpha}(x)) \,\&\, \forall y_{y<x} \neg R(\bar{\alpha}(y))]$.

PROOF OF *26.6a. I. $B(\alpha)$ does indeed restrict α to a non-empty spread. For, we can introduce a function variable σ so that the following formula (a) holds. Specifically, using #22, #D, #E, etc., the right member of (a) is equivalent to $p(a)=0$ for some term $p(a)$ (with $\vdash p(a) \leq 1$). Using Lemma 5.3 (a), assume preparatory to $\exists$-elim. $\forall a[\sigma(a)=p(a)]$. Thence

(a) $\forall a[\sigma(a)=0 \sim \mathrm{Seq}(a) \,\&\, \forall t_{t<\mathrm{lh}(a)} (a)_t \dot{-} 1 \leq \beta(\Pi_{i<t} p_i^{(a)_i})]$.

(The following also proves *26.6a′ in which the "Seq(a)" of *26.6a is replaced by the right side of (a).) By *23.2, *23.4, *23.5 and *6.3,

(b) $\sigma(\bar{\alpha}(x))=0 \sim \forall t_{t<x} \alpha(t) \leq \beta(\bar{\alpha}(t))$.

Thence

(c) $B(\alpha) \sim \alpha \in \sigma$.

Furthermore, the first two hyps **(1)** $\mathrm{Spr}(\sigma)$ (using 0 for the s in the second member) and **(2)** $\sigma(1)=0$ of *26.4a now hold.

II. We shall apply *26.4a with the present σ and R taking A(a) as follows.

$A(a)$: $\exists z \forall \alpha[\forall t \alpha(t) \leq \beta(a * \bar{\alpha}(t)) \supset \exists x_{x \leq z} R(a * \bar{\alpha}(x))]$.

With this A(a), the concl. $\exists z \forall \alpha_{B(\alpha)} \exists x_{x \leq z} R(\bar{\alpha}(x))$ of *26.6a will follow from A(1) by *22.7. So it will suffice, assuming the two hyps. of *26.6a, to deduce the other four hyps. **(3)–(6)** of *26.4a. The next three (3)–(5) we quickly obtain (with 0 for the z in (5)). To deduce (6), assume **(d)** $\sigma(a)=0$ and **(e)** $\forall s\{\sigma(a*2^{s+1})=0 \supset A(a*2^{s+1})\}$; we must deduce A(a). Using (d), (a) and *23.6, we can put (for $\exists$-elims.) **(f)** $a=\bar{\delta}(y)$. Then by (d) and (b): **(g)** $\forall t_{t<y}\delta(t)\leq\beta(\bar{\delta}(t))$. We shall deduce $A(\bar{\delta}(y))$, i.e. $\exists z\forall\alpha[\forall t\alpha(t)\leq\beta(\bar{\delta}(y)*\bar{\alpha}(t)) \supset \exists x_{x\leq z}R(\bar{\delta}(y)*\bar{\alpha}(x))]$.

A. Assume **(h)** $s\leq\beta(\bar{\delta}(y))$. Using *23.6 with *22.5 and *22.8, we can put (for $\exists$-elims.) $\bar{\delta}(y)*2^{s+1}=\overline{\delta'}(u)$, whereupon (by *23.5, *22.8, *20.3) $u=y'$, so $\bar{\delta}(y)*2^{s+1}=\overline{\delta'}(y')$. By *23.2, ×21.1, *19.11 with *23.3 and *19.9, $\delta'(y)=s$; with *23.2 and *19.10, $t<y \supset \delta'(t)=\delta(t)$; hence by *B19 with ×23.1, $t\leq y \supset \overline{\delta'}(t)=\bar{\delta}(t)$. Now (g) and (h) give $\forall t_{t<y'}\delta'(t)\leq\beta(\overline{\delta'}(t))$, whence by (b) $\sigma(\overline{\delta'}(y'))=0$, whence $\sigma(\bar{\delta}(y)*2^{s+1})=0$, whence by (e) and (f) $A(\bar{\delta}(y)*2^{s+1})$, i.e. $\exists z\forall\alpha[\forall t\alpha(t)\leq\beta((\bar{\delta}(y)*2^{s+1})*\bar{\alpha}(t)) \supset \exists x_{x\leq z}R((\bar{\delta}(y)*2^{s+1})*\bar{\alpha}(x))]$; call this formula $\exists zB(s, z)$. By the $\exists$u- and $\exists\delta'$-elim., and $\supset$- and $\forall$-introd., $\forall s_{s\leq\beta(\bar{\delta}(y))}\exists zB(s, z)$. But B(s, z) has the property expressed by the assumption formula of *26.5. Hence $\exists z\forall s_{s\leq\beta(\bar{\delta}(y))}B(s, z)$.

B. Assume (for $\exists$-elim.) $\forall s_{s\leq\beta(\bar{\delta}(y))}B(s, z)$, and $\forall t\alpha(t)\leq\beta(\bar{\delta}(y)*\bar{\alpha}(t))$. Now $\alpha(0)\leq\beta(\bar{\delta}(y))$, and so $B(\alpha(0), z)$, i.e. $\forall\alpha'[\forall t\alpha'(t)\leq\beta((\bar{\delta}(y)*2^{\alpha(0)+1})*\overline{\alpha'}(t)) \supset \exists x_{x\leq z}R((\bar{\delta}(y)*2^{\alpha(0)+1})*\overline{\alpha'}(x))]$. Let "$\alpha'$" abbreviate $\lambda t\alpha(t')$. Then $\alpha'(t) = \alpha(t')$ [×0.1] $\leq \beta(\bar{\delta}(y)*\bar{\alpha}(t')) = \beta(\bar{\delta}(y)*(\bar{\alpha}(1)*\overline{\alpha'}(t)))$ [*23.7] $= \beta((\bar{\delta}(y)*2^{\alpha(0)+1})*\overline{\alpha'}(t))$ [*22.9]; so $\forall t\alpha'(t)\leq\beta((\bar{\delta}(y)*2^{\alpha(0)+1})*\overline{\alpha'}(t))$. So from $B(\alpha(0), z)$, $\exists x_{x\leq z}R((\bar{\delta}(y)*2^{\alpha(0)+1})*\overline{\alpha'}(x))$. Assume $x\leq z$ & $R((\bar{\delta}(y)*2^{\alpha(0)+1})*\overline{\alpha'}(x))$. Thence $x'\leq z'$ & $R(\bar{\delta}(y)*\bar{\alpha}(x'))$, whence $\exists x_{x\leq z'}R(\bar{\delta}(y)*\bar{\alpha}(x))$. By the $\exists$x-elim., $\supset$-, $\forall$- and $\exists$z-introd., and the $\exists$z-elim., $\exists z\forall\alpha[\forall t\alpha(t)\leq\beta(\bar{\delta}(y)*\bar{\alpha}(t)) \supset \exists x_{x\leq z}R(\bar{\delta}(y)*\bar{\alpha}(x))]$.

More generally, the choices permitted for $\alpha(t)$ in a fan need not be a non-empty initial segment of the natural numbers. The choice law is then a function σ satisfying the first two hypotheses of:

*26.7a. $\vdash \mathrm{Spr}(\sigma)$ & $\forall a[\sigma(a)=0 \supset \exists b\forall s\{\sigma(a*2^{s+1})=0 \supset s\leq b\}]$ & $\forall a[\sigma(a)=0 \supset R(a) \vee \neg R(a)]$ & $\forall\alpha_{\alpha\epsilon\sigma}\exists xR(\bar{\alpha}(x))$ $\supset \exists z\forall\alpha_{\alpha\epsilon\sigma}\exists x_{x\leq z}R(\bar{\alpha}(x))$.

PROOF. CASE 1: $\sigma(1)\neq 0$. Then $\neg\alpha\epsilon\sigma$. Use *10a. (The fan is empty and the theorem holds vacuously.)

CASE 2: $\sigma(1)=0$. Assume the four hyps. **(1′)–(4′)** of *26.7a.

I. We have the first two hyps. (1) and (2) of *26.4a, so we can introduce π and γ as in I of the proof there and (α)–(η) will hold.

II. Using $\sigma(a)=0 \vee \sigma(a)\neq 0$ and (2′), $\forall a \exists b[\sigma(a)=0 \supset \forall s\{\sigma(a*2^{s+1})=0 \supset s\leq b\}]$. Applying *2.2, we may assume for ∃-elim.

$$(\theta) \qquad \forall a[\sigma(a)=0 \supset \forall s\{\sigma(a*2^{s+1})=0 \supset s\leq\beta(a)\}].$$

III. We shall apply *26.6a for the β of (θ) (entering into $B(\alpha)$) with $R(\gamma(a))$ (for the γ of I) as the $R(a)$.

A. First we verify that the concl. of *26.7a will then follow from the concl. $\exists z\forall\alpha_{B(\alpha)}\exists x_{x\leq z}R(\gamma(\bar{\alpha}(x)))$ of *26.6a. Assume for ∃-elim. **(ι)** $\forall\alpha_{B(\alpha)}\exists x_{x\leq z}R(\gamma(\bar{\alpha}(x)))$. Assume **($\varkappa$)** $\alpha\in\sigma$, whence $\sigma(\bar{\alpha}(t))=0$ and $\sigma(\bar{\alpha}(t'))=0$. But $\bar{\alpha}(t')=\bar{\alpha}(t)*2^{\alpha(t)+1}$. Applying ($\theta$), $\alpha(t)\leq\beta(\bar{\alpha}(t))$; and by ∀-introd., $B(\alpha)$. Hence by (ι), $\exists x_{x\leq z}R(\gamma(\bar{\alpha}(x)))$. Omitting ∃x for ∃-elim., we have $x\leq z$ and $R(\gamma(\bar{\alpha}(x)))$; by ($\varkappa$), $\sigma(\bar{\alpha}(x))=0$. So by ($\zeta$), $R(\bar{\alpha}(x))$. By &-, ∃- and ⊃-introd., (completing) the ∃x-elim., ∀- and ∃-introd., and the ∃z-elim., $\exists z\forall\alpha_{\alpha\in\sigma}\exists x_{x\leq z}R(\bar{\alpha}(x))$.

B. It remains for us to verify the two hyps. of *26.6a. For the first, assume Seq(a), and put $a=\bar{\alpha}(x)$. By (γ) $\sigma(\gamma(\bar{\alpha}(x)))=0$, so by (3′) $R(\gamma(a)) \vee \neg R(\gamma(a))$. For the second, by ($\varepsilon$) and (4′) $\exists x R(\overline{\alpha_{\gamma}}(x))$, whence by ($\delta$) $\exists xR(\gamma(\bar{\alpha}(x)))$, whence by *11 $B(\alpha) \supset \exists xR(\gamma(\bar{\alpha}(x)))$, whence by ∀-introd. $\forall\alpha_{B(\alpha)}\exists xR(\gamma(\bar{\alpha}(x)))$.

6.11. We now formalize the induction principle in the bar theorem for inferring a property A of any barred sequence number w (cf. 6.6 ¶s 2, 3). This gives us an axiom schema modelled directly on Axiom Schema 13 for ordinary induction. In this the implication of an inductive by an explicit sense of securability, barredness, etc. (i.e. the reversal of direction, 6.5 ¶ 1), which is the kernel of the bar theorem, enters thus: the conclusion of the induction that each barred sequence number w has the property A is formulated using the explicit sense of 'barred'. The intuitionistic restriction on the predicate R with respect to which numbers w are barred we give as a preliminary hypothesis, in two forms. (For a classical result, corresponding to *26.1, we may omit the first hyp. of ˣ26.8a.)

ˣ26.8a. $\forall a[\mathrm{Seq}(a) \supset R(a) \vee \neg R(a)]\ \&$
$\forall a[\mathrm{Seq}(a)\ \&\ R(a) \supset A(a)]\ \&\ \forall a[\mathrm{Seq}(a)\ \&\ \forall sA(a*2^{s+1}) \supset A(a)] \supset$
$\{\mathrm{Seq}(w)\ \&\ \forall\alpha\exists xR(w*\bar{\alpha}(x)) \supset A(w)\}$.

ˣ26.8c. $\forall\alpha\forall x\forall y[R(\bar{\alpha}(x))\ \&\ R(\bar{\alpha}(y)) \supset x=y]\ \&$
$\forall a[\mathrm{Seq}(a)\ \&\ R(a) \supset A(a)]\ \&\ \forall a[\mathrm{Seq}(a)\ \&\ \forall sA(a*2^{s+1}) \supset A(a)] \supset$
$\{\mathrm{Seq}(w)\ \&\ \forall\alpha\exists xR(w*\bar{\alpha}(x)) \supset A(w)\}$.

$^{\times}$26.8d. $\forall\alpha\forall x[R(\bar{\alpha}(x))\ \&\ \forall y_{y<x}\neg R(\bar{\alpha}(y)) \supset A(\bar{\alpha}(x))]\ \&$
$\forall a[\mathrm{Seq}(a)\ \&\ \forall s A(a*2^{s+1}) \supset A(a)] \supset$
$\{\forall\alpha_{\bar{\alpha}(z)=\bar{\beta}(z)}\exists x_{x\geq z}[R(\bar{\alpha}(x))\ \&\ \forall y_{y<x}\neg R(\bar{\alpha}(y))] \supset A(\bar{\beta}(z))\}.$

Derivation of $^{\times}$26.3a from $^{\times}$26.8a. Substitute 1 for w, and use Seq(1), *22.7 and *23.5.

Derivation of $^{\times}$26.8a from $^{\times}$26.3a. Assume the five hyps. of $^{\times}$26.8a. Using *22.8, *22.9 and *22.5, the four hyps. of $^{\times}$26.3a follow for R(w*a), A(w*a) as the R(a), A(a). So by $^{\times}$26.3a A(w*1), whence by *22.6 A(w).

6.12. Finally we consider Brouwer's longer proof of the bar theorem given in the text of **1927**, and slightly differently in **1924** (cf. **1924a**) and **1954**. Brouwer confined his attention (in **1924, 1927**) to an R such that $(\alpha)(E!x)R(\bar{\alpha}(x))$, of which we now assume the uniqueness part $(\alpha)(x)(y)[R(\bar{\alpha}(x))\ \&\ R(\bar{\alpha}(y)) \rightarrow x=y]$. We take the case of the universal spread (though Brouwer was considering any spread), since the theorem for arbitrary spreads is a corollary *26.4.

In this longer proof, Brouwer begins with the following interpretation. Consider any sequence number w securable (i.e. securable$_E$) but not past secured; i.e. assume $\mathrm{Seq}(w)\ \&\ (\alpha)(Ex)R(w*\bar{\alpha}(x))$. (Here we edit Brouwer's proof slightly; he considered the unsecured sequences, which is a bit less convenient.) That w is securable means intuitionistically that there is a "proof (Beweisführung)" of w's being securable. Such a proof must rest "ultimately (in letzter Instanz)" (**1924, 1927**) upon the "atomic" facts in the situation (**1954**), which are only the truth of $R(v)$ for certain sequence numbers v, and the relationships between sequence numbers v and their immediate extensions $v*2^{s+1}$ ($s = 0, 1, 2, \ldots$). So when a proof that w is securable is analyzed into its atomic inferences ("Elementarschlüsse"), these will be of three kinds: η-inferences (from 0 premises) that v is securable because $R(v)$, $\mathcal{F}$-inferences (from $\aleph_0$ premises) that v is securable because all $v*2^{s+1}$ ($s = 0, 1, 2, \ldots$) are securable, and ζ-inferences (from 1 premise) that $v*2^{s+1}$ is securable because v is securable. (Brouwer did not use the term "η-inference".)

Disregarding considerations of the symbolism, such a proof differs from the proofs in metamathematics (IM § 19 especially p. 83 lines 8–10, end § 24 especially bottom p. 106) only in that one of the present three rules of inference (namely $\mathcal{F}$-inference) has infinitely many premises. The logical structure of such a proof is represented directly by taking it in tree form (IM p. 106), but now with infinitely many branches concurrent downward at each F-inference. (Brouwer in **1924**, **1927** uses instead the sequence form IM p. 106, which brings the inferences or the propositions inferred into a linear ordering, which when F-inferences occur is a transfinite well-ordering. Thus he connects with his theory of well-ordered species **1918–9** I § 3, **1924–7** III, **1954** § 4, which Vesley plans to formalize. Brouwer says "species (Spezies)" rather than "set", as he used "set" for what he later called "spread". — Brouwer **1954**, and Heyting **1956** pp. 43–44 dealing with a fan, omit this unnecessary linearization.) Now we can use induction over proofs by η-, F- and ζ-inferences, i.e. the form of induction corresponding to the inductive definition of 'provable formula' (IM p. 83, top p. 260) when used with the present three rules of inference. (Proceeding downward in the trees of IM pp. 106–107 corresponds to proceeding to the left in trees as drawn here, e.g. in 6.5 Figure 1.)

Brouwer's interpretation (just described) of what is entailed in 'w is securable (but not past secured)' taken with the principle of induction over proofs, like his **1927** Footnote 7 (6.5 above), introduces a reversal of direction. We start from the hypothesis $(\alpha)(x)(y)[R(\bar{\alpha}(x)) \,\&\, R(\bar{\alpha}(y)) \rightarrow x=y] \,\&\, \mathrm{Seq}(w) \,\&\, (\alpha)(Ex)R(w*\bar{\alpha}(x))$, with the prefix $(\alpha)(Ex)$ looking forward from w into the tree of diverging paths of growth of a choice sequence α. The interpretation (with the induction principle) then enables us to argue inductively, proceeding backward from many bases at the ends of the branches of a tree along converging paths to a single conclusion. But this time, instead of proceeding in the convergent direction in the original tree of the sequence numbers issuing from w and terminating in the $w*\bar{\alpha}(x)$'s for which $R(w*\bar{\alpha}(x))$, we are doing so in another tree constituting a proof of w's being securable.

By induction over this latter tree, we readily establish that the ζ-inferences are eliminable, so we obtain the bar theorem.

To formalize this proof of the bar theorem, we shall require a postulate expressing Brouwer's presupposition,

(***)
$$\begin{aligned}&(\alpha)(x)(y)[R(\bar{\alpha}(x)) \,\&\, R(\bar{\alpha}(y)) \rightarrow x=y]\\ &\,\&\, \mathrm{Seq}(w) \,\&\, (\alpha)(Ex)R(w*\bar{\alpha}(x))\\ &\rightarrow \{\text{there is a proof of `}w\text{ is securable'}\\ &\quad \text{by } \eta\text{-, } F\text{- and } \zeta\text{-inferences}\}.\end{aligned}$$

To formulate this postulate, we must find a way of rendering the notion of a proof by η-, F- and ζ-inferences into the formal symbolism. Agreement on this must necessarily be reached outside the formalism.

All the propositions in a proof of securability by η-, F- and ζ-inferences are of the form that a number v is securable. So by replacing (the occurrences of) the propositions 'v is securable' in the proof-tree by (occurrences of) the numbers v which those propositions are about, we obtain a tree isomorphic to that proof-tree, while escaping any difficulty from the lack of formal metamathematical symbols in the system. (In **1924** Brouwer dealt at once with the well-ordered species of the finite choice sequences in the order in which their securability is established by the proof. It was in **1924a** that he began first to speak of the proof as a well-ordered species of inferences.)

Now we must find a way to talk about the latter tree of (occurrences of) sequence numbers. But indeed it constitutes simply a mapping of sequence numbers onto the vertices of a certain tree of terminating choice sequences. This, when viewed in the forward (divergent) direction, is precisely a spread, where as in 6.9 ¶ 4 w is to be correlated to the empty sequence of choices, i.e. to the sequence number $\bar{\alpha}(0) = 1$. (Of course here we have no interest in the finite sequences constituting the elements of the spread as such.)

In specifying this spread by a choice law and a correlation law (6.1, 6.9), instead of sometimes terminating sequences of choices or forbidding some choices (while allowing others), it is simpler to correlate 0, which isn't a sequence number, to the choices which would thereby be disallowed (cf. 6.9 ¶ 3 (b)). By this device, the choice law σ becomes trivial, and a correlation law ρ will specify the spread. More explicitly, there are two spreads, the one specified directly by ρ which overlies the universal spread of all the choice sequences and has sequence numbers and 0's as the correlated objects, and the other which is obtainable from that by disallowing the choices to which 0's are correlated and is the spread isomorphic to the proof of 'w is securable'. But we are viewing the latter spread in the backward (convergent) direction, as is essential to Brouwer's argument.

Now we can formulate conditions that ρ thus represent a proof of securability. When $\text{Seq}(a) \& \rho(a)>0$, then $\rho(a)=v$ where 'v is securable' is the result of an η-, F- or ζ-inference as described above; etc. That the proof represented by ρ is of 'w is securable' is expressed by $\rho(1) = w$. Altogether, the "local" requirements $\mathfrak{P}(w, \rho, R)$ on ρ that it represent a proof of 'w is securable' are as follows.

$\mathfrak{P}(w, \rho, R)$: $\rho(1)=w \;\&\; (a)\{\mathrm{Seq}(a) \;\&\; \rho(a)>0 \rightarrow [R(\rho(a)) \;\&\; (s)\rho(a*2^{s+1})=0]$
$\vee\; (s)\rho(a*2^{s+1})=\rho(a)*2^{s+1} \vee [\mathrm{Seq}(\rho(a*2)) \;\&$
$(Es)\rho(a)=\rho(a*2)*2^{s+1} \;\&\; (s)\rho(a*2^{s+2})=0]\} \;\&$
$(a)\{\mathrm{Seq}(a) \;\&\; \rho(a)=0 \rightarrow (s)\rho(a*2^{s+1})=0\}$.

The principle of induction over proofs of 'w is securable' by η-, F- and ζ-inferences, as we have now represented such proofs, is $(A)\mathfrak{J}(w, \rho, A)$ where $\mathfrak{J}(w, \rho, A)$ is as follows.

$\mathfrak{J}(w, \rho, A)$: $(a)\{\mathrm{Seq}(a) \;\&\; \rho(a)>0 \;\&\; (s)[A(\rho(a*2^{s+1})) \vee$
$\rho(a*2^{s+1})=0] \rightarrow A(\rho(a))\} \rightarrow A(w)$.

Let $\mathfrak{P}(w, \rho, R)$ and $\mathfrak{J}(w, \rho, A)$ be the correspondingly constructed formulas, for any formulas R(a) and A(a).

The conclusion of (***) is now $(E\rho)\{\mathfrak{P}(w, \rho, R) \;\&\; (A)\mathfrak{J}(w, \rho, A)\}$. Thus we are led to the following axiom schema, expressing (as nearly as we can in the absence of predicate variables) the presupposition for Brouwer's longer proof.

^{x}26.9. $\forall\alpha\forall x\forall y[R(\bar{\alpha}(x)) \;\&\; R(\bar{\alpha}(y)) \supset x=y] \;\&\; \mathrm{Seq}(w) \;\&\; \forall\alpha\exists x R(w*\bar{\alpha}(x))$
$\supset \exists\rho\{\mathfrak{P}(w, \rho, R) \;\&\; \mathfrak{J}(w, \rho, A)\}$.

With ^{x}26.9 as an axiom schema, we can prove each axiom by ^{x}26.3c, formalizing Brouwer's longer proof, thus.

We take 1 for the w (since in ^{x}26.3c we specialized to 1, though Brouwer didn't), and for the A(a) the following formula.

A′(a): $\exists y_{y<\mathrm{lh}(a)} R(\Pi_{i<y} p_i^{(a)_i}) \vee \forall c\{\mathrm{Seq}(c) \;\&\; \forall\alpha\exists x R((a*c)*\bar{\alpha}(x)) \;\&$
$\forall b[\mathrm{Seq}(b) \;\&\; R(b) \supset A(b)] \;\&$
$\forall b[\mathrm{Seq}(b) \;\&\; \forall s A(b*2^{s+1}) \supset A(b)] \supset A(a*c)\}$.

Assume **(a)** $\forall\alpha\exists! x R(\bar{\alpha}(x))$. If from (a) and the axiom by ^{x}26.9 we deduce A′(1), we will have what we want by taking 1 for the c (using lh(1)=0, Seq(1), *22.7). Using (a), we have the three hyps. of that axiom, so our problem reduces to seeing that from (a), **(b)** $\rho(1)=1$, **(c)** $\forall a\{\mathrm{Seq}(a) \;\&\; \rho(a)>0 \supset [R(\rho(a)) \;\&\; \forall s\rho(a*2^{s+1})=0] \vee \forall s\rho(a*2^{s+1})= \rho(a)*2^{s+1} \vee [\mathrm{Seq}(\rho(a*2)) \;\&\; \exists s\rho(a)=\rho(a*2)*2^{s+1} \;\&\; \forall s\rho(a*2^{s+2})=0]\}$ and **(d)** $\forall a\{\mathrm{Seq}(a) \;\&\; \rho(a)=0 \supset \forall s\rho(a*2^{s+1})=0\}$ we can deduce **(A)** $\forall a\{\mathrm{Seq}(a) \;\&\; \rho(a)>0 \;\&\; \forall s[A'(\rho(a*2^{s+1})) \vee \rho(a*2^{s+1})=0] \supset A'(\rho(a))\}$. By ind. on x, using (b)–(d): **(e)** $\rho(\bar{\alpha}(x))>0 \supset \mathrm{Seq}(\rho(\bar{\alpha}(x)))$. Using (a) in Lemma 5.6: **(f)** $\forall a[\mathrm{Seq}(a) \supset R(a) \vee \neg R(a)]$. Using (for $\rho(a)>0$) *150 with (e), (f), *23.4 etc.: **(g)** $\mathrm{Seq}(a) \supset \exists y_{y<\mathrm{lh}(\rho(a))} R(\Pi_{i<y} p_i^{(\rho(a))_i}) \vee \neg\exists y_{y<\mathrm{lh}(\rho(a))} R(\Pi_{i<y} p_i^{(\rho(a))_i})$. To deduce (A), we assume **(h)** Seq(a), **(i)** $\rho(a)>0$ and **(j)** $\forall s[A'(\rho(a*2^{s+1})) \vee \rho(a*2^{s+1})=0]$, and undertake to deduce $A'(\rho(a))$. Using (g) we have two cases; and using (c) we have three subcases for use in the second case. CASE 2: **(k)** $\neg\exists y_{y<lh(\rho(a))} R(\Pi_{i<y} p_i^{(\rho(a))_i})$. Toward deducing the second disjunctive member of $A'(\rho(a))$, we assume **(ł)** $Seq(c)$, **(m)** $\forall\alpha\exists x R((\rho(a)*c)*\bar{\alpha}(x))$, **(n)** $\forall b[Seq(b) \& R(b) \supset A(b)]$ and **(o)** $\forall b[Seq(b) \& \forall s A(b*2^{s+1}) \supset A(b)]$, and undertake to deduce $A(\rho(a)*c)$. By (h), (i) and (e): **(p)** $Seq(\rho(a))$. SUBCASE 1: $R(\rho(a)) \& \forall s \rho(a*2^{s+1})=0$. Using (m) and (a), $c=1$ (and $x=0$). Using (n) with (p), we obtain $A(\rho(a))$, so $A(\rho(a)*c)$. SUBCASE 2: $\forall s \rho(a*2^{s+1})=\rho(a)*2^{s+1}$. By (j) with (p): **(q)** $\forall s A'(\rho(a*2^{s+1}))$. By (k) with (p), *22.3 etc.: **(r)** $\forall s\{\exists y_{y<lh(\rho(a)*2^{s+1})} R(\Pi_{i<y} p_i^{(\rho(a)*2^{s+1})_i}) \sim R(\rho(a))\}$. By (f) with (p), $R(\rho(a)) \vee \neg R(\rho(a))$. SUB²CASE 1: $R(\rho(a))$. Like Subcase 1. SUB²CASE 2: $\neg R(\rho(a))$. Using (q) with (r), (n) and (o): **(s)** $\forall s \forall d\{Seq(d) \& \forall\alpha\exists x R(\rho(a)*2^{s+1}*d*\bar{\alpha}(x)) \supset \forall s A(\rho(a)*2^{s+1}*d)\}$. SUB³CASE 1: $c=1$. We first deduce $A(\rho(a)*2^{s+1})$ by cases from (f) with (p). CASE A: $R(\rho(a)*2^{s+1})$. By (n), $A(\rho(a)*2^{s+1})$. CASE B: $\neg R(\rho(a)*2^{s+1})$. By (m), $\forall\alpha\exists x R(\rho(a)*2^{s+1}*\bar{\alpha}(x))$. By (s), $A(\rho(a)*2^{s+1})$. — By (o), $A(\rho(a))$, so $A(\rho(a)*c)$. SUB³CASE 2: $c\neq 1$. Using (ł) (and *23.7 etc.), we can put $c=2^{s+1}*d$ where $Seq(d)$. Using (s) and (m), $A(\rho(a)*c)$. SUBCASE 3: $Seq(\rho(a*2)) \& \exists s \rho(a)=\rho(a*2)*2^{s+1} \& \forall s \rho(a*2^{s+2})=0$. Assume **(t)** $\rho(a)=\rho(a*2)*2^{s+1}$. By $Seq(\rho(a*2))$, $\rho(a*2)>0$, so by (j): **(u)** $A'(\rho(a*2))$. By (t) with (k): **(v)** $\neg\exists y_{y<lh(\rho(a*2))} R(\Pi_{i<y} p_i^{(\rho(a*2))_i})$. By (u) with (v), (n) and (o), $\forall c\{Seq(c) \& \forall\alpha\exists x R(\rho(a*2)*c*\bar{\alpha}(x)) \supset A(\rho(a*2)*c)\}$. Thence taking $2^{s+1}*c$ for the c and using (t), (ł) and (m), $A(\rho(a)*c)$.

Conversely, ˣ26.9 is derivable from ˣ26.3. The following derivation by Joan Rand (March 22, 1963) is shorter than the author's original derivation (winter 1958–59) and avoids his direct use (not via *2.2) of ˣ2.1.

Assume **(a)** $\forall\alpha\forall x\forall y[R(\bar{\alpha}(x)) \& R(\bar{\alpha}(y)) \supset x=y]$, **(b)** $Seq(w)$, and **(c)** $\forall\alpha\exists x R(w*\bar{\alpha}(x))$. Then **(d)** $\forall\alpha\exists! x R(w*\bar{\alpha}(x))$. Thence by Lemma 5.6, **(e)** $R(w*\bar{\alpha}(x)) \vee \neg R(w*\bar{\alpha}(x))$. Thence $Seq(a) \& a\neq 1 \supset (R(w*\Pi_{i<lh(a)\dot{-}1} p_i^{(a)_i}) \vee \neg R(w*\Pi_{i<lh(a)\dot{-}1} p_i^{(a)_i}))$. So we can introduce ρ as follows, preparatory to $\exists$-elim. from the result of an application of Lemma 5.5 (c) (with $\bar{\alpha}(y)$), using *23.5 [*23.2 with *6.3, *22.3, *6.7, *B4, *143b etc.] to express a $[\rho(\Pi_{i<lh(a)\dot{-}1} p_i^{(a)_i})]$ in terms of $\bar{\rho}(a)$:

$$\textbf{(f)} \quad \forall a \rho(a) = \begin{cases} w = w*a \text{ if } a=1, \\ w*a \text{ if } Seq(a) \& a\neq 1 \& \neg R(w*\Pi_{i<lh(a)\dot{-}1} p_i^{(a)_i}) \\ \qquad \& \rho(\Pi_{i<lh(a)\dot{-}1} p_i^{(a)_i})>0, \\ 0 \text{ otherwise.} \end{cases}$$

Then **(g)** $\rho(a)>0 \supset \rho(a)=w*a$. (For more detail, cf. (A) in 14.1.)

The first and third conjunctive members **(h)** and **(i)** of $\mathfrak{P}(w, \rho, R)$ are immediate from (f). Toward the second **(j)**, assume $\mathrm{Seq}(a) \,\&\, \rho(a)>0$. Using (e), we have two cases. CASE 1: $R(w*a)$. Then by (g) $R(\rho(a))$, and by (f) $\forall s \rho(a*2^{s+1})=0$. CASE 2: $\neg R(w*a)$. Using this with $\rho(a)>0$ and (f), $\forall s \rho(a*2^{s+1}) = w*(a*2^{s+1}) = (w*a)*2^{s+1} = \rho(a)*2^{s+1}$ [by (g)].

Toward $\mathfrak{J}(w, \rho, A)$, assume **(k)** $\forall a\{\mathrm{Seq}(a) \,\&\, \rho(a)>0 \,\&\, \forall s[A(\rho(a*2^{s+1})) \vee \rho(a*2^{s+1})=0] \supset A(\rho(a))\}$. To deduce $A(w)$ by use of ˣ26.3c, it will suffice to deduce the hypotheses (d), **(A)** and **(B)** of ˣ26.3c with $R(w*a)$, $A(w*a) \,\&\, \rho(a)>0$ as the $R(a)$, $A(a)$. Using *149a: **(ł)** $\rho(\bar{\alpha}(x))=0 \supset \exists y_{y \leq x}[\rho(\bar{\alpha}(y))=0 \,\&\, \forall z_{z<y}\rho(\bar{\alpha}(z))>0]$. DEDUCTION OF (A). Assume $\mathrm{Seq}(a) \,\&\, R(w*a)$. Put $a=\bar{\alpha}(x)$. By (f), $\forall s \rho(a*2^{s+1})=0$. To apply (k), we still need **(m)** $\rho(a)>0$. Assume $\rho(a)=0$. From (ł), assume **(n)** $y \leq x \,\&\, \rho(\bar{\alpha}(y))=0 \,\&\, \forall z_{z<y}\rho(\bar{\alpha}(z))>0$. By $\rho(\bar{\alpha}(y))=0$ with (f) (and $w>0$ from (b)), $y>0$ and $R(w*\bar{\alpha}(y \dot{-} 1)) \vee \rho(\bar{\alpha}(y \dot{-} 1))=0$. But $\rho(\bar{\alpha}(y \dot{-} 1))=0$ is excluded by (n) with $y>0$. Thus $R(w*\bar{\alpha}(y \dot{-} 1)) \,\&\, R(w*\bar{\alpha}(x)) \,\&\, y \dot{-} 1<y \leq x$, contradicting (d) with *172. — Now (k) gives $A(\rho(a))$. Thence by (m) and (g), $A(w*a) \,\&\, \rho(a)>0$. DEDUCTION OF (B). Assume $\mathrm{Seq}(a) \,\&\, \forall s(A(w*(a*2^{s+1})) \,\&\, \rho(a*2^{s+1})>0)$. By (i): **(o)** $\rho(a)>0$. Using (g), $\forall s A(\rho(a*2^{s+1}))$. So by (k), $A(\rho(a))$, whence by (o) and (g), $A(w*a) \,\&\, \rho(a)>0$.

REMARK 6.3. In Brouwer's "bar theorem" **1954** p. 14, he comes out with a (linear) well-ordering, but this entails an induction principle similar to that in our ˣ26.3 etc. At no other place we know in Brouwer's writings is a similar proposition stated explicitly as a theorem, except in **1924** § 1 (not cited from **1954**), where the hypothesis of **1954** p. 14 or ˣ26.3 is replaced by one entailing it via Brouwer's principle § 7 below (making the theorem false classically). However, from **1954** p. 14 he cites the passage pp. 63–65 of **1927** proving that version. We do not consider a proposition to be a version of the bar theorem unless, from an hypothesis giving (or entailing) securability_E or barredness_E, it concludes securability_I or barredness_I or a well-orderedness property or an induction principle; for, that is what the labor in Brouwer's proof in **1954** or **1927** is devoted to obtaining.

§ 7. Postulate on correlation of functions to choice sequences (Brouwer's principle). 7.1. Suppose that to each choice sequence $\alpha(0), \alpha(1), \alpha(2), \ldots$ a natural number b is correlated ("zugeordnet"). Since intuitionistically choice sequences are considered as continually growing by new choices rather than as completed, this correlation can subsist intuitionistically only in such a manner that at some (finite) stage in the growth of the sequence $\alpha(0), \alpha(1), \alpha(2), \ldots$ the correlated number b will be determined (effectively). That is, intuitionistically the b must be determined effectively by the first y choices $\alpha(0), \ldots, \alpha(y-1)$ of α for some y (depending in general on those choices).

We call this "Brouwer's principle (for numbers)". Brouwer maintained it in **1924** § 1, **1927** § 2 and **1954** p. 15, in the course of proving his version of the fan theorem. He also maintained it in **1918–9** I end § 1 and **1924–7** I end § 5, in arguing (without using Cantor's diagonal method, IM § 2 bottom p. 7) that the universal spread C cannot be mapped 1–1 onto a subset of the natural numbers A; taking this with the usual definition of $>$ for cardinal numbers (IM p. 10), and an obvious 1–1 mapping of A onto a subset of C, it is a rather trivial theorem intuitionistically that the cardinal number of C is greater than the cardinal number of A.

Brouwer's principle must be made explicit for our formal development. One might do so by incorporating it explicitly into each statement that a b is correlated to each α. Then the simple statement without the added explanation would not be used. In any case, Brouwer elected rather to consider the determination of the correlated number b by an initial segment $\alpha(0), \ldots, \alpha(y-1)$ of α as implicit in his use of the mode of expression 'a b is correlated to each α'. Then formulas expressing his principle, or something entailing it, must be postulated for the intuitionistic formal system. For the principle is false classically, e.g. in the case of the classically-admissible correlation of 0 to the sequence consisting of all 0's and 1 to all other sequences. So (with 9.2) the principle is independent of the other intuitionistic postulates, which are all true classically.

Brouwer's intention that the determination of the b be effective is attested by his phraseology "the *algorithm* of the correlation law (dem *Algorithmus* des Zuordnungsgesetzes)" (**1924** § 1, **1927** § 2, italics his).

Toward formulating the necessary postulate, consider first what the algorithm must do. It must decide for each initial segment $\alpha(0), \ldots, \alpha(y-1)$ of a choice sequence α whether from that segment it will produce the correlated number b. When this decision is affirmative (as it must be for some y), it must produce the b. It will be convenient, as in § 6, to represent each initial segment $\alpha(0), \ldots, \alpha(y-1)$ of a choice sequence α by the sequence number $\bar{\alpha}(y)$. Now we can combine the two operations the algorithm must perform into one function τ, which operates on sequence numbers $\bar{\alpha}(y)$, thus. As the initial segment $\alpha(0), \ldots, \alpha(y-1)$ of a choice sequence α grows (y increasing), $\tau(\bar{\alpha}(y))$ remains 0 so long as the algorithm does not accept $\alpha(0), \ldots, \alpha(y-1)$ or $\bar{\alpha}(y)$ as a basis for producing the b; but when it first does, then $\tau(\bar{\alpha}(y)) = b+1$. (This representation was selected in January 1956 in the course of the present study. It has meanwhile been used in Kleene **1959a**. Using the terminology of that paper intuitionistically, Brouwer's principle for numbers can be stated: Each type-2 functional is countable.)

We are not saying that, as y increases, $\tau(\bar{\alpha}(y))$ remains 0 only as long as $\alpha(0), \ldots, \alpha(y-1)$ or $\bar{\alpha}(y)$ does not "determine" the b in the sense that the same b is correlated to all α's having $\alpha(0), \ldots, \alpha(y-1)$ as initial segment. Intuitionistically, the correlation is first established by the algorithm itself, i.e. $\tau(\bar{\alpha}(y))$ changes from 0 when $\bar{\alpha}(y)$ determines the b effectively by the algorithm. This does not exclude the possibility that we may be able to prove about the algorithm that, in some case, it puts off producing the b to beyond the first y at which, for all ways of extending the choices $\alpha(0), \ldots, \alpha(y-1)$, it will ultimately produce the same b. (Cf. Brouwer **1924** § 1 end ¶ 1, **1927** § 2 end ¶ 1.)

We give an example (from Kleene **1959a*** p. 83) in which such postponement by the algorithm is essential, if τ to express an algorithm must be general recursive (Church's thesis). Let a b be correlated to each α by the rule that $b = 0$ if $\overline{T}_1(\alpha(0), \alpha(0), \alpha(1))$ and $b = \alpha(1)+1$ otherwise (IM p. 281). If the production of the b is always put off until $y = 2$, a primitive recursive τ suffices, namely

$$\tau(a) = \begin{cases} 1 \text{ if } \overline{T}_1((a)_0 \dot{-} 1, (a)_0 \dot{-} 1, (a)_1 \dot{-} 1) \ \&\ \mathrm{lh}(a)=2, \\ (a)_1+1 \text{ if } T_1((a)_0 \dot{-} 1, (a)_0 \dot{-} 1, (a)_1 \dot{-} 1) \ \&\ \mathrm{lh}(a)=2, \\ 0 \text{ otherwise.} \end{cases}$$

But if the b were always produced as soon as classically it is determined, then we would have $(y)\overline{T}_1(x, x, y) \equiv \tau(2^{x+1})=1$, so by IM p. 283 τ would not be general recursive.

It will be convenient to take $\tau(\bar{\alpha}(y)) = 0$ after the first y for which $\tau(\bar{\alpha}(y)) > 0$, so that we will have $(\alpha)(E!y)\tau(\bar{\alpha}(y))>0$. Had we merely a τ with $(\alpha)(Ey)\tau(\bar{\alpha}(y))>0$, we could get one with $(\alpha)(E!y)\tau(\bar{\alpha}(y))>0$, call it τ_1, by putting

$$\tau_1(a) = \begin{cases} \tau(a) \text{ if } \mathrm{Seq}(a) \;\&\; (z)_{z<\mathrm{lh}(a)}\tau(\Pi_{i<z} p_i^{(a)_i}) = 0, \\ 0 \text{ otherwise.} \end{cases}$$

The formulas by which we shall presently state Brouwer's principle will thus be interderivable with ones expressing it without taking the y to be unique, using Lemma 5.5 (a), *149a (with *159) and *174b.

Next, consider the hypothesis that a natural number b is correlated to each choice sequence α. This asserts the existence of a function from α's to b's, i.e. a type-2 functional F. We might add functional variables to our formal symbolism to formalize the hypothesis of Brouwer's principle. But we wish to keep the symbolism as simple as possible. At least as far as we go in this monograph (including Vesley's Chapter III), all we need is that, for each α, $b = \mathsf{F}(\alpha)$ bear some given relation $A(\alpha, b)$ to α. Then we can take the hypothesis to be simply $(\alpha)(Eb)A(\alpha, b)$. As we understand intuitionism, $(\alpha)(Eb)A(\alpha, b)$ means exactly that there is a process F by which, given any α, one can find a particular b such that $A(\alpha, b)$. Thus an axiom of choice $(\alpha)(Eb)A(\alpha, b) \rightarrow (E\mathsf{F})(\alpha)A(\alpha, \mathsf{F}(\alpha))$ holds intuitionistically. (Other axioms of choice, corresponding to the prefixes $(x)(Ey)$ and $(x)(E\alpha)$, were introduced above as *2.2 and ˣ2.1.) Brouwer's principle says that any F can be represented by a τ in the manner just described. Then the τ can replace the F. So the F has only a transitory role, between the initial hypothesis $(\alpha)(Eb)A(\alpha, b)$ and the appearance of the τ. By combining the principle of choice implicit in the intuitionistic meaning of the prefix $(\alpha)(Eb)$ with Brouwer's principle for numbers as he directly stated it, we obtain a version of Brouwer's principle, (each instance of) which can be expressed in our formal symbolism as it stands. Thus we arrive at *27.2 as our formalization of Brouwer's principle for numbers. (In **1957*** (4) we already stated in like manner a consequence of Brouwer's principle.)

As another case of Brouwer's principle (the case for functions), if to each choice sequence $\alpha(0), \alpha(1), \alpha(2), \ldots$ a function β is correlated, then intuitionistically each value $\beta(t)$ of β must be determined effectively by t and some initial segment $\alpha(0), \ldots, \alpha(y-1)$ of α. This can be handled as the correlation of a number $b = \beta(t)$ to each of the choice sequences $t, \alpha(0), \alpha(1), \alpha(2), \ldots$.

We originally supposed that this would be required in formalizing Brouwer's proof of his uniform continuity theorem (**1923a** Theorem 3,

1924 Theorem 3, **1927** Theorem 3, **1952** p. 145, **1954** § 6). But Vesley in § 15 of Chapter III has managed using only Brouwer's principle for numbers for each value $\beta(t)$ of β separately.

Still, it seems to us that the intuitionistic reasons for accepting the principle for numbers apply equally to the principle for functions (even though we do not know of an explicit affirmation of it in Brouwer's writings). So we elect to postulate the latter, as $^{x}27.1$ (for the intuitionistic, but not the basic, system); cf. 7.15.

7.2. For $^{x}27.1$, α, β and τ are any distinct function variables, t and y are any distinct number variables, and $A(\alpha, \beta)$ is any formula not containing τ free.

$^{x}27.1$. $\forall\alpha\exists\beta A(\alpha, \beta) \supset \exists\tau\forall\alpha\{\forall t\exists! y\tau(2^{t+1}*\bar{\alpha}(y))>0\ \&$
$\forall\beta[\forall t\exists y\tau(2^{t+1}*\bar{\alpha}(y))=\beta(t)+1 \supset A(\alpha, \beta)]\}$.

Thence we can specialize to Brouwer's principle for numbers.

*27.2. $\vdash \forall\alpha\exists b A(\alpha, b) \supset \exists\tau\forall\alpha\exists y\{\tau(\bar{\alpha}(y))>0\ \&$
$\forall x[\tau(\bar{\alpha}(x))>0 \supset y=x]\ \&\ A(\alpha, \tau(\bar{\alpha}(y)) \dot{-} 1)\}$.

PROOF. Assume $\forall\alpha\exists b A(\alpha, b)$. By *0.6, $\forall\alpha\exists\beta A(\alpha, \beta(0))$. Applying $^{x}27.1$, omitting $\exists\tau$ preparatory to $\exists$-elim., and using $\forall$- and &-elim.,

(1) $\forall t\exists! y\tau(2^{t+1}*\bar{\alpha}(y))>0$,
(2) $\forall\beta[\forall t\exists y\tau(2^{t+1}*\bar{\alpha}(y))=\beta(t)+1 \supset A(\alpha, \beta(0))]$.

Applying *2.2 to (1) (unabbreviating $\exists! y$ by IM p. 199), and omitting $\exists\upsilon$ from the result (preparatory to $\exists$-elim.),

(3) $\forall t[\tau(2^{t+1}*\bar{\alpha}(\upsilon(t)))>0\ \&\ \forall x[\tau(2^{t+1}*\bar{\alpha}(x))>0 \supset \upsilon(t)=x]]$.

By $\forall$-elim. from (2) (using $\lambda t\tau(2^{t+1}*\bar{\alpha}(\upsilon(t))) \dot{-} 1$ for β) and $^{x}0.1$,

(4) $\forall t\exists y\tau(2^{t+1}*\bar{\alpha}(y))=(\tau(2^{t+1}*\bar{\alpha}(\upsilon(t))) \dot{-} 1)+1 \supset A(\alpha, \tau(2*\bar{\alpha}(\upsilon(0))) \dot{-} 1)$.

By $\forall$-elim. from (3) and *6.7, $\tau(2^{t+1}*\bar{\alpha}(\upsilon(t)))=(\tau(2^{t+1}*\bar{\alpha}(\upsilon(t))) \dot{-} 1)+1$, whence $\forall t\exists y\tau(2^{t+1}*\bar{\alpha}(y))=(\tau(2^{t+1}*\bar{\alpha}(\upsilon(t))) \dot{-} 1)+1$. Thence by (4), $A(\alpha, \tau(2*\bar{\alpha}(\upsilon(0))) \dot{-} 1)$, and by $^{x}0.1$,

(5) $A(\alpha, \{\lambda t\tau(2*t)\}(\bar{\alpha}(\upsilon(0))) \dot{-} 1)$.

Also by $\forall$-elim. from (3) (with 0 as the t) and $^{x}0.1$,

(6) $\{\lambda t\tau(2*t)\}(\bar{\alpha}(\upsilon(0)))>0\ \&\ \forall x[\{\lambda t\tau(2*t)\}(\bar{\alpha}(x))>0 \supset \upsilon(0)=x]$.

The formula of *27.2 follows from (6) and (5) by &- and $\exists$y-introd., the $\exists\upsilon$-elim., $\forall\alpha$- and $\exists\tau$-introd., the $\exists\tau$-elim., and $\supset$-introd.

A third case of Brouwer's principle (the case for decisions) applies when A and B are classes such that each choice sequence α belongs to A or to B. Then intuitionistically one of A and B to which α belongs must be determinable from a sufficient initial segment $\alpha(0), \ldots, \alpha(y-1)$ of α.

*27.3. $\vdash \forall\alpha(A(\alpha) \vee B(\alpha)) \supset \exists\tau\forall\alpha\exists y\{\forall x[\tau(\bar{\alpha}(x))>0 \supset y=x]$ & $\{(A(\alpha)$ & $\tau(\bar{\alpha}(y))=1) \vee (B(\alpha)$ & $\tau(\bar{\alpha}(y))=2)\}\}$.

PROOF. Assume $\forall\alpha(A(\alpha) \vee B(\alpha))$. By $\forall$-elim., $A(\alpha) \vee B(\alpha)$. CASE 1: $A(\alpha)$. Using *100, $A(\alpha)$ & $0=0$, whence by $\vee$-introd. $(A(\alpha)$ & $0=0) \vee (B(\alpha)$ & $0=1)$, whence by $\exists$-introd. $\exists b\{(A(\alpha)$ & $b=0) \vee (B(\alpha)$ & $b=1)\}$. CASE 2: $B(\alpha)$. Similarly. — By $\forall$-introd., $\forall\alpha\exists b\{(A(\alpha)$ & $b=0) \vee (B(\alpha)$ & $b=1)\}$. By *27.2, $\exists\tau\forall\alpha\exists y\{\tau(\bar{\alpha}(y))>0$ & $\forall x[\tau(\bar{\alpha}(x))>0 \supset y=x]$ & $\{(A(\alpha)$ & $\tau(\bar{\alpha}(y)) \dot{-} 1=0) \vee (B(\alpha)$ & $\tau(\bar{\alpha}(y)) \dot{-} 1=1)\}\}$. Via $\exists$-, $\forall$-, &- and $\vee$-elims. and introds. and *6.7 in proper sequence, and $\supset$-introd., we obtain the formula of *27.3.

Using *2.2 and *6.7, the converse of ˣ27.1 can be proved in the basic system. So in ˣ27.1 the outermost $\supset$ can be strengthened to $\sim$ (analogously to ˣ2.1); call it then *27.1a. Likewise (usually more simply), *27.2–*27.10, *27.15 can be strengthened to *27.2a–*27.10a, *27.15a ($B \supset A$, C & $B \supset A$, D & C & $B \supset A$ becoming $B \sim A$, $C \supset (B \sim A)$, D & $C \supset (B \sim A)$).

7.3. In 7.2 we introduced Brouwer's principle for the universal spread. It applies to other spreads.

*27.4. $\vdash \mathrm{Spr}(\sigma)$ & $\forall\alpha_{\alpha\in\sigma}\exists\beta A(\alpha, \beta) \supset \exists\tau\forall\alpha_{\alpha\in\sigma}\{\forall t\exists! y\tau(2^{t+1}*\bar{\alpha}(y))>0$ & $\forall\beta[\forall t\exists y\tau(2^{t+1}*\bar{\alpha}(y))=\beta(t)+1 \supset A(\alpha, \beta)]\}$.

*27.5. $\vdash \mathrm{Spr}(\sigma)$ & $\forall\alpha_{\alpha\in\sigma}\exists b A(\alpha, b) \supset \exists\tau\forall\alpha_{\alpha\in\sigma}\exists y\{\tau(\bar{\alpha}(y))>0$ & $\forall x[\tau(\bar{\alpha}(x))>0 \supset y=x]$ & $A(\alpha, \tau(\bar{\alpha}(y)) \dot{-} 1)\}$.

*27.6. $\vdash \mathrm{Spr}(\sigma)$ & $\forall\alpha_{\alpha\in\sigma}(A(\alpha) \vee B(\alpha)) \supset \exists\tau\forall\alpha_{\alpha\in\sigma}\exists y\{\forall x[\tau(\bar{\alpha}(x))>0 \supset y=x]$ & $\{(A(\alpha)$ & $\tau(\bar{\alpha}(y))=1) \vee (B(\alpha)$ & $\tau(\bar{\alpha}(y))=2)\}\}$.

PROOFS. *27.4. CASE 1: $\sigma(1)\neq 0$. Then $\neg\alpha\in\sigma$. Use *10a. CASE 2: $\sigma(1)=0$. Assuming also $\mathrm{Spr}(\sigma)$, we can introduce π and γ as in I of the proof of *26.4a and (α)–(η) will hold. Assume $\forall\alpha_{\alpha\in\sigma}\exists\beta A(\alpha, \beta)$. By $\forall$-elim. $\alpha_\gamma\in\sigma \supset \exists\beta A(\alpha_\gamma, \beta)$, whence by (ε) and $\forall$-introd. $\forall\alpha\exists\beta A(\alpha_\gamma, \beta)$.

Using $^{\times}27.1$ for $A(\alpha_\gamma, \beta)$ as the $A(\alpha, \beta)$, and omitting $\exists\tau$ preparatory to $\exists$-elim., $\forall\alpha\{\forall t\exists!y\tau(2^{t+1}*\bar{\alpha}(y))>0 \,\&\, \forall\beta[\forall t\exists y\tau(2^{t+1}*\bar{\alpha}(y)) = \beta(t)+1 \supset A(\alpha_\gamma, \beta)]\}$. Thence, assuming $\alpha\epsilon\sigma$ and using $\forall$-elim. and (η), $\forall t\exists!y\tau(2^{t+1}*\bar{\alpha}(y))>0 \,\&\, \forall\beta[\forall t\exists y\tau(2^{t+1}*\bar{\alpha}(y))=\beta(t)+1 \supset A(\alpha, \beta)]$.

*27.5, *27.6. Similarly from *27.2 and *27.3, respectively, or successively from *27.4 as *27.2 and *27.3 from $^{\times}27.1$.

7.4. Brouwer combined his principle (for numbers) with what we already have in *26.6 (or *26.7) to state his "fan theorem" thus: If to each element δ of a fan a natural number b_δ is correlated, then a natural number z can be determined such that, for each δ, the correlated number b_δ is completely determined by the first z choices of the choice sequence α generating δ. We establish this now. First we deal directly with the fan of the choice sequences α (cf. 6.10).

*27.7. $\vdash \forall\alpha_{B(\alpha)}\exists bA(\alpha, b) \supset$
$\exists z\forall\alpha_{B(\alpha)}\exists b\forall\gamma_{B(\gamma)}\{\forall x_{x<z}\gamma(x)=\alpha(x) \supset A(\gamma, b)\}$

where $B(\alpha)$ is $\forall x\alpha(x)\leq\beta(\bar{\alpha}(x))$.

*27.8. $\vdash \mathrm{Spr}(\sigma) \,\&\, \forall a[\sigma(a)=0 \supset \exists b\forall s\{\sigma(a*2^{s+1})=0 \supset s\leq b\}] \,\&$
$\forall\alpha_{\alpha\epsilon\sigma}\exists bA(\alpha, b) \supset \exists z\forall\alpha_{\alpha\epsilon\sigma}\exists b\forall\gamma_{\gamma\epsilon\sigma}\{\forall x_{x<z}\gamma(x)=\alpha(x) \supset A(\gamma, b)\}$.

PROOFS. *27.7. As in I of the proof of *26.6a, we introduce σ so that (a)–(c), (1) and (2) hold. Assume **(d)** $\forall\alpha_{B(\alpha)}\exists bA(\alpha, b)$. Applying *27.5 with (1) and (c), and omitting the $\exists\tau$ for $\exists$-elim.,

(e) $\forall\alpha_{B(\alpha)}\exists y\{\tau(\bar{\alpha}(y))>0 \,\&\, \forall x[\tau(\bar{\alpha}(x))>0 \supset y=x] \,\&\, A(\alpha, \tau(\bar{\alpha}(y)) \dot{-} 1)\}$.

We have the first hyp. of *26.6a for $\tau(a)>0$ as the R(a) by *159 or by *15.1 and *158, and the second using (e). So, applying *26.6a and omitting $\exists z$ for $\exists$-elim.,

(f) $\forall\alpha_{B(\alpha)}\exists x_{x\leq z}\tau(\bar{\alpha}(x))>0$.

Toward the concl. of *27.7, assume $B(\alpha)$. Using (e), assume for (&- and) $\exists$-elim. **(h_1)** $\forall x[\tau(\bar{\alpha}(x))>0 \supset y_1=x]$. Assume $B(\gamma)$. Using (e), assume **(g_2)** $\tau(\bar{\gamma}(y_2))>0$, **($h_2$)** $\forall x[\tau(\bar{\gamma}(x))>0 \supset y_2=x]$ and **(i_2)** $A(\gamma, \tau(\bar{\gamma}(y_2)) \dot{-} 1)$; and using (f), assume **(j)** $x\leq z$ and **(k)** $\tau(\bar{\gamma}(x))>0$. Assume $\forall x_{x<z}\gamma(x)=\alpha(x)$. Thence using (j) with *B19 and $^{\times}23.1$ $\bar{\gamma}(x)=\bar{\alpha}(x)$. By (k) and ($h_2$) $y_2=x$, so $\bar{\gamma}(y_2) = \bar{\gamma}(x) = \bar{\alpha}(x)$, whence by ($g_2$) and ($h_1$) $y_1=x$, so $\bar{\gamma}(y_2)=\bar{\alpha}(y_1)$. Hence by ($i_2$), $A(\gamma, \tau(\bar{\alpha}(y_1)) \dot{-} 1)$. Now we can carry out the $\supset$-, $\forall$- and $\exists$-introds. and &- and $\exists$-elims. in proper order, with $\tau(\bar{\alpha}(y_1)) \dot{-} 1$ as the b.

*27.8. Similarly from *26.7a. —

Now consider the fan of elements δ determined by a choice law σ (satisfying the first two hypotheses of *27.8) and a correlation law ρ (cf. 6.1, 6.9). That a b (such that $A(\delta, b)$) is correlated to each element δ can be expressed in the formal symbolism by $\forall\delta\{\exists\alpha_{\alpha\in\sigma}\forall t\delta(t)=\rho(\bar{\alpha}(t)) \supset \exists bA(\delta, b)\}$. Via *96, ×0.1 and Lemma 4.2, this is equivalent to $\forall\alpha_{\alpha\in\sigma}\exists bA(\lambda t\rho(\bar{\alpha}(t)), b)$. The fan theorem for such a fan is expressed by *27.8 with $A(\lambda t\rho(\bar{\alpha}(t)), b)$ as the $A(\alpha, b)$.

7.5. The postulates and results thus far obtained place us in a favorable position for following Brouwer's development of his analysis. This is done up to a point by Vesley in Chapter III below. We continue here with investigations concerned more with foundations.

Using Brouwer's principle, we can dispense with the hypothesis $\forall a[\text{Seq}(a) \supset R(a) \vee \neg R(a)]$ ($\forall a[\sigma(a)=0 \supset R(a) \vee \neg R(a)]$) in our previous version *26.6a (*26.7a) of the fan theorem.

*27.9. $\vdash \forall\alpha_{B(\alpha)}\exists xR(\bar{\alpha}(x)) \supset \exists z\forall\alpha_{B(\alpha)}\exists x_{x\leq z}R(\bar{\alpha}(x))$

where $B(\alpha)$ is $\forall t\alpha(t)\leq\beta(\bar{\alpha}(t))$.

*27.10. $\vdash \text{Spr}(\sigma) \;\&\; \forall a[\sigma(a)=0 \supset \exists b\forall s\{\sigma(a*2^{s+1})=0 \supset s\leq b\}] \;\&$
$\forall\alpha_{\alpha\in\sigma}\exists xR(\bar{\alpha}(x)) \supset \exists z\forall\alpha_{\alpha\in\sigma}\exists x_{x\leq z}R(\bar{\alpha}(x))$.

Proofs. *27.9. This will follow from *27.11 as *26.6a from *26.4a; the following proof is more direct. We proceed as in the proof of *27.7 down through (f), for $R(\bar{\alpha}(b))$ as the $A(\alpha, b)$. For $\exists$-elim. from Lemma 5.3 (b), assume

(g) $\gamma(0)=1 \;\&\; \forall k[\gamma(k')=\gamma(k)\cdot(p_k \exp 1+\max_{t\leq\gamma(k)}\beta(t))]$

(cf. Kleene **1956** Footnote 8). By ind. on k as follows,

(h) $B(\alpha) \;\&\; x\leq k \supset \bar{\alpha}(x)\leq\gamma(k)$.

Ind. step. Assume $B(\alpha) \;\&\; x\leq k'$. Case 1: $x\leq k$. Then $\bar{\alpha}(x) \leq \gamma(k)$ [hyp. ind.] $\leq \gamma(k')$ [(g), and *143a with *3.9 and *18.5]. Case 2: $x=k'$. Then $\bar{\alpha}(x) = \bar{\alpha}(k)\cdot p_k^{\alpha(k)+1}$ [*23.8] $\leq \bar{\alpha}(k)\cdot(p_k \exp 1+\max_{t\leq\gamma(k)}\beta(t))$ [for, $\alpha(k) \leq \beta(\bar{\alpha}(k))$ [by $B(\alpha)$] $\leq \max_{t\leq\gamma(k)}\beta(t)$ [hyp. ind., *H7]; now use *145c (in 5.5 ¶ 4 above) with *144b, and *3.12 with *18.5] $\leq \gamma(k)\cdot(p_k \exp 1+\max_{t\leq\gamma(k)}\beta(t))$ [hyp. ind. and *145c again] $=$ $\gamma(k')$ [(g)]. — Toward the concl. of *27.9, assume **(i)** $B(\alpha)$. By (f), $\exists x_{x\leq z}\tau(\bar{\alpha}(x))>0$. Thence assume **(j)** $x\leq z \;\&\; \tau(\bar{\alpha}(x))>0$; and using (e), assume **(k)** $\forall x[\tau(\bar{\alpha}(x))>0 \supset y=x]$ and **(l)** $R(\bar{\alpha}(\tau(\bar{\alpha}(y))\dot{-}1))$. By (j) and (k) y=x, so by (ł): **(m)** $R(\bar{\alpha}(\tau(\bar{\alpha}(x))\dot{-}1))$. By *H7 and (h) with (i) and (j): **(n)** $\tau(\bar{\alpha}(x))\dot{-}1\leq \max_{s\leq\gamma(z)}\tau(s)\dot{-}1$. Using (n) and (m), we obtain the concl. of *27.9 by &-, ∃-, ⊃- and ∀-introds. and &- and ∃-elims. in proper sequence, with $\tau(\bar{\alpha}(x))\dot{-}1$ as the x and $\max_{s\leq\gamma(z)}\tau(s)\dot{-}1$ as the z.

*27.10. From *27.9 as *26.7a from *26.6a.

We can similarly weaken the hyp. of the bar theorem *26.4a in the case the spread is a fan. These results, with those in 6.7–6.9, 7.6 and 7.14 give a survey by proof-theoretic methods of intuitionistic vs. classical forms of the bar theorem.

*27.11. $\vdash \forall\alpha_{B(\alpha)}\exists x R(\bar{\alpha}(x))$ &
$\forall\alpha\forall x[\forall t_{t<x}\alpha(t)\leq\beta(\bar{\alpha}(t))$ & $R(\bar{\alpha}(x))\supset A(\bar{\alpha}(x))]$ &
$\forall\alpha\forall x[\forall t_{t<x}\alpha(t)\leq\beta(\bar{\alpha}(t))$ &
$\forall s\{s\leq\beta(\bar{\alpha}(x))\supset A(\bar{\alpha}(x)*2^{s+1})\}\supset A(\bar{\alpha}(x))]\supset A(1)$

where $B(\alpha)$ is $\forall t\alpha(t)\leq\beta(\bar{\alpha}(t))$.

*27.12. $\vdash \mathrm{Spr}(\sigma)$ & $\sigma(1)=0$ &
$\forall a[\sigma(a)=0\supset\exists b\forall s\{\sigma(a*2^{s+1})=0\supset s\leq b\}]$ &
$\forall\alpha_{\alpha\in\sigma}\exists x R(\bar{\alpha}(x))$ & $\forall a[\sigma(a)=0$ & $R(a)\supset A(a)]$ &
$\forall a[\sigma(a)=0$ & $\forall s\{\sigma(a*2^{s+1})=0\supset A(a*2^{s+1})\}\supset A(a)]\supset A(1)$.

Proofs. *27.11. We proceed as in the proof of *27.9 through (h). Let $R'(a)$ be $\exists u_{u\leq\gamma(z)}[\sigma(u)=0$ & $\mathrm{lh}(a)+1=\tau(u)$ & $\forall i_{i<\min(\mathrm{lh}(a),\mathrm{lh}(u))}(a)_i=(u)_i]$. Now assume the remaining two hyps. **(o)** and **(p)** of *27.11. We shall apply *26.4a with $R'(a)$ as the $R(a)$. Of the remaining hyps. **(3′)–(6′)** of *26.4a, we have (3′) by Remark 4.1 (or #D, #E etc.). Toward (4′), assume $\alpha\in\sigma$, whence (c) gives (i) $B(\alpha)$, so we may further assume (j). Using (h)–(j), (b), *6.7, *23.2 etc. we get $R'(\bar{\alpha}(\tau(\bar{\alpha}(x))\dot{-}1))$ with $\bar{\alpha}(x)$ as the u. To deduce (5′) from (o) and (a), (b) etc., it will suffice (using (a), *23.6) to deduce $\sigma(\overline{\alpha_1}(x_1))=0$ & $R'(\overline{\alpha_1}(x_1))\supset R(\overline{\alpha_1}(x_1))$. So assume $\sigma(\overline{\alpha_1}(x_1))=0$ & $R'(\overline{\alpha_1}(x_1))$, and for ∃-elim. (simplifying): **(q)** $u\leq\gamma(z)$ & $\sigma(u)=0$ & $x_1+1=\tau(u)$ & $\forall i_{i<\min(x_1,\mathrm{lh}(u))}\alpha_1(i)+1=(u)_i$. By $\sigma(u)=0$ with (a), Seq(u). Using Lemma 5.3 (a), let $\forall i\alpha(i)=\max((\overline{\alpha_1}(x_1))_i, (u)_i)\dot{-}1$. Using the last part of (q), *23.2, *23.3, *22.2 etc., $i<x_1\supset \alpha(i)=\alpha_1(i)$, $i<\mathrm{lh}(u)\supset\alpha(i)=(u)_i\dot{-}1$ and **(r)** $i\geq\max(x_1, \mathrm{lh}(u))\supset\alpha(i)=0$. So using first *B19 and ×23.1, then also *22.3, *6.7 with *22.1, etc.: **(s)** $\bar{\alpha}(x_1)=\overline{\alpha_1}(x_1)$ & $\bar{\alpha}(\mathrm{lh}(u))=u$. Thence using also $\sigma(\overline{\alpha_1}(x_1))=0$ and $\sigma(u)=0$ with (b), (c) and (r): **(i)** $B(\alpha)$. So as in the proof of *27.9 we can assume (k) and (ł). By (s) with (q): **(t)** $x_1+1=\tau(\bar{\alpha}(\mathrm{lh}(u)))$. So by

(k) y=lh(u), whence by (ł) $R(\bar{\alpha}(\tau(\bar{\alpha}(lh(u))) \dot{-} 1))$, whence by (t) and *6.3 with (s), $R(\overline{\alpha_1}(x_1))$. — Finally (6′) follows from (p) and (b).

*27.12. Assuming the six hyps. **(1″)–(6″)** of *27.12, we take over I and II of the proof of *26.7a Case 2.

III. We shall apply *27.11 for the β of (θ) with R(γ(a)) as the R(a) and σ(a)=0 ⊃ A(γ(a)) as the A(a). B. The first hyp. of *27.11 (= the second of *26.6a) is verified as before. For the second, assume $R(\gamma(\bar{\alpha}(x)))$. By (γ) and (5″) $A(\gamma(\bar{\alpha}(x)))$, whence $\sigma(\bar{\alpha}(x))=0 \supset A(\gamma(\bar{\alpha}(x)))$. For the third it will suffice, assuming

(ι) $\forall s\{s \leq \beta(\bar{\alpha}(x)) \supset [\sigma(\bar{\alpha}(x)*2^{s+1})=0 \supset A(\gamma(\bar{\alpha}(x)*2^{s+1}))]\}$,

to deduce $\sigma(\bar{\alpha}(x))=0 \supset A(\gamma(\bar{\alpha}(x)))$. Assume **(κ)** $\sigma(\bar{\alpha}(x))=0$. By (γ) and (6″) it will suffice, assuming **(λ)** $\sigma(\gamma(\bar{\alpha}(x))*2^{s+1})=0$, to deduce $A(\gamma(\bar{\alpha}(x))*2^{s+1})$. From (λ) by (κ) and (ζ): **(μ)** $\sigma(\bar{\alpha}(x)*2^{s+1})=0$. By (θ) with (κ) and (μ): **(ν)** $s \leq \beta(\bar{\alpha}(x))$. By (ι) with (ν) and (μ) $A(\gamma(\bar{\alpha}(x)*2^{s+1}))$, whence by (ζ) with (μ) $A(\bar{\alpha}(x)*2^{s+1})$, whence by (ζ) with (κ) $A(\gamma(\bar{\alpha}(x))*2^{s+1})$.

7.6. Using (ˣ27.1 via) *27.2, we can establish a fifth intuitionistic version of the bar theorem (cf. ˣ26.3a–ˣ26.3d).

*27.13. $\vdash \forall\alpha\forall x[R(\bar{\alpha}(x)) \supset \forall y_{y>x}R(\bar{\alpha}(y))]\ \&$
$\forall\alpha\exists xR(\bar{\alpha}(x))\ \&\ \forall a[Seq(a)\ \&\ R(a) \supset A(a)]\ \&$
$\forall a[Seq(a)\ \&\ \forall sA(a*2^{s+1}) \supset A(a)] \supset A(1)$.

Proof. Assume the four hyps. **(1)–(4)** of *27.13. Using (2) with *27.2, and omitting ∃τ for ∃-elim.,

(a) $\forall\alpha\exists y\{\tau(\bar{\alpha}(y))>0\ \&\ \forall x[\tau(\bar{\alpha}(x))>0 \supset y=x]\ \&\ R(\bar{\alpha}(\tau(\bar{\alpha}(y)) \dot{-} 1))\}$.

Let R′(a) be $lh(a)=\langle(lh(a))_0, (lh(a))_1\rangle\ \&\ \tau(\Pi_{i<(lh(a))_0}p_i^{(a)_i})=(lh(a))_1+1$, so

(b) $R'(\bar{\alpha}(x)) \sim x=\langle(x)_0, (x)_1\rangle\ \&\ \tau(\bar{\alpha}((x)_0))=(x)_1+1$

[*23.4 with *19.2 etc.]. We shall apply ˣ26.3a with R′(a) as the R(a). We have the first hyp. of ˣ26.3a by Remark 4.1 or #D. The second with $\langle y, \tau(\bar{\alpha}(y)) \dot{-} 1\rangle$ as the x we obtain from (a) and *25.1, *6.7. Toward the third, assume Seq(a) & R′(a), and $a=\bar{\alpha}(x)$, so we have the right side **(c)** of (b). Using (a), assume for ∃-elim. **(d)** $\forall x[\tau(\bar{\alpha}(x))>0 \supset y=x]$ and **(e)** $R(\bar{\alpha}(\tau(\bar{\alpha}(y)) \dot{-} 1))$. By (c) with (d) $y=(x)_0$, so $\tau(\bar{\alpha}(y)) \dot{-} 1 = \tau(\bar{\alpha}((x)_0)) \dot{-} 1 = ((x)_1+1) \dot{-} 1 = (x)_1$ [*6.3] $\leq x$ [*19.2,

*19.6]. Hence by (e) and (l), $R(\bar{\alpha}(x))$, i.e. $R(a)$. So by (3), $A(a)$. The fourth is (4).

DERIVATION OF ×26.3a FROM *27.13 (as postulate in place of ×26.3, without ×27.1). Assume the four hyps. **(1)–(4)** of ×26.3a. Pick $R'(a)$, $A'(a)$ so that via *23.4

(a) $R'(\bar{\alpha}(x)) \sim \exists t_{t \leq x} R(\bar{\alpha}(t))$,
(b) $A'(\bar{\alpha}(x)) \sim A(\bar{\alpha}(x)) \vee \exists t_{t<x} R(\bar{\alpha}(t))$.

We shall apply *27.13 with $R'(a)$, $A'(a)$ as the $R(a)$, $A(a)$. The first hyp. of *27.13 is immediate. The second follows from (2). Toward the third, assume $\text{Seq}(a) \& R'(a)$, and put $a=\bar{\alpha}(x)$. Using (1), $R(a) \vee \neg R(a)$. CASE 1: $R(a)$. Then $A(a)$ by (3), whence $A'(a)$ by (b). CASE 2: $\neg R(a)$. Then by $R'(a)$ with (a) $\exists t_{t<x} R(\bar{\alpha}(t))$, whence $A'(a)$ by (b). — Toward the fourth, assume $\text{Seq}(a) \& \forall s A'(a*2^{s+1})$, and put $a=\bar{\alpha}(x)$. By (1) with *150, $\exists t_{t<x'} R(\bar{\alpha}(t)) \vee \neg \exists t_{t<x'} R(\bar{\alpha}(t))$. CASE 1: $\exists t_{t<x'} R(\bar{\alpha}(t))$. SUBCASE 1: $\exists t_{t<x} R(\bar{\alpha}(t))$. Then $A'(a)$ by (b). SUBCASE 2: $R(\bar{\alpha}(x))$. Then $A'(a)$ by (3) and (b). CASE 2: $\neg \exists t_{t<x'} R(\bar{\alpha}(t))$. But $\forall s A'(a*2^{s+1})$ gives $\forall s[A(a*2^{s+1}) \vee \exists t_{t<x'} R(\bar{\alpha}(t))]$, whence by the case $\forall s A(a*2^{s+1})$. By (4) and (b), $A'(a)$.

*27.14. $\vdash \text{Spr}(\sigma) \& \sigma(1)=0 \& \forall\alpha\forall x[\sigma(\bar{\alpha}(x))=0 \& R(\bar{\alpha}(x)) \supset \forall y_{y>x}\{\sigma(\bar{\alpha}(y))=0 \supset R(\bar{\alpha}(y))\}] \& \forall\alpha_{\alpha\in\sigma}\exists x R(\bar{\alpha}(x)) \& \forall a[\sigma(a)=0 \& R(a) \supset A(a)] \& \forall a[\sigma(a)=0 \& \forall s\{\sigma(a*2^{s+1})=0 \supset A(a*2^{s+1})\} \supset A(a)] \supset A(1)$.

7.7. Brouwer's principle for numbers can be stated (similarly to the fan theorem in *27.7) without mentioning the algorithm τ explicitly.

*27.15. $\vdash \forall\alpha\exists b A(\alpha, b) \supset \forall\alpha\exists y\exists b\forall\gamma\{\forall x_{x<y}\gamma(x)=\alpha(x) \supset A(\gamma, b)\}$.

This formula appears in Kreisel **1962** Remark 9, where it is misnamed "the bar theorem"; cf. Remark 6.3. (Kreisel seems to be suggesting it as a postulate to replace the postulate *27.7 of Kleene **1957**; but by Corollary 9.9 below, *27.7 would not then be provable lacking ×26.3.) It is Heyting 1930a 12.22, allowing for considerable differences in the symbolism. In the hypothesis Heyting explicitly uses a function from choice sequences to natural numbers, which our symbolism doesn't provide. In *27.15 our practice (initiated in **1957**) is followed under which the existence of the correlation of b to α is implied by the simple prefix $\forall\alpha\exists b$, or for a fan $\forall\alpha_{B(\alpha)}\exists b$ etc. (cf. 7.1). — The proof of *27.15

from ${}^{\times}27.1$ via $*27.2$ (without ${}^{\times}26.3$) is straightforward by reasoning used in the proof of $*27.7$. We do not see how $*27.2$ could be proved from $*27.15$ as a postulate replacing ${}^{\times}27.1$.

7.8. In 7.1 we explained Brouwer's principle on the basis of the intuitionistic conception of choice sequences as continually growing rather than as completed objects. This applies in talking about the totality of choice sequences α making up the universal spread or any other spread. It is true that the property $\alpha \in \sigma$ of being a choice sequence α of a spread involves all the values of α (cf. 6.9). But it involves them only in such a way that, whenever at a given stage $\bar{\alpha}(x)$ in the growth of α the property $\alpha \in \sigma$ is satisfied "thus far", i.e. $\sigma(\bar{\alpha}(x))=0$, the growth can continue so that the property will be satisfied at every subsequent finite stage, and thus so that $\alpha \in \sigma$ itself will be satisfied. But a species (not a spread) of choice sequences α can be characterized by a property $C(\alpha)$ involving all the values of α not merely in that manner (cf. 6.12 ¶ 3). In reasoning about the members α of such a species, $C(\alpha)$ functions as a non-constructive hypothesis, which so to speak augments what the intuitionist can do "by himself". (This theme will be developed in Chapter II, especially 8.6.)

We now show that Brouwer's principle as formulated in $*27.4$–$*27.6$ for any spread (characterized by $\alpha \in \sigma$) cannot be extended consistently with the basic system to arbitrary species (characterized by $C(\alpha)$). Indeed, using only the intuitionistic Postulate Groups A–D:

$*27.16$. $\vdash \neg[\forall\alpha_{C(\alpha)}(A(\alpha) \vee B(\alpha)) \supset \exists\tau\forall\alpha_{C(\alpha)}\exists y\{\forall x[\tau(\bar{\alpha}(x))>0 \supset y=x]$
$\quad\& \{(A(\alpha) \,\&\, \tau(\bar{\alpha}(y))=1) \vee (B(\alpha) \,\&\, \tau(\bar{\alpha}(y))=2)\}\}]$

when $A(\alpha)$ is $\forall x\alpha(x)=0$, $B(\alpha)$ is $\neg\forall x\alpha(x)=0$, and $C(\alpha)$ is $A(\alpha) \vee B(\alpha)$.

PROOF. $\forall\alpha_{C(\alpha)}(A(\alpha) \vee B(\alpha))$ holds by the principle of identity $*1$. So via $\exists$-elim. it will suffice to deduce a contradiction from

(a) $\forall\alpha_{C(\alpha)}\exists y\{\forall x[\tau(\bar{\alpha}(x))>0 \supset y=x] \,\&\, \{(A(\alpha) \,\&\, \tau(\bar{\alpha}(y))=1) \vee$
$(B(\alpha) \,\&\, \tau(\bar{\alpha}(y))=2)\}\}$.

Assume $\forall x\alpha_1(x)=0$ (Lemma 5.3 (a)), so $A(\alpha_1)$, hence $C(\alpha_1)$. Applying (a) and omitting $\exists y_1$ for $\exists$-elim.,

(b) $\tau(\overline{\alpha_1}(y_1))=1$.

Now assume $\forall x\alpha_2(x)=x' \dot{-} y_1$, so (by $*6.11$ etc.)

(c) $x<y_1 \supset \alpha_2(x)=0$, (d) $x \geq y_1 \supset \alpha_2(x)>0$.

By (d), $\alpha_2(y_1) \neq 0$, so $\neg\forall x \alpha_2(x)=0$, i.e. B($\alpha_2$), hence C($\alpha_2$). Applying (a) and omitting $\exists y_2$ for $\exists$-elim.,

(e) $\forall x[\tau(\overline{\alpha_2}(x))>0 \supset y_2=x]$, (f) $\tau(\overline{\alpha_2}(y_2))=2$.

But by (c) (and *B19, ^{x}23.1), $\overline{\alpha_2}(y_1)=\overline{\alpha_1}(y_1)$, so by (b):

(g) $\tau(\overline{\alpha_2}(y_1))=1$.

By (g) and (e) $y_2=y_1$, so by (f) and (g) $2=1$, contradicting $2\neq 1$.

Thus, with a particular choice of A(α), B(α), C(α), we have refuted the modification of *27.6 obtained by suppressing Spr(σ) as hyp. and changing $\alpha\in\sigma$ to C(α). The corresponding modifications of *27.4 and *27.5 are likewise refutable, since that of *27.6 is deducible from each of them in the way that *27.3 was deduced from *27.2 and thence from ^{x}27.1.

7.9. Now (through subsection 7.14), we shall explore consequences of Brouwer's principle which contradict classical results.

We begin in the next subsection with refutations of laws of classical logic. Demonstrations that various laws of classical logic are not provable in intuitionistic logic have been given in a number of ways.

For the propositional calculus, cf. Gödel 1932, Gentzen 1934–5 (or IM pp. 479–486), Jaśkowski 1936 (and Pil'čak **1952**, G. F. Rose **1953**), Wajsberg 1938, Stone **1937–8**, Tarski **1938**, McKinsey-Tarski **1946**, 1948 (which uses ideas that appeared in Skolem 1919; cf. Skolem **1958**, Scott **1960**), Scott **1957**, Kreisel-Putnam **1957**, Schmidt **1958**, Harrop **1956**, **1960**, Kleene **1962a**.

For the predicate calculus, the methods of Gentzen 1934–5 were applied to this end by Curry 1950 and Kleene 1948 and IM § 80. Mostowski 1948 used a topological interpretation of the predicate calculus (extending Stone **1937–8**, Tarski **1938**). Kleene 1945 used unrealizable number-theoretic formulas constituting (free-)substitution instances of the predicate letter formulas, from which the unprovability of the latter formulas follows by Nelson's theorem 1947 that all formulas provable in the predicate calculus are realizable (also in IM § 82). An elegant new method of Beth **1956**, related to Gentzen 1934–5, is available on the basis of an outline in Kreisel **1958b** p. 381 of how to correct errors in Beth's proof (cf. the reviews Kleene **1957a** and Kreisel **1960**) and a report Dyson-Kreisel **1961** carrying out these corrections. Other criteria are in Rasiowa **1954**, **1954a**, Harrop **1960**, Kleene **1962a**.

Brouwer in **1924–7** I, **1925**, **1927**, 1928 and Heyting 1930 p. 50, 1930a p. 65 used contradicting propositions in intuitionistic analysis to refute formulas of the classical propositional and predicate calculi. We shall return to this earliest method here. But we shall be using explicitly given formation and transformation rules, while Brouwer's and Heyting's examples were given informally. (Heyting had such a formalism later in 1930a, but he didn't restate those examples in it.) Also their examples presuppose that intuitionistic analysis provides a model for intuitionistic predicate calculus, which hardly had been demonstrated explicitly then. Likewise our exhibiting a contradicting formula in our system of intuitionistic analysis will not show unprovability in the intuitionistic predicate calculus, until we have given a consistency proof for the system. Such a proof, employing a realizability notion for intuitionistic analysis, will be given in Chapter II. By Gödel's second theorem (1931; IM pp. 210–213), no proof of the consistency can be elementary. The only demonstrations of the intuitionistic unprovability of classically provable formulas of the predicate (not merely the propositional) calculus which seem to us really elementary are those based on Gentzen's Hauptsatz 1934–5 (IM p. 453), including Beth's demonstrations. Simply to demonstrate unprovability in the intuitionistic predicate calculus, there would be slight point to the examples here. But having for other reasons gone through the work of reaching the position in which we now stand, and granting what we will also do in Chapter II, the examples are simple and sweeping. Moreover, as we shall see in a moment, the existence of a contradicting formula in intuitionistic analysis is a stronger result than unprovability in the intuitionistic predicate calculus. Also, it rules out provability in any extension of the intuitionistic predicate calculus compatible with intuitionism. The examples are also applied in analysis.

There is no effective method or "decision procedure" (IM §§ 30, 60, 61) to decide in general whether a predicate letter formula A (IM p. 143) already known to be provable in the classical predicate calculus is provable in the intuitionistic predicate calculus, as the following argument (due to Kleene, and published in Beth **1955a** p. 341) shows. Let C be a fixed (say closed) predicate letter formula provable in the classical predicate calculus, but unprovable in the intuitionistic predicate calculus (examples in the literature or below). Let A be any (closed) predicate letter formula. Then A ∨ C is provable in the classical predicate calculus. But by Gentzen 1934–5 p. 407 (or IM Theorem 57 (a) p. 486 but for the predicate calculus), A ∨ C is provable in the intuitionistic predicate calculus exactly if A or C is so provable, i.e. since C is not, exactly if A is. So, were there a decision procedure for the intuitionistic provability of classically provable formulas, there would be one for the intuitionistic provability of arbitrary formulas. This is contrary to the version for the intuitionistic predicate calculus of a theorem of Church 1936a and Turing 1936–7 (IM Theorem 54 p. 432).

We should not seek to "refute" formulas of the predicate calculus in intuitionistic analysis simply by proving the negations of substitution instances. Thus $\mathcal{A} \vee \neg\mathcal{A}$ is unprovable in the intuitionistic propositional (and predicate) calculus. But no formula of the form ¬(A ∨ ¬A) is provable in intuitionistic analysis, if that is consistent. For $\neg\neg(\mathcal{A} \vee \neg\mathcal{A})$ is provable in the intuitionistic propositional calculus (*51a), so ¬¬(A ∨ ¬A) is in intuitionistic analysis. In general, we must seek rather to prove negations of closures of (free-)substitution instances (cf. Kleene 1945 §§ 10, 16). For example, we will "refute" $\mathcal{A} \vee \neg\mathcal{A}$ by proving in intuitionistic analysis $\neg\forall\alpha(A(\alpha) \vee \neg A(\alpha))$ for a suitable $A(\alpha)$ (*27.17). This of course does show $\mathcal{A} \vee \neg\mathcal{A}$ unprovable in intuitionistic propositional (and predicate) calculus, if intuitionistic analysis is consistent, since its provability in intuitionistic propositional (or predicate) calculus would entail that of $A(\alpha) \vee \neg A(\alpha)$, and by ∀-introd. $\forall\alpha(A(\alpha) \vee \neg A(\alpha))$, in intuitionistic analysis.

Not every predicate letter formula provable in the classical predicate calculus but not in the intuitionistic can be refuted in this way, using our formal system of intuitionistic analysis or any other given formalism *F* consistent with the intuitionistic predicate calculus. For if all could be, we would have a decision procedure for the provability in the intuitionistic predicate calculus of a predicate letter formula A provable in the classical predicate calculus thus: search through an enumeration of the provable formulas of the intuitionistic predicate calculus for A itself, and simultaneously through an enumeration of the provable formulas of *F* for the negation of the closure of a substitution instance of A.

7.10. However, with one possible exception, all the predicate letter formulas mentioned in IM as being classically provable but intuitionistically unprovable can be thus refuted in our intuitionistic analysis.

*27.17. $\vdash \neg\forall\alpha(\forall x\alpha(x)=0 \vee \neg\forall x\alpha(x)=0)$.

PROOF. Assume $\forall\alpha(\forall x\alpha(x)=0 \vee \neg\forall x\alpha(x)=0)$. Writing $A(\alpha)$ for $\forall x\alpha(x)=0$ and $B(\alpha)$ for $\neg\forall x\alpha(x)=0$, and applying *27.3,

(a) $$\forall\alpha\exists y\{\forall x[\tau(\bar{\alpha}(x))>0 \supset y=x] \;\&\; \{(A(\alpha) \;\&\; \tau(\bar{\alpha}(y))=1) \vee (B(\alpha) \;\&\; \tau(\bar{\alpha}(y))=2)\}\}.$$

We continue as in the proof of *27.16, except that $C(\alpha_1)$ and $C(\alpha_2)$ aren't required for using (a).

*27.18. $\vdash \neg\forall\alpha(\neg\exists x\alpha(x)\neq 0 \vee \neg\neg\exists x\alpha(x)\neq 0)$.

PROOF. $\forall x\alpha(x)=0 \vee \neg\forall x\alpha(x)=0$ comes from $\neg\exists x\alpha(x)\neq 0 \vee \neg\neg\exists x\alpha(x)\neq 0$ by *86, *158 and *49c.

In *27.17 we directly refute (a closure of a substitution instance of) $\mathcal{A} \vee \neg\mathcal{A}$, $\forall x(\mathcal{A}(x) \vee \neg\mathcal{A}(x))$ and $\neg\neg\forall x(\mathcal{A}(x) \vee \neg\mathcal{A}(x))$ (cf. *49b), and *27.18 adds $\neg\mathcal{A} \vee \neg\neg\mathcal{A}$ etc. to this list.

There must be a like refutation of any predicate letter formula B from a (free-)substitution instance B_1 of which a formula A thus refutable is deducible in the intuitionistic predicate calculus. PROOF. Given: in intuitionistic analysis $\vdash \neg\forall A^*$ where A^* is a substitution instance of A, and in intuitionistic predicate calculus $B_1 \vdash A$. By $\forall$-elim. and $\supset$-introd. $\vdash \forall B_1 \supset A$, whence by substitution (IM Theorem 15 p. 159) and *69 (and perhaps *75) $\vdash \forall B_1^* \supset \forall A^*$ in intuitionistic analysis, whence by contraposition *12 $\vdash \neg\forall A^* \supset \neg\forall B_1^*$. By $\supset$-elim. $\vdash \neg\forall B_1^*$. But B_1^* is a substitution instance of B.

For example, we can hence refute also $\neg\neg\mathcal{A} \supset \mathcal{A}$ (by IM Remark 1 p. 120), $\neg\neg(\forall x\neg\neg\mathcal{A}(x) \supset \neg\neg\forall x\mathcal{A}(x))$ (which is $\neg\neg(Ic_1 \supset Ib)$ IM pp. 166, 491), $\neg\forall x\neg\mathcal{A}(x) \supset \exists x\mathcal{A}(x)$ (since $\neg\neg\mathcal{A} \supset \mathcal{A}$ follows after substituting $\mathcal{A}$ for $\mathcal{A}(x)$), etc. Altogether, using these examples and deductions noted in IM pp. 486 and 491, we can from *27.17 and *27.18 thus refute all the examples in IM for the propositional calculus (Example 4 p. 485, and those listed in Theorem 57 (b) p. 486), and all of those for the predicate calculus (listed in Theorem 58 p. 487) except (b) (i) (or *92), (b) (iii), IIc $\supset$ IIb, IIIb $\supset$ IIIa and *97.

We now refute (b) (i) (or *92).

*27.19. $\vdash \neg\forall\alpha\{\forall x(\alpha(x)=0 \vee \neg\forall x\alpha(x)=0) \supset \forall x\alpha(x)=0 \vee \neg\forall x\alpha(x)=0\}$.

PROOF. By *158, $\alpha(x)=0 \vee \neg\alpha(x)=0$. Thence by cases (using *85a in the second case) and $\forall$-introd., **(a)** $\forall x(\alpha(x)=0 \vee \neg\forall x\alpha(x)=0)$.

Assume $\forall\alpha\{\forall x(\alpha(x)=0 \vee \neg\forall x\alpha(x)=0) \supset \forall x\alpha(x)=0 \vee \neg\forall x\alpha(x)=0\}$. Thence by (a) and *41 $\forall\alpha(\forall x\alpha(x)=0 \vee \neg\forall x\alpha(x)=0)$, contradicting *27.17.

Next we refute *97.

*27.20. $\vdash \neg\forall\alpha\{(\exists x\alpha(x)=0 \supset \exists x\alpha(x)=0) \supset \exists x(\exists x\alpha(x)=0 \supset \alpha(x)=0)\}$.

Proof. Assume $\forall\alpha\{(\exists x\alpha(x)=0 \supset \exists x\alpha(x)=0) \supset \exists x(\exists x\alpha(x)=0 \supset \alpha(x)=0)\}$. Using *1 and *41, $\forall\alpha\exists x(\exists x\alpha(x)=0 \supset \alpha(x)=0)$. Applying *27.2, and omitting $\exists\tau$ for $\exists$-elim.,

(a) $\forall\alpha\exists y\{\tau(\bar{\alpha}(y))>0 \;\&\; \forall x[\tau(\bar{\alpha}(x))>0 \supset y=x] \;\&\; [\exists x\alpha(x)=0 \supset \alpha(\tau(\bar{\alpha}(y)) \dot{-} 1)=0]\}$.

Assume $\forall x\alpha_1(x)=1$. Using $\forall$-elim. from (a), and omitting $\exists y_1$ for $\exists$-elim.,

(b) $\tau(\overline{\alpha_1}(y_1))>0$.

Assume $\forall x\alpha_2(x)=\mathrm{sg}(\max(y_1, \tau(\overline{\alpha_1}(y_1))) \dot{-} x)$, so

(c) $x<\max(y_1, \tau(\overline{\alpha_1}(y_1))) \supset \alpha_2(x)=1$,
(d) $x\geq\max(y_1, \tau(\overline{\alpha_1}(y_1))) \supset \alpha_2(x)=0$.

Applying (a), and omitting $\exists y_2$ for $\exists$-elim.,

(e) $\tau(\overline{\alpha_2}(y_2))>0$, (f) $\forall x[\tau(\overline{\alpha_2}(x))>0 \supset y_2=x]$,
(g) $\exists x\alpha_2(x)=0 \supset \alpha_2(\tau(\overline{\alpha_2}(y_2)) \dot{-} 1)=0$.

By (c): **(h)** $\overline{\alpha_2}(y_1)=\overline{\alpha_1}(y_1)$. So by (b) and (f): **(i)** $y_2=y_1$. By (c) with (h) and (b): **(j)** $\alpha_2(\tau(\overline{\alpha_2}(y_1)) \dot{-} 1)=1$. By (d) $\exists x\alpha_2(x)=0$, so by (g): **(k)** $\alpha_2(\tau(\overline{\alpha_2}(y_2)) \dot{-} 1)=0$. By (j), (k) and (i) $1=0$, contradicting $1\neq0$.

By *158 and *49c $\alpha(x)=0 \sim \neg\neg\alpha(x)=0$, so we also refute $(\mathcal{A} \supset \exists x\mathcal{B}(x)) \supset \exists x(\mathcal{A} \supset \neg\neg\mathcal{B}(x))$. But this we can deduce from a substitution instance of IIc $\supset$ IIb, thus. Assume **(a)** $\mathcal{A} \supset \exists x\mathcal{B}(x)$ and **(b)** $\forall x\neg(\mathcal{A} \supset \mathcal{B}(x))$. From (b) by $\forall$-elim. and *60d, $\neg\neg\mathcal{A}$ and $\neg\mathcal{B}(x)$; from the latter by $\forall$-introd. and *86, $\neg\exists x\mathcal{B}(x)$, whence from (a) by *12, $\neg\mathcal{A}$, contrad. $\neg\neg\mathcal{A}$. So by $\neg$-introd. (discharging the assumption (b)), $\neg\forall x\neg(\mathcal{A} \supset \mathcal{B}(x))$. Thence by a substitution instance of $\mathrm{IIc}_3 \supset$ IIb, $\exists x\neg\neg(\mathcal{A} \supset \mathcal{B}(x))$, and by *60g, h $\exists x(\mathcal{A} \supset \neg\neg\mathcal{B}(x))$. By $\supset$-introd. (discharging (a)), $(\mathcal{A} \supset \exists x\mathcal{B}(x)) \supset \exists x(\mathcal{A} \supset \neg\neg\mathcal{B}(x))$.

Finally, $\mathrm{IIc}_2 \supset$ IIb is a substitution instance of $\mathrm{IIIb}_1 \supset$ IIIa, which for the predicate calculus leaves only (b) (iii) unrefuted.

7.11. We refute the closure of an instance of the least number principle *149 (whence *148 is refutable, like IM p. 513 (vii)).

*27.21. $\vdash \neg \forall \alpha \{\exists x C(\alpha, x) \supset \exists y [C(\alpha, y) \& \forall z (z < y \supset \neg C(\alpha, z))]\}$

when $C(\alpha, y)$ is $y = 1 \vee (\forall x \alpha(x) = 0 \& y = 0)$.

PROOF. Call the formula $\neg \forall \alpha D(\alpha)$. From $\forall \alpha D(\alpha)$ by the method of IM (vi) pp. 512–513 we deduce $\forall \alpha (\forall x \alpha(x) = 0 \vee \neg \forall x \alpha(x) = 0)$, contradicting *27.17.

7.12. We refute the classically-provable duals of *2.1a, *2.2a and *25.9, beginning with the last. (In each of these duals, one implication is provable by the predicate calculus simply.)

*27.22. $\vdash \neg \forall \alpha \{\forall x \exists i_{i<2} A(\alpha, i, (x)_i) \supset \exists i_{i<2} \forall x A(\alpha, i, x)\}$

when $A(\alpha, x)$ is $\alpha(x) = 0$, $B(\alpha)$ is $\neg \forall x \alpha(x) = 0$ and $A(\alpha, i, x)$ is $(A(\alpha, x) \& i = 0) \vee (B(\alpha) \& i = 1)$.

PROOF. Assume

(a) $\forall \alpha \{\forall x \exists i_{i<2} A(\alpha, i, (x)_i) \supset \exists i_{i<2} \forall x A(\alpha, i, x)\}$.

Assume $\forall x (A(\alpha, x) \vee B(\alpha))$. By $\forall$-elim., $A(\alpha, (x)_0) \vee B(\alpha)$. CASE 1: $A(\alpha, (x)_0)$. Then $A(\alpha, (x)_0) \& 0 = 0$, hence $A(\alpha, 0, (x)_0)$, hence $0 < 2 \& A(\alpha, 0, (x)_0)$, hence $\exists i_{i<2} A(\alpha, i, (x)_i)$. CASE 2: $B(\alpha)$. Similarly. Completing the case argument ($\vee$-elim.) and using $\forall$-introd., $\forall x \exists i_{i<2} A(\alpha, i, (x)_i)$. By (a) $\exists i_{i<2} \forall x A(\alpha, i, x)$. Assume $i < 2 \& \forall x A(\alpha, i, x)$ for $\exists$-elim. CASE 1: $i = 0$. Then $\forall x A(\alpha, 0, x)$, hence (using $0 \neq 1$, *47, *48, *89) $\forall x A(\alpha, x)$, hence $\forall x A(\alpha, x) \vee B(\alpha)$. CASE 2: $i = 1$. Similarly. Thus, completing the $\vee$- and $\exists$-elim. and using $\supset$- and $\forall$-introd., $\forall \alpha \{\forall x (A(\alpha, x) \vee B(\alpha)) \supset \forall x A(\alpha, x) \vee B(\alpha)\}$, contradicting *27.19.

This refutation extends to the duals of *2.2a and *2.1a (or their implications beginning with $\forall \alpha \exists x$), since the dual of *25.9 follows from that of *2.2a and thence of *2.1a in the same manner as *25.9 itself from *2.2a and thence from *2.1a.

7.13. We take one illustration from Brouwer's theory of species. In **1924–7** I p. 246 ¶ 9 he gives a pair of species *M* and *N* which are "congruent" but not "identical". As a newcomer to this field in 1941, the present author found this example hard to decipher from Brouwer's austere text. (A different example is in Brouwer **1954** p. 6 ¶ 4.) We now give the **1924–7** example rearranged and simplified. Let

"$\alpha \in M$" abbreviate $\forall x \alpha(x) \neq 0 \vee \neg \forall x \alpha(x) \neq 0$,
"$\alpha \in N$" abbreviate $\forall x \alpha(x) = 0 \vee \neg \forall x \alpha(x) = 0$.

That the species M is "congruent" to the species N is expressed by

(i) $$\neg \exists \alpha(\alpha \in M \ \& \ \neg \alpha \in N) \ \& \ \neg \exists \alpha(\alpha \in N \ \& \ \neg \alpha \in M).$$

This formula is provable, since by *51a $\neg\neg \alpha \in N$ and $\neg\neg \alpha \in M$ are both provable. That M and N are not "identical" is expressed, after a simplification permissible because the elements of M and N are already choice sequences (and not merely spread elements overlying choice sequences by a correlation law), by

(ii) $$\neg\{\forall \alpha(\alpha \in M \supset \alpha \in N) \ \& \ \forall \alpha(\alpha \in N \supset \alpha \in M)\}.$$

Both $\neg \forall \alpha(\alpha \in M \supset \alpha \in N)$ and $\neg \forall \alpha(\alpha \in N \supset \alpha \in M)$, a fortiori (ii), are provable. To prove the first, assume $\forall \alpha[\alpha \in M \supset \alpha \in N]$, i.e.

(a) $\forall \alpha[\forall x \alpha(x) \neq 0 \vee \neg \forall x \alpha(x) \neq 0 \supset \forall x \alpha(x) = 0 \vee \neg \forall x \alpha(x) = 0]$.

Assume $\forall x \alpha_1(x) = sg(x) \cdot \alpha(x \dot{-} 1)$, so $\alpha_1(0) = 0$, $\alpha_1(x') = \alpha(x)$. Then $\neg \forall x \alpha_1(x) \neq 0$, hence **(b)** $\forall x \alpha_1(x) \neq 0 \vee \neg \forall x \alpha_1(x) \neq 0$. Also **(c)** $\forall x \alpha_1(x) = 0 \sim \forall x \alpha(x) = 0$. From (a) using $\forall$-elim. (with α_1 for α), (b), (c), $\exists \alpha_1$-elim. and $\forall$-introd. we deduce $\forall \alpha(\forall x \alpha(x) = 0 \vee \neg \forall x \alpha(x) = 0)$, contradicting *27.17.

7.14. Finally, we refute the classical version *26.1 of the bar theorem.

*27.23. $\vdash \neg \forall \beta\{\forall \alpha \exists x R(\beta, \bar{\alpha}(x)) \ \& \ \forall a[\mathrm{Seq}(a) \ \& \ R(\beta, a) \supset A(\beta, a)] \ \& \ \forall a[\mathrm{Seq}(a) \ \& \ \forall s A(\beta, a * 2^{s+1}) \supset A(\beta, a)] \supset A(\beta, 1)\}$

when $R(\beta, a)$ is $(a = 1 \ \& \ \neg \forall x \beta(x) = 0) \vee (\mathrm{lh}(a) = 1 \ \& \ \beta((a)_0 \dot{-} 1) = 0)$ and $A(\beta, a)$ is $R(\beta, a) \vee \forall x R(\beta, a * 2^{x+1})$.

Proof. Call the formula $\neg \forall \beta B(\beta)$. Assume $\forall \beta B(\beta)$.

I. We establish the three hypotheses of the implication $B(\beta)$. (A) For the first, by *158 $\beta(\alpha(0)) = 0 \vee \beta(\alpha(0)) \neq 0$. Case 1: $\beta(\alpha(0)) = 0$. Then by *23.5, *23.2 and *6.3, $\mathrm{lh}(\bar{\alpha}(1)) = 1 \ \& \ \beta((\bar{\alpha}(1))_0 \dot{-} 1) = 0$. So $R(\beta, \bar{\alpha}(1))$, hence $\exists x R(\beta, \bar{\alpha}(x))$. Case 2: $\beta(\alpha(0)) \neq 0$. Then $\exists x \beta(x) \neq 0$, whence by *85a $\neg \forall x \beta(x) = 0$. Also $\bar{\alpha}(0) = 1$. So $R(\beta, \bar{\alpha}(0))$, hence $\exists x R(\beta, \bar{\alpha}(x))$. Completing the case argument ($\vee$-elim.), and using $\forall$-introd., $\forall \alpha \exists x R(\beta, \bar{\alpha}(x))$. (B) The second hyp. of $B(\beta)$ is immediate, because $R(\beta, a)$ is a disjunctive member of $A(\beta, a)$. (C) Assume $\mathrm{Seq}(a)$ and $\forall s A(\beta, a*2^{s+1})$, i.e. $\forall s[R(\beta, a*2^{s+1}) \vee \forall x R(\beta, (a*2^{s+1})*2^{x+1})]$. But $\neg\forall x R(\beta, (a*2^{s+1})*2^{x+1})$; for by *22.8 with *22.5, $\mathrm{lh}((a*2^{s+1})*2^{x+1}) = \mathrm{lh}(a)+2$, which with $\mathrm{lh}(1)=0$ contradicts both disjunctive members of $R(\beta, (a*2^{s+1})*2^{x+1})$. So $\forall s R(\beta, a*2^{s+1})$, whence $A(\beta, a)$.

II. Now by $\forall$- and $\supset$-elim. from $\forall\beta B(\beta)$ we infer $A(\beta, 1)$, whence (using *22.7 with *22.5) $R(\beta, 1) \vee \forall x R(\beta, 2^{x+1})$. Case 1: $R(\beta, 1)$. Then since $\mathrm{lh}(1)=0$, $\neg\forall x\beta(x)=0$, whence $\forall x\beta(x)=0 \vee \neg\forall x\beta(x)=0$. Case 2: $\forall x R(\beta, 2^{x+1})$, i.e. $\forall x[(2^{x+1}=1 \;\&\; \neg\forall x\beta(x)=0) \vee (\mathrm{lh}(2^{x+1})=1 \;\&\; \beta((2^{x+1})_0 \dot{-} 1)=0)]$. But by *3.10, $2^{x+1}\neq 1$. So $\forall x\beta((2^{x+1})_0 \dot{-} 1)=0$, whence by *19.9 and *6.3 $\forall x\beta(x)=0$, whence $\forall x\beta(x)=0 \vee \neg\forall x\beta(x)=0$. Completing the case argument and using $\forall$-introd., we contradict *27.17.

7.15. In **1957** Kleene proposed *2.2 as a postulate. Subsequently he thought there would be need for the apparently stronger ^{x}2.1, which likewise is acceptable intuitionistically (as well as classically). Then Kleene (in February 1963) obtained a result *R15.1 bypassing the only direct use (not via *2.2) of ^{x}2.1 in Vesley's Chapter III; Joan Rand (in March 1963) eliminated Kleene's direct use of ^{x}2.1 in deriving ^{x}26.9 from ^{x}26.3; and finally Vesley (in July 1963) obtained the following:

Derivation of ^{x}2.1 from *2.2. Assume $\forall x\exists\beta A(x, \beta)$. By *0.5, $\forall\alpha\exists\beta A(\alpha(0), \beta)$. By ^{x}27.1 (omitting $\exists\tau$ prior to $\exists$-elim.), $\forall\alpha\{\forall t\exists! y\, \tau(2^{t+1}*\bar{\alpha}(y))>0 \;\&\; \forall\beta[\forall t\exists y\tau(2^{t+1}*\bar{\alpha}(y))=\beta(t)+1 \supset A(\alpha(0), \beta)]\}$. Using $\forall$-elim. with $\lambda z x$ for α, ^{x}0.1 and $\forall$-introd.: **(a)** $\forall x\forall t\exists y\tau(2^{t+1}*\overline{\lambda z x}(y))>0$, **(b)** $\forall x\forall\beta[\forall t\exists y\tau(2^{t+1}*\overline{\lambda z x}(y))=\beta(t)+1 \supset A(x, \beta)]$. Using (a) in *25.7 (cf. Remark 5.4), assume (omitting $\exists\delta$ prior to $\exists$-elim.): **(c)** $\forall x\forall t\tau(2^{t+1}*\overline{\lambda z x}(\delta(\langle x, t\rangle)))>0$. Assume (prior to $\exists$-elim. from Lemma 5.3 (a)): **(d)** $\forall w\gamma(w)=\tau(2^{(w)_1+1}*\overline{\lambda z(w)_0}(\delta(\langle (w)_0, (w)_1\rangle)))\dot{-}1$. Now $(\lambda t\gamma(2^x\cdot 3^t))(t)+1 = \gamma(2^x\cdot 3^t)+1 = (\tau(2^{t+1}*\overline{\lambda z x}(\delta(\langle x, t\rangle)))\dot{-}1)+1$ [(d), *25.1] $= \tau(2^{t+1}*\overline{\lambda z x}(\delta(\langle x, t\rangle)))$ [*6.7 with (c)]. By $\exists$- and $\forall$-introd., $\forall t\exists y\tau(2^{t+1}*\overline{\lambda z x}(y))=(\lambda t\gamma(2^x\cdot 3^t))(t)+1$. Thence using (b), $A(x, \lambda t\gamma(2^x\cdot 3^t))$, whence $\exists\gamma\forall x A(x, \lambda t\gamma(2^x\cdot 3^t))$. —

This derivation uses ^{x}27.1, directly (not via *27.2). So replacing ^{x}2.1 as postulate by *2.2 would leave it uncertain whether ^{x}2.1 and *25.8 (our sole remaining result other than *2.1a using ^{x}2.1 directly) would hold in the basic system (or even in the present classical system); cf. end § 2. Also ×27.1 is in a similar status to ×2.1, as a postulate that is acceptable, but could for our essential purposes be replaced by its specialization from functions to numbers; it is used directly only here (and for *27.4).

The modification *27.1′ of ×27.1 with $\forall\alpha\exists\beta$ replaced by $\forall\alpha\exists!\beta$ (where $\exists!\beta$ is like $\exists!x$ IM p. 199, using the $=$ for functors 4.5) is derivable from *27.2. HINT: $\forall\alpha\exists!\beta A(\alpha, \beta) \vdash \forall\alpha\exists b A'(\alpha, b)$ where $A'(\alpha, b)$ is $\exists\beta\{\beta(\alpha(0))=b \;\&\; A(\lambda t\alpha(t+1), \beta) \;\&\; \forall\gamma[A(\lambda t\alpha(t+1), \gamma) \supset \beta=\gamma]\}$.

CHAPTER II

VARIOUS NOTIONS OF REALIZABILITY

by S. C. KLEENE

§ 8. Definition of realizability. 8.1. In the introductory § 1 we proposed to relate intuitionistic analysis and the theory of general recursive functions. But in setting up the formal system in Chapter I we did not use the general recursive functions. In particular, we did not carry out the early proposal by Beth and ourselves to take the "laws" or "algorithms" in Brouwer's definition of a spread to be general recursive functions; those laws, and also the one in Brouwer's principle, we expressed simply by function variables, the σ in *26.4, the ρ in ˣ26.9 and the τ in ˣ27.1. (Cf. Remark 6.1.)

A non-classical meaning of the prefixes $\forall\alpha\exists\beta$ and $\forall\alpha\exists b$ is incorporated into the intuitionistic system through the postulate ˣ27.1 expressing Brouwer's principle. But can one give a special meaning to just these prefixes, without having to consider the effect in the presence of the predicate calculus on all other forms of composition of formulas by the logical connectives? Initiates to intuitionism may ask for an explicit interpretation that applies to all formulas while satisfying Brouwer's principle for those beginning with $\forall\alpha\exists\beta$ or $\forall\alpha\exists b$. Indeed, a classical mathematician might question the consistency of the intuitionistic system with Brouwer's principle. He can be assured of its consistency by the interpretation we now give, which is to be based on only principles acceptable classically as well as intuitionistically.

If our semantical (or model-theoretic) arguments using this interpretation should be formalized in the basic formal system (end § 2), a metamathematical consistency proof for the intuitionistic system relative to the basic system would result. We shall discuss later the possibility of such a formalization (cf. 9.2 ¶ 5).

8.2. We begin by recasting some results in the theory of general and partial recursive functions of number and (one-place) function variables (end 3.2) into a form convenient for the present application.

Let Ψ be a list of variables, number or function or both. We write e.g. $\varphi[\Psi]$, with square brackets instead of parentheses, for a function of the variables Ψ with (partial or total) one-place number-theoretic functions as values. We say $\varphi[\Psi]$ is *primitive* (*general*) [*partial*] *recursive*, absolutely or in Θ, if $\varphi[\Psi] = \lambda t\, \varphi(\Psi, t)$ where the function $\varphi(\Psi, t)$ with natural numbers as values is such.

As in § 7 where we formalized Brouwer's principle, a certain kind of functional which correlates a one-place number-theoretic function β to a one-place number-theoretic function α (in fact, a "countable" functional Kleene **1959a** § 5) can be represented by a one-place number-theoretic function τ such that, for each t and α, $\tau(2^{t+1}*\bar{\alpha}(y)) > 0$ for exactly one y, and $\tau(2^{t+1}*\bar{\alpha}(y)) = \beta(t)+1$ for that y. We now write $\{\tau\}[\alpha]$ for the function β. In order to be able to construe $\{\tau\}[\alpha]$ as a partial recursive function of τ and α, we define it in general by

$$\{\tau\}[\alpha] = \lambda t\, \tau(2^{t+1}*\bar{\alpha}(y_t)) \dot{-} 1 \tag{8.1}$$

where $y_t \simeq \mu y \tau(2^{t+1}*\bar{\alpha}(y)) > 0$. But we shall say that $\{\tau\}[\alpha]$ is *properly defined* if $(t)(E!y)\tau(2^{t+1}*\bar{\alpha}(y)) > 0$.

LEMMA 8.1. *To each partial recursive function $\varphi[\Theta, \alpha]$, there is a primitive recursive function $\psi[\Theta]$ such that, for each Θ, α: $\{\psi[\Theta]\}[\alpha] = \varphi[\Theta, \alpha]$, and if $\varphi[\Theta, \alpha]$ is completely defined then $\{\psi[\Theta]\}[\alpha]$ is properly defined.*

(Proof follows.) We shall write

$$\Lambda\alpha\, \varphi[\Theta, \alpha] = \psi[\Theta], \tag{8.2}$$

i.e. $\Lambda\alpha\, \varphi[\Theta, \alpha]$ shall be a notation for some primitive recursive function $\psi[\Theta]$ with the properties in the lemma, so that *for each Θ, α*:

$$\{\Lambda\alpha\, \varphi[\Theta, \alpha]\}[\alpha] = \varphi[\Theta, \alpha], \tag{8.3}$$

and $\{\Lambda\alpha\, \varphi[\Theta, \alpha]\}[\alpha]$ is properly defined if completely defined.

If $\varphi[z, \Theta, \alpha]$ is partial recursive and $\psi[z, \Theta] = \Lambda\alpha\, \varphi[z, \Theta, \alpha]$, and we put $\varphi[\Theta, \alpha] = \varphi[e, \Theta, \alpha]$ for a fixed e, then $\psi[e, \Theta] = \Lambda\alpha\, \varphi[e, \Theta, \alpha]$ is a $\Lambda\alpha\, \varphi[\Theta, \alpha]$, i.e. it has the properties.

PROOF OF LEMMA 8.1. Consider e.g. $\varphi[b, \beta, \alpha]$ with b, β as the Θ. Write $\varphi[b, \beta, \alpha] = \lambda t\, \varphi(b, \beta, \alpha, t)$ where the latter φ is partial recursive. By the normal form theorem IM pp. 292, 330, there is a number e

such that, for each b, β, α, t:

(i) $\varphi(b, \beta, \alpha, t) \simeq U(\mu y T_2^{1,1}(\tilde{\beta}(y), \tilde{\alpha}(y), e, b, t, y))$,
(ii) $T_2^{1,1}(\tilde{\beta}(y), \tilde{\alpha}(y), e, b, t, y)$ for at most one y.

Let $\psi[b, \beta] = \lambda s\, \psi(b, \beta, s)$ where

$$\psi(b, \beta, s) = \begin{cases} U(y)+1 \text{ if } \mathrm{lh}(s)>0 \,\&\, T_2^{1,1}(\tilde{\beta}(y), \tilde{\alpha}(y), e, b, t, y), \\ 0 \text{ otherwise,} \end{cases}$$

where the y, t, α on the right are to be expressed in terms of s by

$$y = \mathrm{lh}(s) \dot{-} 1,\ t = (s)_0 \dot{-} 1,\ \alpha(i) = (s)_{i+1} \dot{-} 1\ (i < y). —$$

We shall generalize these notations to allow other lists of arguments $a_1, \ldots, a_k, \alpha_1, \ldots, \alpha_l$ in place of a single function α (the case $(k, l) = (0, 1)$). First, we write $p_0^{a_0} \cdot \ldots \cdot p_m^{a_m}$ as $\langle a_0, \ldots, a_m\rangle$ ($= 1$ when $m = -1$); $\lambda t \langle a_1, \ldots, a_k, \alpha_1(t), \ldots, \alpha_l(t)\rangle$ as $\langle a_1, \ldots, a_k, \alpha_1, \ldots, \alpha_l\rangle^1$, omitting the superscript ^{1} when the context makes it clear that $l > 0$ or that the whole is a function; $\lambda t\, (\alpha(t))_i$ as $(\alpha)_i$, and $((\alpha)_i)_j$ as $(\alpha)_{i,j}$ etc. (cf. 5.7, Kleene **1959** § 2).

Now we define

(8.1a) $\{\tau\}[a] = \{\tau\}[\lambda t\, a]$,
(8.1b) $\{\tau\} = \{\tau\}[0] = \{\tau\}[\lambda t\, 0]$,
(8.1c) $\{\tau\}[a_1, \ldots, a_k, \alpha_1, \ldots, \alpha_l]$
$= \{\tau\}[\langle a_1, \ldots, a_k, \alpha_1, \ldots, \alpha_l\rangle] \qquad (k+l > 1)$

and say the results are *properly defined* if the expressions under (8.1) to which they reduce are properly defined. In conjunction with this notation, we shall avoid using curly brackets as simply marks of inclusion.

Furthermore, we write

(8.2a) $\Lambda a\, \varphi[\Theta, a] = \Lambda \alpha\, \varphi[\Theta, \alpha(0)]$,
(8.2b) $\Lambda\, \varphi[\Theta] = \Lambda a\, \varphi[\Theta] = \Lambda \alpha\, \varphi[\Theta]$,
(8.2c) $\Lambda a_1 \ldots a_k \alpha_1 \ldots \alpha_l\, \varphi[\Theta, a_1, \ldots, a_k, \alpha_1, \ldots, \alpha_l]$
$= \Lambda \alpha\, \varphi[\Theta, (\alpha(0))_0, \ldots, (\alpha(0))_{k-1}, (\alpha)_k, \ldots, (\alpha)_{k+l-1}] \ (k+l > 1)$.

Now

(8.3a) $\{\Lambda a\, \varphi[\Theta, a]\}[a] = \varphi[\Theta, a]$,
(8.3b) $\{\Lambda\, \varphi[\Theta]\} = \varphi[\Theta]$,
(8.3c) $\{\Lambda a_1 \ldots a_k \alpha_1 \ldots \alpha_l\, \varphi[\Theta, a_1, \ldots, a_k, \alpha_1, \ldots, \alpha_l]\}[a_1, \ldots, a_k, \alpha_1, \ldots, \alpha_l]$
$= \varphi[\Theta, a_1, \ldots, a_k, \alpha_1, \ldots, \alpha_l] \qquad (k+l > 1)$,

each being properly defined when completely defined.

The notations of this subsection are analogous to those of IM pp. 341–342, 344, which we shall also use occasionally but writing the functions on the line as in **1957**, namely: We shall write $\{z\}(\Psi)$ for the $\{z\}^{\Psi_1}(\Psi_2)$ of p. 341 where Ψ_1 are the functions, and Ψ_2 the numbers, in order among Ψ. We write $\Lambda\Psi\,\varphi(\Theta, \Psi)$ for the $S_n^{m,1,\ldots,1}(e, \Theta)$ of p. 342 or $\Lambda^{1,\ldots,1}\Psi_2\,\varphi(\Theta, \Psi)$ of p. 344 when Θ consists of numbers only (m of them), where e is a Gödel number of $\lambda\Theta\Psi_2\,\varphi(\Theta, \Psi)$ uniform in Ψ_1 (Ψ_2 being n numbers, and Ψ_1 as many functions as there are superscripts ^{1}); thus, for each Θ, $\Lambda\Psi\,\varphi(\Theta, \Psi)$ is a Gödel number of $\Lambda\Psi_2\,\varphi(\Theta, \Psi)$ uniform in Ψ_1, or briefly a *Gödel number of* $\lambda\Psi\,\varphi(\Theta, \Psi)$.

The T-predicates of IM pp. 291–292 can be reformulated to use $\bar{\alpha}$ instead of $\tilde{\alpha}$, etc.; e.g. $T_1^{1,1}(\bar{\beta}(y), \bar{\alpha}(y), e, a) \equiv T_1^{1,1}(\tilde{\beta}(y), \tilde{\alpha}(y), e, a, y)$ when we put $T_1^{1,1}(u, v, e, a) \equiv T_1^{1,1}(\Pi_{i<\mathrm{lh}(u)} p_i^{(u)_i \dot{-} 1}, \Pi_{i<\mathrm{lh}(u)} p_i^{(v)_i \dot{-} 1}, e, a, \mathrm{lh}(u))$ (Kleene **1955** Footnote 2).

Using IM top p. 235 and the normal form theorem IM pp. 292, 330: A total (total) [partial] function $\varphi(\Psi)$ is primitive (general) [partial] recursive in total one-place functions Σ, if and only if there is a primitive (partial) [partial] recursive function $\lambda\Psi\Sigma\,\varphi(\Psi, \Sigma)$ such that, for the Σ in question, $\lambda\Psi\,\varphi(\Psi) = \lambda\Psi\,\varphi(\Psi, \Sigma)$; and similarly with $\varphi[\Psi]$, $\varphi[\Psi, \Sigma]$.

Since $\lambda a_0 \ldots a_m\,\sigma(a_0, \ldots, a_m)$ is primitive recursive in $\lambda a\,\sigma((a)_0, \ldots, (a)_m)$ and inversely: In 'φ is primitive (general) [partial] recursive in Σ', many-place functions in Σ can be replaced equivalently by one-place functions.

8.3. In our former interpretation of intuitionistic number theory 1945 (found in 1941) or IM § 82, we began with the idea that intuitionistically each statement, except of the most elementary kind, constitutes an incomplete communication of information, asserting that further information could be given effectively to complete it. For example, an existential statement "$(Ex)A(x)$" is an incomplete communication, asserting that an x could be given such that $A(x)$ together with information which would complete the statement "$A(x)$" for that x. A generality statement "$(x)A(x)$" is an incomplete communication, asserting that an effective general process could be given by which, to each x, information which would complete the statement "$A(x)$" for that x could be found. A statement containing no logical connective is a value of one of the fundamental predicates, and requires no completion, assuming those predicates have been given effectively.

Here "$(Ex)A(x)$" and "$(x)A(x)$" to be "statements" must either have no "free" variables, or each free variable in them must be understood as having a specified value; then in turn we talk about "$A(x)$" for a specified value of x as a statement.

In applying these ideas to the number-theoretic formalism it was convenient, instead of specifying natural numbers as values of the free variables, to substitute for the variables the numerals expressing those values. Thereby the formal counterpart of "statements" became closed formulas.

Furthermore, the information by which incomplete communications can be completed admits of being codified in natural numbers, using for effective processes Gödel numbers of recursive functions. We said these numbers "realize" the respective closed formulas. Thus we said a natural number e *realizes* a closed formula $\exists$xA(x), if $e = 2^x 3^a$ where a ($= (e)_1$) realizes the (likewise closed) formula A($\boldsymbol{x}$), where $\boldsymbol{x}$ is the numeral for the natural number x ($= (e)_0$). A number e *realizes* a closed formula $\forall$xA(x), if e is the Gödel number of a general recursive function φ such that, for each x, $\varphi(x)$ realizes A($\boldsymbol{x}$). A number e *realizes* a closed prime formula P, if $e = 0$ and P is true under the usual interpretation of its individual, function and predicate symbols. (For the other four clauses of that definition of 'realizes', see Kleene 1945 or IM § 82. Some minor improvements in details appear in **1960** § 5.)

Finally we said a closed formula A is *realizable* if some number realizes it. An open formula A($y_1, \ldots, y_m$) containing free only the distinct variables $y_1, \ldots, y_m$ is *realizable*, if its closure is realizable (1945), or equivalently (1948 and IM § 82) if there is a general recursive function φ such that, for each $y_1, \ldots, y_m$, the number $\varphi(y_1, \ldots, y_m)$ realizes A($\boldsymbol{y}_1, \ldots, \boldsymbol{y}_m$).

8.4. We shall use these ideas with some modifications to obtain an interpretation of intuitionistic analysis. We now say a statement "$(E\beta)A(\beta)$" is an incomplete communication asserting that a β could be given such that $A(\beta)$ together with information which would complete the statement "$A(\beta)$" for that β. Here from the constructive point of view the β should be a general recursive function, unless "$(E\beta)A(\beta)$" contains free function variables having non-recursive functions as their values. The latter situation arises e.g. in considering the interpretation of "$(\alpha)(E\beta)A(\alpha, \beta)$"; say this contains no free variables. This is to be completed by giving an effective process by which, to each α, information which completes "$(E\beta)A(\alpha, \beta)$" for that α could be found. Here α ranges over all one-place number-theoretic functions. (By Kleene 1950a § 3 or Lemma 9.8 below, the fan theorem would not hold if the functions were restricted to be general recursive.) The β to complete the communication "$(E\beta)A(\alpha, \beta)$" for a given α will in general depend on what that α is, and we should not restrict the β to be a general recursive function. In the simple example that "$A(\alpha, \beta)$" is "$(x)\alpha(x)=\beta(x)$", the β must be α itself. In general, the β to complete "$(E\beta)A(\beta)$", when "$(E\beta)A(\beta)$" has free function variables having specified functions as values, should be a function general recursive in those functions, unless "$(E\beta)A(\beta)$" is contingent (cf. 8.6).

Proceeding now to the formal symbolism, we cannot in general avoid specifying functions as values of the free function variables by substituting for those variables functors expressing their values. For the formal system does not have a functor to express each particular function; no formal system (with only countably many symbols) can. This being the case, we might as well specify the values of function and number variables alike. (In our **1957** definition of realizability, found in 1951, we specified values of the function variables, while substituting numerals for the number variables.)

We find it advantageous now to codify the information, by which communications are to be completed, not in natural numbers but in one-place number-theoretic functions. Before any functions have been introduced, as to evaluate function variables, the "realizing" functions will be general recursive; later they will be general recursive in the functions which have been introduced. Actually functions primitive recursive absolutely or in those functions will suffice, in consequence of our using via the notation $\{\}[\,]$ a form of the normal form theorem (cf. 8.8). (In **1957**, we still used numbers, while adding function arguments to the recursive functions represented by Gödel numbers. The definition in 8.5 is equivalent to the **1957** one, but is more manageable in the proofs.)

8.5. Consider any formula E, and let Ψ be a list of distinct variables of both types including all which occur free in E. We define when a one-place number-theoretic function ϵ 'realizes' E for a given assignment of numbers and functions Ψ as values of Ψ, or in brief when ϵ 'realizes-Ψ' E. The definition, as before (1945, IM § 82), is by induction on the number of (occurrences of) logical symbols in E. We have 9 cases according to the form of E.

1. ϵ *realizes-*Ψ a prime formula P, if P is *true-*Ψ, i.e. if P is true when Ψ have the respective values Ψ.

2. ϵ *realizes-*Ψ A & B, if $(\epsilon)_0$ *realizes-*Ψ A and $(\epsilon)_1$ *realizes-*Ψ B.

3. ϵ *realizes-*Ψ A ∨ B, if $(\epsilon(0))_0 = 0$ and $(\epsilon)_1$ *realizes-*Ψ A, or $(\epsilon(0))_0 \neq 0$ and $(\epsilon)_1$ *realizes-*Ψ B.

In Clause 4 (next) it is to be understood that, when α realizes-Ψ A, $\{\epsilon\}[\alpha]$ is properly defined; and similarly in Clauses 6, 8 and 9.

4. ϵ *realizes-*Ψ A ⊃ B, if, for each α, if α *realizes-*Ψ A then $\{\epsilon\}[\alpha]$ *realizes-*Ψ B.

5. ϵ *realizes-*Ψ ¬A, if, for each α, not α *realizes-*Ψ A. (Equivalently, since by Clause 1 α *realizes-*Ψ $1=0$ for no α, Ψ: ϵ *realizes-*Ψ ¬A, if ϵ *realizes-*Ψ A ⊃ $1=0$ under Clause 4.)

In Clause 6 it is to be understood that by first changing bound variables if necessary it has been arranged that x not occur in the list Ψ, and that Ψ, x are the values of Ψ, x, respectively. Similarly in Clauses 7, 8, 9 (in Clause 7, $(\epsilon(0))_0$ is the value of x; etc.).

6. ϵ *realizes-*Ψ ∀xA, if, for each x, $\{\epsilon\}[x]$ *realizes-*Ψ, x A.

7. ϵ *realizes-*Ψ ∃xA, if $(\epsilon)_1$ *realizes-*Ψ, $(\epsilon(0))_0$ A.

8. ϵ *realizes-*Ψ ∀αA, if, for each α, $\{\epsilon\}[\alpha]$ *realizes-*Ψ, α A.

9. ϵ *realizes-*Ψ ∃αA, if $(\epsilon)_1$ *realizes-*Ψ, $\{(\epsilon)_0\}$ A.

We say a closed formula E is *realizable*, if a general recursive function ϵ realizes E; an open formula, if its closure is.

Realizability can be relativized to a class (or list) T of (total) number-theoretic functions, thus. A closed formula E is *realizable*/T, if a function ϵ general recursive in (some finitely-many functions of) T realizes E; an open formula, if its closure is.

Consider, now in the formal symbolism, the example $\forall\alpha\exists\beta\forall x\alpha(x)=\beta(x)$ discussed in 8.4. This formula is provable by elementary logical reasoning. It is realizable, if there is a general recursive function ϵ which realizes it, i.e. by Clause 8, such that, for each α, $\{\epsilon\}[\alpha]$ realizes-α $\exists\beta\forall x\alpha(x)=\beta(x)$. Write $\zeta = \{\epsilon\}[\alpha]$. By Clause 9, $(\zeta)_1$ must realize-α, $\{(\zeta)_0\}$ $\forall x\alpha(x)=\beta(x)$. The obvious choice for $\{(\zeta)_0\}$ is α; then by Clause 6, for each x, $\{(\zeta)_1\}[x]$ must realize-α, α, x $\alpha(x)=\beta(x)$. But, for each x, $\alpha(x)=\beta(x)$ is true-α, α, x; so $\{(\zeta)_1\}[x]$ can be any function, e.g. $\lambda t\, 0$. Using Lemma 8.1 etc., it suffices to take $\zeta = \langle \Lambda\, \alpha, \Lambda x\, \lambda t\, 0\rangle$,

$\epsilon = \Lambda\alpha \langle \Lambda\, \alpha, \Lambda x\, \lambda t\, 0 \rangle$. That is, this ϵ is general (actually, primitive) recursive, and realizes $\forall\alpha\exists\beta\forall x \alpha(x)=\beta(x)$.

As a second illustration, take $\forall\alpha\forall x(\alpha(x)=0 \vee \alpha(x)\neq 0)$ (cf. Remark 6.1). To realize it, ϵ must be general recursive and, for each α, $\{\epsilon\}[\alpha] = \zeta$ where, for each x, $\{\zeta\}[x] = \eta$ where η realizes-α, x $\alpha(x)=0 \vee \alpha(x)\neq 0$. We can take $\eta = \langle \alpha(x), \lambda t\, 0 \rangle$, $\zeta = \Lambda x \langle \alpha(x), \lambda t\, 0\rangle$, $\epsilon = \Lambda\alpha\, \Lambda x \langle \alpha(x), \lambda t\, 0 \rangle$. Notice how the choice whether to realize $\alpha(x)=0$ or $\alpha(x)\neq 0$ (by an arbitrary function, e.g. $\lambda t\, 0$) depends on α as well as on x.

As a third illustration, $\exists\alpha[\alpha(0)=1\ \&\ \forall x\alpha(x')=x'\cdot\alpha(x)]$ (cf. Lemma 5.3 (b)) is realized by $\langle \Lambda\, \lambda t\, t!, \langle \lambda t\, 0, \Lambda x\, \lambda t\, 0 \rangle\rangle$.

8.6. The new notion of realizability, besides departing from the earlier one 1945 in obvious respects connected with the presence of the function variables, alters the treatment of implication and (thence under the alternative Clause 5) of negation.

Consider a closed implication A ⊃ B, taken by itself. We said in 1945: e realizes A ⊃ B, if, for each a, if a realizes A then $\{e\}(a)$ realizes B. Now we say: ϵ realizes A ⊃ B, if, for each α, if α realizes A then $\{\epsilon\}[\alpha]$ realizes B. The range of α is all number-theoretic functions, not just the general recursive ones. So the new interpretation of implication requires in the general recursive ϵ a process by which, from even a highly non-constructive completion α of the incomplete communication expressed by A, one can get constructively a completion of the communication expressed by B. In other words, the new realizability interpretation treats A ⊃ B as "true constructively" whenever, for A "true" but not necessarily "constructively" so, B will be likewise "true" with "degree of non-constructiveness" not greater than that of A. This enforces the intuitionistic demand for constructiveness in a less drastic form than before; we now allow non-trivial "contrary-to-fact" conditionals, instead of placing all B's on a par as consequences of a non-(intuitionistically-true) antecedent A. Under the 1945 notion, if A was unrealizable, then A ⊃ B was always realizable. That would leave no place in intuitionistic mathematics for the theory of relative recursiveness. (The change in question, which enters rather unobtrusively here with functions used as the realizing objects, was effected in the **1957** version of the new notion of realizability by departing from what in that notation is the direct generalization of the 1945 definition.)

We were forced to make this change, since without it we failed to extend Nelson's 1947 Theorem 1 (IM Theorem 62 (a) p. 504) from intuitionistic one-sorted predicate calculus with number variables to the intuitionistic two-sorted predicate calculus with number and function variables.

But the change alters the notion of realizability for *number-theoretic formulas*, i.e. those not containing function variables or λ.

Consider a closed such formula A. Formerly, $\neg$A was realizable if and only if no number a realized A, i.e. if and only if A was unrealizable. Now $\neg$A is realizable (and realizable/η for any given η), if and only if no function α realizes A, i.e. if and only if A is unrealizable/θ for every θ.

In IM Theorem 63 (ii) p. 511 we gave an example A of a closed formula such that A was unrealizable, so $\neg$A was realizable, so $\neg\neg$A was unrealizable. The formula A was $\forall$x(A(x) $\vee$ $\neg$A(x)) where A(x) is $\exists$zA(x, z) where A(x, z) numeralwise expresses $T_1(x, x, z)$; A was unrealizable because, if a number a realized it, $(\{a\}(x))_0$ would be a general recursive representing function of the predicate $(Ez)T_1(x, x, z)$, which is not general recursive (IM Theorem V p. 283). Now similarly no general recursive function α realizes A (details in a moment), so A is unrealizable; but a function α primitive recursive in the representing function τ of $(Ez)T_1(x, x, z)$ can be defined classically which realizes A, so A is realizable/τ. Hence now $\neg$A is also unrealizable, and unrealizable/η for every η, and $\neg\neg$A is realizable.

To simplify the details (which otherwise should follow Corollary 9.6) we may either suppose the list of function symbols (introduced as described in 5.1) extended sufficiently to include one f for the representing function of $T_1(x, x, z)$ and let A(x, z) be f(x, z)=0, or without extending the symbolism we may choose the A(x, z) to numeralwise express $T_1(x, x, z)$ by the method of Lemma 8.5 below. In either case (using Lemma 8.4a in the second), $(E\epsilon)\{\epsilon$ realizes-x, z A(x, z)$\} \rightarrow T_1(x, x, z)$, and there is a primitive recursive function $\epsilon_{A(x,z)}$ (arbitrary in the first case) such that $T_1(x, x, z) \rightarrow \{\epsilon_{A(x,z)}$ realizes-x, z A(x, z)$\}$. The proof that A is realized in the former sense by no number a is as before (IM p. 512 (i), after applying Clause 7 p. 503). Similarly, A is realized in the present sense by no general recursive function α, because then the general recursive function $\mathrm{sg}((\{\alpha\}[x](0))_0)$ would represent $(Ez)T_1(x, x, z)$. Let $\varphi[\tau, x] = \langle \overline{\mathrm{sg}}\ \varepsilon z T_1(x, x, z), \langle \varepsilon z T_1(x, x, z), \epsilon_{A(x,z)}\rangle\rangle$ (cf. IM p. 317). Then $\varphi[\tau, x]$ is general, a fortiori partial, recursive; so using Lemma 8.1 via (8.2a), and 8.2 next to last ¶, the function $\alpha = \Lambda x\, \varphi[\tau, x]$ is primitive recursive in τ, and realizes A (using $(x)\overline{T}_1(x, x, 0)$ and cases from $(Ez)T_1(x, x, z) \vee \overline{(Ez)}T_1(x, x, z)$).

Although we now consider that the new notion of realizability gives a more faithful interpretation of intuitionistic number theory, the old one remains of interest. It is simpler for establishing unprovability results for intuitionistic formal number theory as in IM Theorem 63 p. 511. Also it lends itself to investigations of intuitionistic number theory which depend on the realizability of each formula being expressible by another formula of the same system or the system inessentially extended (cf. IM top p. 406), as in Kleene 1945 §§ 12–16 and Nelson 1947 §§ 2, 11, 12. This property of realizability is lost under the new notion as applied to the formal system of number theory, though we have it again for the intuitionistic formal system of analysis. Kleene in 1945 used this property of 1945-realizability for intuitionistic number theory to set up a system of number theory corresponding to the starker form of constructiveness which that realizability interpretation represents, by adjoining to intuitionistic formal number theory certain realizable but classically false formulas (also cf. IM p. 514). Such a form of constructivism has been favored by Markov and Šanin, if we correctly understand their position. A different-appearing interpretation by Šanin **1958**, **1958a** is shown in Kleene **1960** to be equivalent to 1945-realizability.

8.7. Lemma 8.2. *Let* Ψ_1 *be a list of those of the variables* Ψ *which actually occur free in* E, *not necessarily in the same order as in* Ψ; *and let* $\boldsymbol{\Psi}_1$ *be the list of the corresponding ones of the numbers and functions* $\boldsymbol{\Psi}$. *Then* ϵ *realizes-*$\boldsymbol{\Psi}_1$ E, *if and only if* ϵ *realizes-*$\boldsymbol{\Psi}$ E.

Proof. By induction on the number of logical symbols in E. —

Equivalently to the definition in 8.5 (as we show in a moment), a formula E containing free only Ψ is *realizable* (*realizable*/T), if there is a function φ, general recursive (general recursive in T), such that, for each $\boldsymbol{\Psi}$, $\varphi[\boldsymbol{\Psi}]$ realizes-$\boldsymbol{\Psi}$ E. Any function φ with the latter property we call a *realization function for* E (*in* the list Ψ); and any function ϵ which realizes-$\boldsymbol{\Psi}$ E, e.g. $\varphi[\boldsymbol{\Psi}]$, we call a *realizing-*$\boldsymbol{\Psi}$ *function for* E (*in* the list Ψ). This notion of realizability is independent of the choice of the list Ψ, as we see thus (similarly for realizability/T). Suppose φ is a general recursive realization function for E in a given list Ψ, and let $\varphi_1[\boldsymbol{\Psi}_1] = \varphi[\boldsymbol{\Psi}^*]$ where Ψ_1 is a minimal list (as in

Lemma 8.2) and Ψ^* comes from Ψ by replacing each number (function) correlated to a variable in Ψ not in Ψ_1 by 0 (by $\lambda t\, 0$). Then by Lemma 3.2 (with 8.2 ¶ 2) φ_1 is general recursive, and by Lemma 8.2 is a realization function for E in the list Ψ_1. Inversely, if ψ_1 is a general recursive realization function for E in Ψ_1, and $\psi[\Psi] = \psi_1[\Psi_1]$, then ψ is a general recursive realization function for E in Ψ.

To prove the equivalence to the definition in 8.5, take Ψ to be the free variables of E in order of first free occurrence. For E closed, Ψ is empty; and if ϵ is general recursive $\varphi[\Psi] = \epsilon$ is, and inversely. For E open, say e.g. its closure is $\forall\alpha\forall a$E. If φ is a general recursive realization function for E in α, a, then using Clauses 8 and 6 and Lemma 8.1 $\Lambda\alpha\, \Lambda a\, \varphi[\alpha, a]$ is a primitive recursive realizing function for $\forall\alpha\forall a$E. (If φ is a realization function for E general recursive in T, then using end 8.2 $\varphi[\alpha, a] = \varphi_1[\alpha, a, \Sigma]$ for some partial recursive function φ_1 and list Σ of one-place functions primitive recursive in T, and $\Lambda\alpha\Lambda a\, \varphi_1[\alpha, a, \Sigma]$ is a realizing function for $\forall\alpha\forall a$E primitive recursive in T.) Inversely, if ϵ is a general recursive realizing function for $\forall\alpha\forall a$E, then, using Clauses 8 and 6, $\lambda\alpha a\, \{\{\epsilon\}[\alpha]\}[a]$ is a general recursive realization function for E.

In the foregoing definition of 'E is realizable/T' the functions in the class or list T are not assigned as values to respective function variables. Now say as before that E contains free at most the variables Ψ (of either type), suppose Ψ is (Φ, Θ), and let $\Psi = (\Phi, \Theta)$ be values of Ψ. We say E is *realizable-Θ/T* if there is a function φ general recursive in (functions of) Θ, T such that, for each Φ, $\varphi[\Phi]$ realizes-Ψ E; or equivalently, if a function ϵ general recursive in Θ, T realizes-Θ the closure $\forall\Phi$E of E with respect to Φ. Similarly, without the T (cf. Kleene 1957). The notion 'E is realizable-Θ/T' differs from 'E is realizable/T' in that Θ are assigned as values to Θ, and the universal quantification 'for each Φ' (or $\forall\Phi$) applies only to the rest of Ψ (or Ψ). In view of Lemma 8.2, functions among Θ correlated to function variables among Θ not actually occurring free in E can equivalently be considered as part of the T. For the same reason, one-place functions of T can equivalently be included among the Θ by correlating them to function variables not free in E; and so only in the case T is an infinite class of functions is the notion 'E is realizable-Θ/T' more general that 'E is realizable-Θ'.

The reader may elect to start § 9 next (filling in later).

8.8. From a realization function φ general recursive (general recursive in T), we obtain realizing-Ψ functions $\varphi[\Psi]$ general recursive in Ψ (in Ψ, T). However, by the next lemma, with $\Theta = \Psi$, then φ_{E} is another such realization function with realizing-Ψ functions $\varphi_{\mathrm{E}}[\Psi]$ primitive recursive in Ψ (in Ψ, T).

LEMMA 8.3. *To any formula* E *containing free only* Ψ, *any* $\Theta \subset \Psi$, *and any function* χ *general recursive absolutely (in* T*), there is another such function* χ_{E} *such that: For each* Ψ, $\chi_{\mathrm{E}}[\Theta]$ *is primitive recursive in* Θ *(in* Θ, T*), and if* $\chi[\Theta]$ *realizes-*Ψ E *so does* $\chi_{\mathrm{E}}[\Theta]$ *(where* Θ *are those of* Ψ *which correspond to* Θ*).*

PROOF, by induction on the number of logical symbols in E.

CASES 1 AND 5: E is prime or E is $\neg$A. Let $\chi_{\mathrm{E}}[\Theta] = \lambda t\, 0$.

CASE 2: E is A & B. Let $\chi_{\mathrm{E}}[\Theta] = \langle(\chi[\Theta])_{0\mathrm{A}}, (\chi[\Theta])_{1\mathrm{B}}\rangle$ where $(\chi[\Theta])_{0\mathrm{A}}$ is $(\lambda\Theta\, (\chi[\Theta])_0)_{\mathrm{A}}[\Theta]$ etc.

CASE 3: E is A $\vee$ B. Let $\chi_{\mathrm{E}}[\Theta] = \langle(\chi[\Theta](0))_0,$
$\lambda t\, \overline{\mathrm{sg}}((\chi[\Theta](0))_0)\cdot(\chi[\Theta])_{1\mathrm{A}}(t) + \mathrm{sg}((\chi[\Theta](0))_0)\cdot(\chi[\Theta])_{1\mathrm{B}}(t)\rangle$.

CASES 4 AND 8: A $\supset$ B or $\forall\alpha$A. $\Lambda\alpha\,\{\chi[\Theta]\}[\alpha]$. (For χ general recursive in T, $\Lambda\alpha\,\{\chi_1[\Theta, \Sigma]\}[\alpha]$; cf. 8.7 ¶ 4.)

CASE 6: $\forall x$A. $\Lambda x\,\{\chi[\Theta]\}[x]$.

CASE 7: $\exists x$A. $\langle(\chi[\Theta](0))_0, (\chi[\Theta])_{1\mathrm{A}}\rangle$.

CASE 9: $\exists\alpha$A. $\langle\Lambda\,\{(\chi[\Theta])_0\}, (\chi[\Theta])_{1\mathrm{A}}\rangle$.

8.9. Brief indications of the following results were given in Kleene **1960** (in the last sentence of § 1 with Lemma 2.1a and 2.1b and Footnote 9, and Footnote 1). That E in Lemma 8.4a is true-Ψ is simply the proposition expressed upon translating E directly into the informal language under the usual reading of the symbols with Ψ as the values of Ψ. (In Clause 1 of 8.5 we did not have any logical symbols to translate. On IM p. 499 we had no free variables, while on p. 500 we gave the free variables the generality interpretation p. 149 instead of values under the predicate interpretation p. 146.)

LEMMA 8.4a. *To each formula* E *containing free only the variables* Ψ *and not containing* $\vee$ *or* $\exists$, *there is a primitive recursive function* ϵ_{E} *such that, for each* Ψ:

(i) *If* $(E\epsilon)[\epsilon$ *realizes-*Ψ E], *then* E *is true-*Ψ.
(ii) *If* E *is true-*Ψ, *then* ϵ_{E} *realizes-*Ψ E.

PROOF, by induction on the number of logical symbols in E. According as E is of the form P (a prime formula), A & B, A $\supset$ B, $\neg$A,

$\forall$xA or $\forall\alpha$A, let $\epsilon_{\mathrm{E}} = \lambda t\, 0$, $\langle\epsilon_{\mathrm{A}}, \epsilon_{\mathrm{B}}\rangle$, $\Lambda\alpha\, \epsilon_{\mathrm{B}}$, $\lambda t\, 0$, $\Lambda x\, \epsilon_{\mathrm{A}}$ or $\Lambda\alpha\, \epsilon_{\mathrm{A}}$, respectively. One case will suffice to illustrate the reasoning.

Case 4: E is A $\supset$ B. Then $\epsilon_{\mathrm{E}} = \Lambda\alpha\, \epsilon_{\mathrm{B}}$. (i) Suppose ϵ realizes-Ψ A $\supset$ B. Suppose A is true-Ψ; then by hyp. ind. (ii), ϵ_{A} realizes-Ψ A, so $\{\epsilon\}[\epsilon_{\mathrm{A}}]$ realizes-Ψ B, and by hyp. ind. (i), B is true-Ψ. (ii). Suppose A $\supset$ B is true-Ψ. Suppose α realizes-Ψ A; then by hyp. ind. (i), A is true-Ψ, so B is true-Ψ, and by hyp. ind. (ii), ϵ_{B} realizes-Ψ B. Thus $\Lambda\alpha\, \epsilon_{\mathrm{B}}$ realizes-Ψ A $\supset$ B.

Lemma 8.4b. *To each formula* E *containing free only the variables* Ψ, *and containing no* $\exists$, *other than in parts of the form* $\exists$xP(x) *with* P(x) *prime or of the form* $\exists\alpha$P(α) *with* P(α) *prime, and no* $\vee$, *there is a partial recursive function* $\epsilon_{\mathrm{E}}[\Psi]$ *such that, for each* Ψ:

(i) *If* $(E\epsilon)[\epsilon$ *realizes*-Ψ E], *then* E *is true*-Ψ.

(ii) *If* E *is true*-Ψ, *then* $\epsilon_{\mathrm{E}}[\Psi]$ $(= \lambda t\, \epsilon_{\mathrm{E}}(\Psi, t)$ *is completely defined and*) *realizes*-Ψ E.

Proof, by induction. According as E is of the form P, A & B, A $\supset$ B, $\neg$A, $\forall$xA(x), $\exists$xP(x) or $\forall\alpha$A(α), let $\epsilon_{\mathrm{E}}[\Psi] = \lambda t\, 0$, $\langle\epsilon_{\mathrm{A}}[\Psi], \epsilon_{\mathrm{B}}[\Psi]\rangle$, $\Lambda\alpha\, \epsilon_{\mathrm{B}}[\Psi]$, $\lambda t\, 0$, $\Lambda x\, \epsilon_{\mathrm{A(x)}}[\Psi, x]$, $\langle\mu x P(\Psi, x), \lambda t\, 0\rangle$ (where $P(\Psi, x)$ is the primitive recursive predicate expressed by P(x)) or $\Lambda\alpha\, \epsilon_{\mathrm{A}(\alpha)}[\Psi, \alpha]$.

Case 9: E is $\exists\alpha$P(α). Say e.g. Ψ is a, β. Now P(α) expresses a primitive recursive predicate $P(a, \beta, \alpha)$. Using IM Theorem VI* (a) with its proof from Theorem IV* (6) (IM pp. 292 etc.) with $\bar{\beta}$, $\bar{\alpha}$ instead of $\tilde{\beta}$, $\tilde{\alpha}$ (cf. end 8.2), there is a number e such that $(E\alpha)P(a, \beta, \alpha) \equiv (E\alpha)(Ey)T_1^{1,1}(\bar{\beta}(y), \bar{\alpha}(y), e, a) \equiv (Es)[\mathrm{Seq}(s)\ \&\ T_1^{1,1}(\bar{\beta}(\mathrm{lh}(s)), s, e, a)]$ and $\mathrm{Seq}(s)\ \&\ T_1^{1,1}(\bar{\beta}(\mathrm{lh}(s)), s, e, a) \rightarrow P(a, \beta, \lambda t\, (s)_t \dot{-} 1)$. So we take $\epsilon_{\mathrm{E}}[a, \beta] = \langle\Lambda\, \lambda t\, (\mu s[\mathrm{Seq}(s)\ \&\ T_1^{1,1}(\bar{\beta}(\mathrm{lh}(s)), s, e, a)])_t \dot{-} 1, \lambda t\, 0\rangle$. — Cf. Remark 9.16.

Lemma 8.5. *To each general recursive function* $\varphi(\Psi)$ [*predicate* $P(\Psi)$], *there is a formula* P(Ψ, w) [P(Ψ)] *not containing* $\vee$ *or* $\exists$ *which numeralwise represents* [*expresses*] *it* (5.3 [5.4]) *in the intuitionistic formal system of analysis* (*or any subsystem including Postulate Groups* A *and* B *and Axiom* ^{x}1.1), *besides representing* [*expressing*] *it under the usual interpretation of the symbolism.*

Proof. We modify the proofs of IM Theorems 27 p. 243 and 32 (a) p. 295 and Corollaries as extended on p. 298 (lines 13–16), and material contributing thereto, as follows. In Case (IV) for Theorem I p. 241, we change to $\varphi(x_1, \ldots, x_n, \alpha_1, \ldots, \alpha_l) = w \equiv (y_1)\ldots(y_m)[\chi_1(x_1, \ldots, x_n, \alpha_1, \ldots, \alpha_l) = y_1\ \&\ \ldots\ \&\ \chi_m(x_1, \ldots, x_n, \alpha_1, \ldots, \alpha_l) = y_m \rightarrow \psi(y_1, \ldots,$

$y_m, \alpha_1, \ldots, \alpha_l)=w]$. In Case (Vb) for Theorem 27 (and similarly in Case (Va)), we make several changes. First (also on top p. 296), we take $a<b$ to be an abbreviation for $\forall c a \neq b+c$ (not $\exists c c'+a=b$) or to be prime (cf. above preceding *15.1). Then, instead of using $B(c, d, i, w)$ as defined on p. 203, we take it to abbreviate $\neg\forall v c \neq (i'\cdot d)'\cdot v+w \;\&\; w<(i'\cdot d)'$ (cf. *180b p. 204). Finally, the displayed formula middle p. 243 becomes

$$\forall c\forall d\{\forall u[B(c, d, 0, u) \supset Q(x_2, \ldots, x_n, \alpha_1, \ldots, \alpha_l, u)] \;\&\; \forall i_{i<y}\forall u\forall v [B(c,d,i',u) \;\&\; B(c,d,i,v) \supset R(i,v,x_2,\ldots,x_n,\alpha_1,\ldots,\alpha_l,u)] \supset B(c,d,y,w)\},$$

abbreviated $P(y, x_2, \ldots, x_n, \alpha_1, \ldots, \alpha_l, w)$. Instead of proving directly that this works, we can establish its equivalence to the previous formula, as we shall do in Lemma 8.7.

Remark 8.6. IM pp. 244–245 Remark 1, concerning the formula $P(\Psi, w)$ given by the method of proof of IM Theorem 27 to numeralwise represent a primitive recursive function $\varphi(\Psi)$, extends to the case Ψ may include function variables.

Lemma 8.7. *For each primitive (general) recursive function* $\varphi(\Psi)$, *the formulas* $P(\Psi, w)$ *and* $P_1(\Psi, w)$ *given by the proofs of* IM *Theorem* 27 (32 (a)) *and the present Lemma* 8.5 *respectively to numeralwise represent* $\varphi(\Psi)$ *are equivalent (even in the subsystem of Lemma* 8.5), *i.e.* $\vdash P(\Psi, w) \sim P_1(\Psi, w)$.

Proof, for $\varphi(\Psi)$ primitive recursive, by induction on the length k of a given primitive recursive description of $\varphi(\Psi)$. In Cases (Va) and (Vb), the proposition will follow from I, II, IV below and the hyp. ind.; in these cases, we write (y, Θ) for Ψ, and in (Va) $Q(w)$ is $w=\boldsymbol{q}$.

I. $\vdash a<b \sim a<_1 b$. For, $a<b \sim \neg(a\geq b)$ [*140, *141, *63] $\sim \neg\exists c a=b+c \sim \forall c a\neq b+c$ [*86].

II. $\vdash B(c, d, i, w) \sim B_1(c, d, i, w)$. For, write the right member of *180b as $\exists v(A \;\&\; B)$. Then $\exists vA \sim \exists v_{v\leq c}A$. By *150 with *158, $\exists v_{v\leq c}A \vee \neg\exists v_{v\leq c}A$. Hence $\exists vA \vee \neg\exists vA$. Hence by *49c, $\exists vA \sim \neg\neg\exists vA \sim \neg\forall v\neg A$ [IM p. 166 II]. So $B(c, d, i, w) \sim \exists v(A \;\&\; B)$ [*180b] $\sim \exists vA \;\&\; B$ [*91] $\sim \neg\forall v\neg A \;\&\; B$.

III. *In Cases* (Va) *and* (Vb) *for* IM *Theorem* 27,

$$\vdash \exists c\exists d\{\exists u[B(c, d, 0, u) \;\&\; Q(\Theta, u)] \;\&\; \forall i_{i<y}\exists u\exists v[B(c, d, i', u) \;\&\; B(c, d, i, v) \;\&\; R(i, v, \Theta, u)] \;\&\; B(c, d, y, w)\} \sim$$
$$\forall c\forall d\{\exists u[B(c, d, 0, u) \;\&\; Q(\Theta, u)] \;\&\; \forall i_{i<y}\exists u\exists v[B(c, d, i', u) \;\&\; B(c, d, i, v) \;\&\; R(i, v, \Theta, u)] \supset B(c, d, y, w)\}.$$

IIIa. Assume, preparatory to ∃-elims.,

(a) $\exists u[B(c_0, d_0, 0, u) \& Q(\Theta, u)] \& \forall i_{i<y} \exists u \exists v[B(c_0, d_0, i', u) \& B(c_0, d_0, i, v) \& R(i, v, \Theta, u)] \& B(c_0, d_0, y, w)$.

Assume, preparatory to ⊃- and ∀-introds.,

(b) $\exists u[B(c, d, 0, u) \& Q(\Theta, u)] \& \forall i_{i<y} \exists u \exists v[B(c, d, i', u) \& B(c, d, i, v) \& R(i, v, \Theta, u)]$.

We aim to deduce B(c, d, y, w). But first we shall deduce, by ind. on i,

(c) $i \leq y \supset \forall u_0 \forall u[B(c_0, d_0, i, u_0) \& B(c, d, i, u) \supset u_0 = u]$.

Basis. Similar to the: Ind. step. Assume **(1)** $i' \leq y$, **(2)** $B(c_0, d_0, i', u_0)$ and **(3)** B(c, d, i′, u). From (a), assume for &- and ∃-elim., **(4)** $B(c_0, d_0, i', u_1)$, **(5)** $B(c_0, d_0, i, v_1)$ and **(6)** $R(i, v_1, \Theta, u_1)$. From (b), assume **(7)** $B(c, d, i', u_2)$, **(8)** $B(c, d, i, v_2)$ and **(9)** $R(i, v_2, \Theta, u_2)$. By hyp. ind. with (1), (5) and (8): **(10)** $v_1 = v_2$. By IM p. 245 top line (with Remark 8.6), $\exists! w R(i, v_1, \Theta, w)$. So by (6) and (9) with *172: **(11)** $u_1 = u_2$. By *180c, $\exists! w B(c_0, d_0, i', w)$. So by (2) and (4): **(12)** $u_0 = u_1$. Similarly by (7) and (3): **(13)** $u_2 = u$. Combining (12), (11) and (13): $u_0 = u$. — By (a), $B(c_0, d_0, y, w)$. Case 1: $y > 0$ (or Case 2: $y = 0$). From (b), assume B(c, d, y, u). By (c), w=u. So B(c, d, y, w).

IIIb. Write the converse (to be proved) as

$$\forall c \forall d\{A(c, d) \supset B(c, d, w)\} \supset \exists c \exists d\{A(c, d) \& B(c, d, w)\}.$$

Assume **(a)** $\forall c \forall d\{A(c, d) \supset B(c, d, w)\}$. Using IM p. 245 (3), $\exists w \exists c \exists d\{A(c, d) \& B(c, d, w)\}$. Thence assume **(b)** $A(c, d) \& B(c, d, w_0)$. By (a), B(c, d, w). So $\exists c \exists d\{A(c, d) \& B(c, d, w)\}$.

IV. *In Cases* (Va) *and* (Vb) *for* IM *Theorem* 27,

$$\vdash P(y, \Theta, w) \sim \forall c \forall d\{\forall u[B(c, d, 0, u) \supset Q(\Theta, u)] \& \forall i_{i<y} \forall u \forall v[B(c, d, i', u) \& B(c, d, i, v) \& R(i, v, \Theta, u)] \supset B(c, d, y, w)\}.$$

Use successively: III; *91; *181 p. 408 with *180c; *95; *4, *5.

Lemma 8.8. *Let* P(Ψ, w) [P(Ψ)] *be picked by the method of proof of Lemma* 8.5 (*or of* IM *Theorem* 27 [*Corollary Theorem* 27]) *to numeralwise represent* [*express*] *a primitive recursive function* φ(Ψ) [*predicate* P(Ψ)]. *Then* (*even in the subsystem*)

$$\vdash P(\Psi, w) \vee \neg P(\Psi, w) \qquad [\vdash P(\Psi) \vee \neg P(\Psi)].$$

PROOF, for Lemma 8.5. Then $P(\Psi)$ is $P(\Psi, 0)$, where by IM p. 245 (3) (with Remark 8.6 above) and Lemma 8.7 (with $P(\Psi, w)$ as the $P_1(\Psi, w)$), $\vdash \exists! w P(\Psi, w)$. Hence by Lemma 5.6, $\vdash P(\Psi, w) \vee \neg P(\Psi, w)$, whence by substitution, $\vdash P(\Psi) \vee \neg P(\Psi)$.

§ 9. Realizability under deduction in the intuitionistic formal system. 9.1. LEMMA 9.1. (a) *Let* Ψ *be a list of distinct variables not including* x; *let* A(x) *be a formula containing free only* Ψ, x; *and let* t *be a term containing free only* Ψ, x, *free for* x *in* A(x), *and* (*for given values* Ψ, *x of* Ψ, x) *expressing the number* $t(\Psi, x)$. *Then* ϵ *realizes-*Ψ, $t(\Psi, x)$ A(x) *if and only if* ϵ *realizes-*Ψ, *x* A(t). (b) *With similar stipulations* (u *a functor*), ϵ *realizes-*Ψ, $u[\Psi, \alpha]$ A(α) *if and only if* ϵ *realizes-*Ψ, α A(u).

With this we may combine uses of Lemma 8.2, hereafter tacit. For example, if A(a) contains free only a (distinct from x), and α and x are free for a in A(a): $\{\epsilon$ realizes-$\bar{\alpha}(x)$ A(a)$\} \equiv \{\epsilon$ realizes-α, x, $\bar{\alpha}(x)$ A(a)$\}$ [Lemma 8.2] $\equiv \{\epsilon$ realizes-α, x, a A($\bar{\alpha}$(x))$\}$ [Lemma 9.1 (a), with α, x as the Ψ and a as the x] $\equiv \{\epsilon$ realizes-α, x A($\bar{\alpha}$(x))$\}$ [Lemma 8.2].

LEMMA 9.2. ϵ *realizes-*Ψ E *if and only if* ϵ *realizes-*Ψ *the result of replacing each part of* E *of the form* $\neg$A *where* A *is a formula by* $A \supset 1=0$.

THEOREM 9.3. (a) *If* $\Gamma \vdash E$ *in the intuitionistic formal system of analysis, and the formulas* Γ *are realizable, then* E *is realizable.* (b) *Similarly, reading "realizable/T" in place of "realizable".*

(c) *If* Γ, E *contain free only* $\Psi = (\Phi, \Theta)$, *and in the intuitionistic formal system of analysis* $\Gamma \vdash E$ *with* Θ *held constant, and* Γ *are realizable-*Θ (end 8.7), *then* E *is realizable-*Θ. (d) *Similarly, reading "realizable-*Θ*/T" in place of "realizable-*Θ*".*

PROOF. (a) (Cf. the proof of IM Theorem 62 (a).)

AXIOMS (except by $^{\times}$26.3, $^{\times}$27.1). For each particular axiom E (For most of the axiom schemata), we give a particular primitive recursive function ϵ such that, for (for any axiom E by the schema,) any list Ψ of variables including all that occur free in E, and any assignment Ψ of values to Ψ, the function ϵ realizes-Ψ E; so taking $\varphi[\Psi] = \epsilon$ (i.e. $\varphi[\Psi] = \lambda t\, \varphi(\Psi, t)$ where $\varphi(\Psi, t) = \epsilon(t)$), φ is a primitive recursive realization function for E (cf. 8.7). For the rest of these axiom schemata, the ϵ may depend on some of the Ψ as parameters (e.g. for Axiom Schema 13 on x).

1a. $A \supset (B \supset A)$ is realized-Ψ by $\Lambda\alpha\Lambda\beta\,\alpha$. For, suppose **(1)**

α realizes-Ψ A; by Clause 4 in 8.5, we must infer from (1) that $\{\Lambda\alpha\, \Lambda\beta\, \alpha\}[\alpha]$ realizes-Ψ B ⊃ A. But by (8.3), $\{\Lambda\alpha\, \Lambda\beta\, \alpha\}[\alpha] = \Lambda\beta\, \alpha$. Suppose **(2)** β realizes-Ψ B; we must infer from (1) and (2) that $\{\Lambda\beta\, \alpha\}[\beta]$ realizes-Ψ A. By (8.3), $\{\Lambda\beta\, \alpha\}[\beta] = \alpha$; so what we need is (1).

1b. (A ⊃ B) ⊃ ((A ⊃ (B ⊃ C)) ⊃ (A ⊃ C)). Also 7, via Lemma 9.2. $\Lambda\pi\, \Lambda\rho\, \Lambda\alpha\, \{\{\rho\}[\alpha]\}[\{\pi\}[\alpha]]$.

3. A ⊃ (B ⊃ A & B). $\Lambda\alpha\, \Lambda\beta\, \langle\alpha, \beta\rangle$.

4a. A & B ⊃ A. $\Lambda\gamma\, (\gamma)_0$. 4b. A & B ⊃ B. $\Lambda\gamma\, (\gamma)_1$.

5a. A ⊃ A ∨ B. $\Lambda\alpha\, \langle 0, \alpha\rangle$. 5b. B ⊃ A ∨ B. $\Lambda\beta\, \langle 1, \beta\rangle$.

6. (A ⊃ C) ⊃ ((B ⊃ C) ⊃ (A ∨ B ⊃ C)). $\Lambda\pi\, \Lambda\rho\, \Lambda\sigma$ $\lambda t\, \overline{\mathrm{sg}}((\sigma(0))_0)\cdot(\{\pi\}[(\sigma)_1])(t) + \mathrm{sg}((\sigma(0))_0)\cdot(\{\rho\}[(\sigma)_1])(t)$.

8^{I}. ¬A ⊃ (A ⊃ B). $\Lambda\pi\, \lambda t\, 0$. Suppose π realizes-Ψ ¬A. Then no function α realizes-Ψ A. So any function, e.g. $\lambda t\, 0$, realizes-Ψ A ⊃ B.

10N. ∀xA(x) ⊃ A(t), where A(x), t are as in Lemma 9.1 (a), so the free variables of the axiom are only Ψ, x. $\Lambda\pi\, \{\pi\}[t(\Psi, x)]$. For, suppose π realizes-Ψ, x ∀xA(x). Then by Lemma 8.2, π realizes-Ψ ∀xA(x). So $\{\pi\}[t(\Psi, x)]$ realizes-$\Psi, t(\Psi, x)$ A(x), whence by Lemma 9.1 (a) $\{\pi\}[t(\Psi, x)]$ realizes-Ψ, x A(t).

10F. ∀αA(α) ⊃ A(u). $\Lambda\pi\, \{\pi\}[u[\Psi, \alpha]]$.

11N. A(t) ⊃ ∃xA(x). $\Lambda\pi\, \langle t(\Psi, x), \pi\rangle$.

11F. A(u) ⊃ ∃αA(α). $\Lambda\pi\, \langle\Lambda\, u[\Psi, \alpha], \pi\rangle$.

13. A(0) & ∀x(A(x) ⊃ A(x′)) ⊃ A(x). $\Lambda\alpha\, \rho[x, \alpha]$, where ρ is defined by the "functional recursion"

$$\rho[0, \alpha] = (\alpha)_0,$$
$$\rho[x', \alpha] = \{\{(\alpha)_1\}[x]\}[\rho[x, \alpha]].$$

Writing $\rho[x, \alpha] = \lambda t\, \rho(x, \alpha, t)$, this takes the form

$$\rho(0, \alpha, t) = \psi(\alpha, t),$$
$$\rho(x', \alpha, t) \simeq \chi(x, \alpha, \lambda t\, \rho(x, \alpha, t), t)$$

where ψ is primitive, and χ is partial, recursive. To prove this ρ partial recursive, we apply the recursion theorem IM p. 353 for α as the Ψ ($l = 1$) with uniformity to solve for z the equation

$$\{z\}(x, \alpha, t) \simeq \begin{cases} \psi(\alpha, t) \text{ if } x = 0, \\ \chi(x \dot{-} 1, \alpha, \lambda t\, \{z\}(x \dot{-} 1, \alpha, t), t) \text{ if } x \neq 0. \end{cases}$$

Call the solution e, and put $\rho(x, \alpha, t) = \{e\}(x, \alpha, t)$. (Cf. Lemma 3.2, Kleene **1956** § 4, **1959** XXIV.)

14, 17, ×1.1: $\Lambda\pi\,\lambda t\,0$. 16: $\Lambda\pi\,\Lambda\rho\,\lambda t\,0$.

15, and all prime axioms (namely, 18–21, ×0.1, the axioms of Group D): $\lambda t\,0$.

×2.1. $\forall x \exists \alpha A(x, \alpha) \supset \exists \alpha \forall x A(x, \lambda y \alpha(\langle x, y\rangle))$. $\Lambda\pi\,\langle \Lambda\,\lambda t\,\{(\{\pi\}[(t)_0])_0\}((t)_1), \Lambda x\,(\{\pi\}[x])_1\rangle$. Suppose π realizes-Ψ $\forall x \exists \alpha A(x, \alpha)$. Then, for each x, $(\{\pi\}[x])_1$ realizes-Ψ, x, $\{(\{\pi\}[x])_0\}$ $A(x, \alpha)$. Hence by Lemma 9.1 (b), $(\{\pi\}[x])_1$ realizes-Ψ, x, $\lambda t\,\{(\{\pi\}[(t)_0])_0\}((t)_1)$ $A(x, \lambda y\,\alpha(\langle x, y\rangle))$.

Rules of inference. 2. A, A ⊃ B / B. Noting 8.7, we choose Ψ to include all variables free in A ⊃ B. By hyp. ind., there are general recursive functions α and ψ such that, for each Ψ, $\alpha[\Psi]$ realizes-Ψ A and $\psi[\Psi]$ realizes-Ψ A ⊃ B. Let $\varphi[\Psi] = \{\psi[\Psi]\}[\alpha[\Psi]]$. Then φ is partial recursive, and, for each Ψ, $\varphi[\Psi]$ realizes-Ψ B; hence φ is general recursive.

9N. C ⊃ A(x) / C ⊃ ∀xA(x). Say, for each Ψ and x, $\psi[\Psi, x]$ realizes-Ψ, x C ⊃ A(x). Then, for each Ψ, $\Lambda\gamma\,\Lambda x\,\{\psi[\Psi, x]\}[\gamma]$ realizes-Ψ C ⊃ ∀xA(x). 9F. $\Lambda\gamma\,\Lambda\alpha\,\{\psi[\Psi, \alpha]\}[\gamma]$.

12F. A(α) ⊃ C / ∃αA(α) ⊃ C. $\Lambda\pi\,\{\psi[\Psi, \{(\pi)_0\}]\}[(\pi)_1]$ is a realization function for the conclusion, if $\psi[\Psi, \alpha]$ is one for the premise. 12N. $\Lambda\pi\,\{\psi[\Psi, (\pi(0))_0]\}[(\pi)_1]$.

Axiom Schema ×26.3c. $\forall\alpha\exists! x R(\bar\alpha(x))$ & $\forall a[\mathrm{Seq}(a)$ & $R(a) \supset A(a)]$ & $\forall a[\mathrm{Seq}(a)$ & $\forall s A(a*2^{s+1}) \supset A(a)] \supset A(1)$.

Assume that π realizes-Ψ the antecedent of the main implication of an axiom by this schema containing free only Ψ; all the definitions and inferences below are under this assumption until the final step, except that, when we say a predicate or function with π as a variable is partial recursive, π ranges over all functions.

Now $(\pi)_{0,0}$ realizes-Ψ $\forall\alpha\exists! x R(\bar\alpha(x))$, i.e. $\forall\alpha\exists x[R(\bar\alpha(x))$ & $\forall y(R(\bar\alpha(y)) \supset x{=}y)]$; $(\pi)_{0,1}$ realizes-Ψ $\forall a[\mathrm{Seq}(a)$ & $R(a) \supset A(a)]$; and $(\pi)_1$ realizes-Ψ $\forall a[\mathrm{Seq}(a)$ & $\forall s A(a*2^{s+1}) \supset A(a)]$. Hence: **(1)** For each α, $(\{(\pi)_{0,0}\}[\alpha])_1$ realizes-Ψ, α, x $R(\bar\alpha(x))$ & $\forall y(R(\bar\alpha(y)) \supset x{=}y)$ for $x = (\{(\pi)_{0,0}\}[\alpha](0))_0$. **(2)** For each a, ρ_0, ρ_1, if ρ_0 realizes-a Seq(a) and ρ_1 realizes-Ψ, a R(a), then $\{\{(\pi)_{0,1}\}[a]\}[\rho_0, \rho_1]$ realizes-Ψ, a A(a) (cf. (8.1c)). **(3)** For each a, ρ_0, ρ_1, if ρ_0 realizes-a Seq(a), and, for each s, $\{\rho_1\}[s]$ realizes-Ψ, a, s $A(a*2^{s+1})$, then $\{\{(\pi)_1\}[a]\}[\rho_0, \rho_1]$ realizes-Ψ, a A(a).

Furthermore: **(4)** For each σ, α, y, if σ realizes-Ψ, α, y $R(\bar\alpha(y))$, then $y = x$ for the x of (1). For, by (1) $\{\{(\{(\pi)_{0,0}\}[\alpha])_{1,1}\}[y]\}[\sigma]$ realizes-x, y x=y, so x=y is true-x, y.

By (1), $\{(\pi)_{0,0}\}[\alpha](0)$ is defined for each α. Let

$$R(\pi, a) \equiv (E\alpha)[a = \bar{\alpha}(x) \text{ for } x = (\{(\pi)_{0,0}\}[\alpha](0))_0].$$

Clearly **(5)** $(\beta)(Ex)R(\pi, \bar{\beta}(x))$.

We define a partial recursive predicate R_1 thus.

$$R_1(\pi, a) \simeq [a = \overline{\alpha_1}(x_1) \text{ for } \alpha_1 = \lambda t\,(a)_t \dot{-} 1, x_1 = (\{(\pi)_{0,0}\}[\alpha_1](0))_0].$$

We show that **(6)** $R(\pi, a) \equiv R_1(\pi, a)$. Assume $R(\pi, a)$, and put $a = \bar{\alpha}(x)$ with $x = (\{(\pi)_{0,0}\}[\alpha](0))_0$. By (1), $(\{(\pi)_{0,0}\}[\alpha])_{1,0}$ realizes-Ψ, α, x $\mathrm{R}(\bar{\alpha}(\mathrm{x}))$, whence by Lemma 9.1 (a) it realizes-$\Psi, \bar{\alpha}(\mathrm{x})$ $\mathrm{R}(\mathrm{a})$. Let $\alpha_1 = \lambda t\,(a)_t \dot{-} 1$. Then α_1 agrees with α in its first x values, so $\overline{\alpha_1}(x) = \bar{\alpha}(x)$, and by Lemma 9.1 (a) $(\{(\pi)_{0,0}\}[\alpha])_{1,0}$ realizes-Ψ, α_1, x $\mathrm{R}(\bar{\alpha}(\mathrm{y}))$. By (4) with α_1, x and $x_1 = (\{(\pi)_{0,0}\}[\alpha_1](0))_0$ as the α, y and x, $x = x_1$. So $a = \overline{\alpha_1}(x_1)$. Conversely, $a = \overline{\alpha_1}(x_1)$ implies $R(\pi, a)$.

We shall find a partial recursive function η with the following property. Let S_1^π be the set of the sequence numbers barred with respect to $\lambda a\, R_1(\pi, a)$ (cf. 6.3, 6.5, 6.6). **(7)** $a \in S_1^\pi \rightarrow \{\eta[\pi, a]$ realizes-Ψ, a $\mathrm{A}(\mathrm{a})\}$. To prove this, we use an intuitive application of the bar theorem, i.e. we use an induction over S_1^π (the informal analog of $^{\times}$26.8a in 6.11, with the recursiveness of $\lambda a\, R_1(\pi, a)$ providing the first hypothesis). We begin by giving the basis and induction step. In each we derive a specification for $\eta[\pi, a]$ that will suffice there. Afterwards we show that a partial recursive η can be chosen to satisfy both specifications.

BASIS: $R_1(\pi, a)$. Then $a = \overline{\alpha_1}(x_1)$ etc. By (1), $(\{(\pi)_{0,0}\}[\alpha_1])_{1,0}$ realizes-Ψ, α_1, x_1 $\mathrm{R}(\bar{\alpha}(\mathrm{x}))$, whence by Lemma 9.1 (a) it realizes-Ψ, a $\mathrm{R}(\mathrm{a})$. Also $\mathrm{Seq}(a)$, so $\lambda t\, 0$ realizes-a $\mathrm{Seq}(\mathrm{a})$. So using (2), $\eta[\pi, a]$ will realize-Ψ, a $\mathrm{A}(\mathrm{a})$ if $\eta[\pi, a] = \{\{(\pi)_{0,1}\}[a]\}[\lambda t\, 0, (\{(\pi)_{0,0}\}[\alpha_1])_{1,0}]$.

IND. STEP: $\mathrm{Seq}(a)$ & $(s)[a{*}2^{s+1} \in S_1^\pi]$. By hyp. ind., for each s, $\eta[\pi, a{*}2^{s+1}]$ realizes-$\Psi, a{*}2^{s+1}$ $\mathrm{A}(\mathrm{a})$, whence using Lemma 9.1 (a) it realizes-Ψ, a, s $\mathrm{A}(\mathrm{a}{*}2^{\mathrm{s}+1})$. So, using (3), $\eta[\pi, a]$ will realize-Ψ, a $\mathrm{A}(\mathrm{a})$ if $\eta[\pi, a] = \{\{(\pi)_1\}[a]\}[\lambda t\, 0, \Lambda s\, \eta[\pi, a{*}2^{s+1}]]$.

DEFINITION OF η. It will suffice to have $\eta[\pi, a] = \lambda u\, \eta(\pi, a, u)$ where

$$\eta(\pi, a, u) \simeq \begin{cases} \{\{(\pi)_{0,1}\}[a]\}[\lambda t\, 0, (\{(\pi)_{0,0}\}[\lambda t\,(a)_t \dot{-} 1])_{1,0}](u) & \text{if } R_1(\pi, a), \\ \{\{(\pi)_1\}[a]\}[\lambda t\, 0, \Lambda s\, \lambda t\, \eta(\pi, a{*}2^{s+1}, t)](u) & \text{otherwise.} \end{cases}$$

Upon replacing η by $\{z\}$, this equation assumes the form $\{z\}(\pi, a, u) \simeq \psi(z, \pi, a, u)$ with a partial recursive ψ. A solution e for z is given by the recursion theorem IM p. 353 for π as the Ψ with uniformity. We take $\eta(\pi, a, u) \simeq \{e\}(\pi, a, u)$; as remarked after Lemma 8.1 (with (8.2a)), the specialization of z to e under the operation Λs is valid.

By (5) and (6): **(8)** $1 \in S_1^{\pi}$. Hence by (7), $\eta[\pi, 1]$ realizes-Ψ, 1 A(a), whence by Lemma 9.1 (a): **(9)** $\eta[\pi, 1]$ realizes-Ψ A(1).

Finally, $\Lambda\pi\, \eta[\pi, 1]$ realizes-Ψ the axiom.

Axiom Schema ˣ27.1. $\forall\alpha\exists\beta A(\alpha, \beta) \supset \exists\tau\forall\alpha\{\forall t\exists!y\tau(2^{t+1}*\bar{\alpha}(y))>0\ \&\ \forall\beta[\forall t\exists y\tau(2^{t+1}*\bar{\alpha}(y))=\beta(t)+1 \supset A(\alpha, \beta)]\}$.

Assume (outside the definitions of the recursive functions and the final step) that π realizes-Ψ $\forall\alpha\exists\beta A(\alpha, \beta)$.

Consider any α. Now **(1)** for each α, $(\{\pi\}[\alpha])_1$ realizes-Ψ, α, β_1 $A(\alpha, \beta)$ for $\beta_1 = \{(\{\pi\}[\alpha])_0\}$. Let $\tau = \Lambda\alpha\, \beta_1 = \Lambda\alpha\, \{(\{\pi\}[\alpha])_0\}$. By (1) and Lemma 8.1, $\{\tau\}[\alpha]$ is properly defined, i.e. **(2)** $(t)(E!y)\tau(2^{t+1}*\bar{\alpha}(y))>0$, and $\{\tau\}[\alpha] = \beta_1$, whence by (8.1): **(3)** $(t)\tau(2^{t+1}*\bar{\alpha}(y_t))=\beta_1(t)+1$ where $y_t = \mu y\tau(2^{t+1}*\bar{\alpha}(y))>0$.

Now we seek a function ρ_0 to realize-τ, α $\forall t\exists!y\tau(2^{t+1}*\bar{\alpha}(y))>0$, i.e. $\forall t\exists y[\tau(2^{t+1}*\bar{\alpha}(y))>0\ \&\ \forall z(\tau(2^{t+1}*\bar{\alpha}(z))>0 \supset y=z)]$, taking the inequality as a prime formula (cf. preceding *15.1 in 5.5). Consider any t. Using (3), $\tau(2^{t+1}*\bar{\alpha}(y))>0$ is true-τ, α, t, y_t and hence is realized-τ, α, t, y_t by $\lambda s\, 0$. If σ realizes-τ, α, t, z $\tau(2^{t+1}*\bar{\alpha}(z))>0$, then $\tau(2^{t+1}*\bar{\alpha}(z))>0$ is true-τ, α, t, z, hence by (2) $z = y_t$, and hence $\lambda s\, 0$ realizes-z, y_t $z=y$. Combining these results, $\forall t\exists!y\tau(2^{t+1}*\bar{\alpha}(y))>0$ is realized-τ, α by $\rho_0 = \Lambda t\, \langle \mu y\tau(2^{t+1}*\bar{\alpha}(y))>0, \langle \lambda s\, 0, \Lambda z\, \Lambda\sigma\, \lambda s\, 0\rangle\rangle$.

Next we seek a function ρ_1 to realize-Ψ, τ, α $\forall\beta[\forall t\exists y\tau(2^{t+1}*\bar{\alpha}(y))=\beta(t)+1 \supset A(\alpha, \beta)]$. Consider any β. Suppose σ realizes-τ, α, β $\forall t\exists y\tau(2^{t+1}*\bar{\alpha}(y))=\beta(t)+1$. Then, for each t, $(\{\sigma\}[t])_1$ realizes-$\tau, \alpha, \beta, \bar{y}_t$ $\tau(2^{t+1}*\bar{\alpha}(y))=\beta(t)+1$ for $\bar{y}_t = (\{\sigma\}[t](0))_0$; thus $(t)\tau(2^{t+1}*\bar{\alpha}(\bar{y}_t))=\beta(t)+1$. Hence by (2) and (3), $\beta = \beta_1$, so by (1) $(\{\pi\}[\alpha])_1$ realizes-Ψ, α, β $A(\alpha, \beta)$. So we take $\rho_1 = \Lambda\beta\, \Lambda\sigma\, (\{\pi\}[\alpha])_1$.

Altogether, the axiom in question is realized-Ψ by $\Lambda\pi\, \langle \Lambda\tau, \Lambda\alpha \langle \rho_0, \rho_1\rangle\rangle$ for τ, ρ_0, ρ_1 as above.

(b) Say Γ is $D_1, \ldots, D_l$. Now each D_j has a realization function $\varphi_j[\Psi_j]$ general recursive in some finitely-many functions of T; let Σ be a list of all the functions of T thus used for $j = 1, \ldots, l$, reduced (if necessary) to one-place functions by end 8.2. By the penultimate remark in 8.2, it now suffices to construct, for each formula A_i of the given deduction, a realization function $\lambda\Psi_i\, \varphi_i[\Psi_i]$ of the form $\lambda\Psi_i\, \varphi_i[\Psi_i, \Sigma]$ with $\lambda\Psi_i\Sigma\, \varphi_i[\Psi_i, \Sigma]$ partial recursive; e.g. for Rule 9N we now use $\Lambda\gamma\, \Lambda x\, \{\psi[\Psi, x, \Sigma]\}[\gamma]$.

(c) By an application of IM Lemma 8b p. 104 and changes of bound variables, we can replace the given deduction of E from Γ by one of $\tilde{\mathrm{E}}$ from $\tilde{\Gamma}$ in which Θ do not occur as bound variables, where $\tilde{\Gamma}$, $\tilde{\mathrm{E}}$ are congruent to Γ, E. Then also $\tilde{\Gamma}$ are realizable-Θ, and E is such if $\tilde{\mathrm{E}}$ is. Now we can reconstrue Θ to be individual and function symbols rather than number and function variables; in the resulting formal system, contrary to Lemma 3.3, terms express under the intended interpretation functions primitive recursive in Θ instead of absolutely. So now we adapt the proof of (b) to construct realization functions $\lambda\Psi_i\, \varphi_i[\Psi_i]$ of the form $\lambda\Psi_i\, \varphi_i[\Psi_i, \Theta]$ with $\lambda\Psi_i\Theta\, \varphi_i[\Psi_i, \Theta]$ partial recursive.

(d) The realization functions are of the form $\lambda\Psi_i\, \varphi_i[\Psi_i, \Theta, \Sigma]$.

9.2. COROLLARY 9.4. *If* A *is realizable, and* B *is unrealizable, then* A ⊃ B *is unrealizable. If* A *is realizable or realizable*/η *for any* η, *then* ¬A *is unrealizable and unrealizable*/θ *for every* θ. *Similarly reading "realizable-*Θ*" in place of "realizable"; etc.*

COROLLARY 9.5. *The intuitionistic formal system of analysis is simply consistent, i.e. for no formula* A *are both* A *and* ¬A *provable.*

PROOF. For no A are both A and ¬A realizable, by Corollary 9.4. —

This consistency proof uses informally only methods corresponding to the common portion of the classical and intuitionistic formal systems.

It is of course not a metamathematical consistency proof, as a non-elementary interpretation is used.[6] But presumably it can be formalized in the basic formal system (i.e. intuitionistic analysis minus ˣ27.1) to give a strictly finitary metamathematical consistency proof for the intuitionistic formal system of analysis relative to the basic system, just as in Nelson 1947 with Kleene 1945 § 14 the proof by the old realizability interpretation of the consistency of a certain non-classical extension of intuitionistic number theory was formalized to provide a metamathematical consistency proof for the extended system relative to the unextended one. Such a formalization must be quite laborious, as it must begin with a formalization of the presupposed theory of partial recursive functionals. Our confidence that it can be carried out is based on a careful review of everything which went into the above interpretative consistency proof, and on a part

[6] The "N" of IM pp. 500 ff. applies to the theorems, corollaries, lemmas and remarks here (not already labelled "C") which use realizability notions.

of the work of formalization already carried out. (We shall perhaps say more in a later publication.)

COROLLARY 9.6. *If* $\mathrm{P}(\mathrm{a}_1, \ldots, \mathrm{a}_k, \alpha_1, \ldots, \alpha_l)$ *numeralwise expresses a general recursive predicate* $P(a_1, \ldots, a_k, \alpha_1, \ldots, \alpha_l)$ *in the intuitionistic formal system of analysis (or any subsystem thereof), then, for each* $a_1, \ldots, a_k, \alpha_1, \ldots, \alpha_l$, $\mathrm{P}(\mathrm{a}_1, \ldots, \mathrm{a}_k, \alpha_1, \ldots, \alpha_l)$ *is realizable-*$a_1, \ldots, a_k, \alpha_1, \ldots, \alpha_l$ *if and only if* $P(a_1, \ldots, a_k, \alpha_1, \ldots, \alpha_l)$.

PROOF, adapting that of IM Lemma 47 p. 512. Suppose $P(a_1, \ldots, a_k, \alpha_1, \ldots, \alpha_l)$. Then by IM § 41 (i) p. 195 with p. 298 (cf. 5.4), $\mathrm{E}^{\alpha_1 \ldots \alpha_l}_{\alpha_1 \ldots \alpha_l} \vdash \mathrm{P}(\boldsymbol{a}_1, \ldots, \boldsymbol{a}_k, \alpha_1, \ldots, \alpha_l)$ with $\alpha_1, \ldots, \alpha_l$ held constant. But $\mathrm{E}^{\alpha_1 \ldots \alpha_l}_{\alpha_1 \ldots \alpha_l}$ are prime and true-$\alpha_1, \ldots, \alpha_l$ and hence realizable-$\alpha_1, \ldots, \alpha_l$. So by (c) of the theorem, $\mathrm{P}(\boldsymbol{a}_1, \ldots, \boldsymbol{a}_k, \alpha_1, \ldots, \alpha_l)$ is realizable-$\alpha_1, \ldots, \alpha_l$, and hence by Lemma 9.1 (a), $\mathrm{P}(\mathrm{a}_1, \ldots, \mathrm{a}_k, \alpha_1, \ldots, \alpha_l)$ is realizable-$a_1, \ldots, a_k, \alpha_1, \ldots, \alpha_l$. Conversely, suppose $\mathrm{P}(\mathrm{a}_1, \ldots, \mathrm{a}_k, \alpha_1, \ldots, \alpha_l)$ is realizable-$a_1, \ldots, a_k, \alpha_1, \ldots, \alpha_l$. Values of $\alpha_1, \ldots, \alpha_l$ being available as required, $P(a_1, \ldots, a_k, \alpha_1, \ldots, \alpha_l) \vee \overline{P}(a_1, \ldots, a_k, \alpha_1, \ldots, \alpha_l)$; etc. as before. —

Corollary 9.6 may be used in conjunction with IM Corollaries to Theorems 27 and 32. A different approach is provided by Lemmas 8.5 and 8.4a.

9.3. In the definitions of 'realizes-Ψ' and 'realizable', let the range of the informal function variables ϵ, Ψ, α be confined to the functions belonging to a class C closed under general recursiveness (i.e., whenever $\Sigma \in C$ and φ is general recursive in Σ, then $\varphi \in C$), or general recursive in a function ξ or in section $\bar{\mathbf{c}}$ (Kleene **1963** p. 133), and hence containing all general recursive functions. There result notions *C/realizes-*Ψ and *C/realizable*, or *ξ/realizes-*Ψ etc.

We can simultaneously use recursiveness in T instead of recursiveness in the definition of 'realizable', obtaining for a $T \subset C$ a notion *C/realizable/T* (and similarly for Θ, $T \subset C$, *C/realizable-*Θ*/T*; etc.). The C acts as a ceiling on all the functions considered, the T as a threshold below which constructivity is not demanded.

THEOREM 9.7. *For any class C of functions closed under general recursiveness (e.g. the general recursive functions): If* $\Gamma \vdash \mathrm{E}$ *in the intuitionistic formal system of analysis without Axiom Schema* ×26.3 *(the bar theorem), and the formulas* Γ *are C/realizable [C/realizable/T, where* $T \subset C$*], then* E *is C/realizable [C/realizable/T] (cf. Theorem* 9.3 (a), (b)). *With corresponding changes, Lemmas* 8.2, 8.3, 8.4a, 8.4b,

9.1, 9.2, *Theorem* 9.3 (c) *and* (d), *and Corollaries* 9.4 *and* 9.6 *hold* (call them LEMMA $C/8.2$ etc.).

PROOF. We reëxamine the former proofs, omitting the case of ×26.3 in that of Theorem 9.3, to verify that the reasoning holds good when the universe of functions is C.

LEMMA 9.8, toward Corollary 9.9. (Kleene 1950a § 3.) *There is a primitive recursive predicate* $R(a)$ *such that, writing* $\alpha \in \mathbf{0} \equiv \{\alpha$ *is general recursive*$\}$ (Kleene-Post **1954**) *and* $B(\alpha) \equiv (t)\alpha(t) \leq 1$:

(a) $(\alpha)_{\alpha \in \mathbf{0} \& B(\alpha)}(Ex)R(\bar{\alpha}(x))$;

(b) $(z)(E\alpha)_{\alpha \in \mathbf{0} \& B(\alpha)}(x)_{x \leq z}\bar{R}(\bar{\alpha}(x))$,

whence

(c) $\overline{(Ez)}(\alpha)_{\alpha \in \mathbf{0} \& B(\alpha)}(Ex)_{x \leq z}R(\bar{\alpha}(x))$,

whence by the fan theorem (the informal analog of *26.6a)

(d) $\overline{(\alpha)}_{B(\alpha)}(Ex)R(\bar{\alpha}(x))$.

PROOF. Using the W_0, W_1 of IM p. 308, let

$$W(i, t, y) \equiv \begin{cases} W_0(t, y) \text{ if } i = 0, \\ W_1(t, y) \text{ if } i \neq 0, \end{cases}$$

$$R(a) \equiv (Et)_{t < \text{lh}(a)}(Ey)_{y < \text{lh}(a) \dot{-} t}W((a)_t \dot{-} 1, t, y).$$

Then, for each α with $B(\alpha)$,

(1) $R(\bar{\alpha}(x)) \equiv (Et)_{t<x}(Ey)_{y<x \dot{-} t}W_{\alpha(t)}(t, y)$.

(a) Consider any general recursive α with $B(\alpha)$. Using IM Theorem IV p. 281,

(2) $\alpha(t) = 1 \equiv (Ey)T_1(f_0, t, y) \equiv (Ey)T_1((f)_0, t, y)$,

(3) $\alpha(t) = 0 \equiv (Ey)T_1(f_1, t, y) \equiv (Ey)T_1((f)_1, t, y)$,

for suitable numbers $f_0, f_1, f = \langle f_0, f_1 \rangle$. CASE 1: $\alpha(f) = 1$. Then $(Ey)T_1((f)_0, f, y)$; and $\overline{(Ey)}T_1((f)_1, f, y)$, whence $(z)\bar{T}_1((f)_1, f, z)$. So $(Ey)W_1(f, y)$, i.e. $(Ey)W_{\alpha(f)}(f, y)$. CASE 2: $\alpha(f) = 0$. Similarly. — By (1) (with $x = f+y+1$, $t = f$), $(Ex)R(\bar{\alpha}(x))$.

(b) Consider any z. Let

$$\alpha(t) = \begin{cases} 1 \text{ if } t<z \ \& \ (Ey)_{y<z \dot{-} t}W_0(t, y) \ (\text{CASE A}), \\ 0 \text{ if } t<z \ \& \ (Ey)_{y<z \dot{-} t}W_1(t, y) \ (\text{CASE B}), \\ 0 \text{ otherwise.} \end{cases}$$

Then $\alpha \in \mathbf{0}$ & $B(\alpha)$. Consider any $x \leq z$. Suppose $R(\bar{\alpha}(x))$. By (1), there is a $t < x \leq z$ and a $y < x \dot{-} t \leq z \dot{-} t$ such that $W_{\alpha(t)}(t, y)$. Thence we obtain a contradiction, by cases. CASE 1: $\alpha(t) = 1$. Then $W_1(t, y)$, and by Case B of the definition of α, $\alpha(t) = 0$. CASE 2: $\alpha(t) = 0$. Similarly.

COROLLARY 9.9. *The bar theorem* ˣ26.3 *and the fan theorem* *26.6 *or* *27.7 *do not hold in the intuitionistic system without the former as axiom schema, i.e. some formulas of the forms* ˣ26.3, *26.6, *27.7 (*and via deducibility relationships*, *26.4, *26.7–ˣ26.9, *27.8–*27.14) *are unprovable in it.*

Also, by Corollary 9.5, the negation of no instance of ˣ26.3 etc. is unprovable. So ˣ26.3 etc. are "independent" of the other postulates of the intuitionistic system.

PROOF OF COROLLARY 9.9. Taking $C = \{$the general recursive functions$\} = \mathbf{0}$ in Theorem 9.7, all formulas provable in the system in question are $\mathbf{0}$/realizable.

ˣ26.3, *26.6. We shall show that the following substitution instance of the fan theorem *26.6a (deducible in this system from an instance of ˣ26.3a) is not $\mathbf{0}$/realizable: $\forall a[\mathrm{Seq}(a) \supset \mathrm{R}(a) \vee \neg \mathrm{R}(a)]$ & $\forall \alpha_{\mathrm{B}(\alpha)} \exists x \mathrm{R}(\bar{\alpha}(x)) \supset \exists z \forall \alpha_{\mathrm{B}(\alpha)} \exists x_{x \leq z} \mathrm{R}(\bar{\alpha}(x))$, where $\mathrm{B}(\alpha)$ is $\forall t \alpha(t) \leq 1$ ($\lambda t 1$ being substituted for β in *26.6), $\mathrm{R}(a)$ is a formula numeralwise expressing the primitive recursive predicate $R(a)$ of Lemma 9.8 obtained by the method of proof of Lemma 8.5, and for simplicity $x \leq z$ and $\alpha(t) \leq 1$ are prime. Suppose a general recursive function ϵ $\mathbf{0}$/realizes it. By Lemma $\mathbf{0}$/8.4a (ii) in Theorem 9.7, $R(\bar{\alpha}(x)) \rightarrow \{\epsilon_{\mathrm{R}(\bar{\alpha}(x))}$ $\mathbf{0}$/realizes-α, x $\mathrm{R}(\bar{\alpha}(x))\}$. By Lemma 8.8, $\vdash \mathrm{R}(a) \vee \neg \mathrm{R}(a)$, whence $\vdash \forall a[\mathrm{Seq}(a) \supset \mathrm{R}(a) \vee \neg \mathrm{R}(a)]$. By Theorem $\mathbf{0}$/9.3 (a), the latter formula is $\mathbf{0}$/realized by a general recursive function ζ_0. Consider any $\alpha \in \mathbf{0}$; by Lemma $\mathbf{0}$/8.4a (i), a function $\rho \in \mathbf{0}$ $\mathbf{0}$/realizes-α $\mathrm{B}(\alpha)$ only when $B(\alpha)$, and in this case by Lemma 9.8 (a) $x_1 = \mu x R(\bar{\alpha}(x))$ is defined and $R(\bar{\alpha}(x_1))$. So $\zeta_1 = \Lambda \alpha \Lambda \rho \langle x_1, \epsilon_{\mathrm{R}(\bar{\alpha}(x))} \rangle$ $\mathbf{0}$/realizes $\forall \alpha_{\mathrm{B}(\alpha)} \exists x \mathrm{R}(\bar{\alpha}(x))$. Hence $\eta = (\{\epsilon\}[\zeta_0, \zeta_1])_1$ $\mathbf{0}$/realizes-z $\forall \alpha_{\mathrm{B}(\alpha)} \exists x_{x \leq z} \mathrm{R}(\bar{\alpha}(x))$ for $z = (\{\epsilon\}[\zeta_0, \zeta_1](0))_0$. Now consider any general recursive α such that $B(\alpha)$. By Lemma $\mathbf{0}$/8.4a (ii), $\mathrm{B}(\alpha)$ is $\mathbf{0}$/realized-α by $\epsilon_{\mathrm{B}(\alpha)}$; so $x \leq z$ is $\mathbf{0}$/realized-x, z by $(\{\{\eta\}[\alpha]\}[\epsilon_{\mathrm{B}(\alpha)}])_{1,0}$, and $\mathrm{R}(\bar{\alpha}(x))$ is $\mathbf{0}$/realized-α, x by $(\{\{\eta\}[\alpha]\}[\epsilon_{\mathrm{B}(\alpha)}])_{1,1}$, for $x = (\{\{\eta\}[\alpha]\}[\epsilon_{\mathrm{B}(\alpha)}](0))_0$. Then $x \leq z$, and by Lemma $\mathbf{0}$/8.4a (i), $R(\bar{\alpha}(x))$. Thus $(\alpha)_{\alpha \in \mathbf{0} \& B(\alpha)} (Ex)_{x \leq z} R(\bar{\alpha}(x))$, contradicting Lemma 9.8 (c).

*27.7. Proceeding similarly, for $\mathrm{R}(\bar{\alpha}(\mathrm{b}))$ as the $\mathrm{A}(\alpha, \mathrm{b})$ of *27.7, we obtain η and z such that η **0**/realizes-z $\forall\alpha_{B(\alpha)}\exists b\forall\gamma_{B(\gamma)}[\forall x_{x<z}$ $\gamma(x)=\alpha(x) \supset R(\bar{\gamma}(b))]$, whence we infer that

(i) $(\alpha)_{\alpha\in\mathbf{0}\&B(\alpha)}(Eb)(\gamma)_{\gamma\in\mathbf{0}\&B(\gamma)}[(x)_{x<z}\gamma(x)=\alpha(x) \to R(\bar{\gamma}(b))]$.

Let S be the finite set of the numbers a such that $\mathrm{Seq}(a)$ & $\mathrm{lh}(a)=z$ & $(t)[(a)_t \leq 2]$, and let z_1 be the maximum of the b's given by (i) for $\alpha = \lambda t\ (a)_t \dot{-} 1$ for $a \in S$. Then $(\alpha)_{\alpha\in\mathbf{0}\&B(\alpha)}(Eb)_{b\leq z_1}R(\bar{\alpha}(b))$, contradicting Lemma 9.8 (c).

9.4C. In this and the next subsection we use classical reasoning.

The "jump" operation $'$ of Kleene-Post **1954** takes a predicate $\lambda a\ A(a)$ into the predicate $\lambda a\ (Ex)T_1^A(a, a, x)$, or more generally a function $\lambda a\ \alpha(a)$ into (the representing function of) the predicate $\lambda a\ (Ex)T_1^\alpha(a, a, x)$ (cf. IM p. 292).

A (number-theoretic) function $\varphi(a_1, \ldots, a_n)$ is *arithmetical* if its representing predicate $\varphi(a_1, \ldots, a_n)=w$ is arithmetical IM pp. 239, 285.

LEMMA 9.10C. *The arithmetical functions constitute the least class of functions closed under general recursiveness and the jump operation* $'$. (Kleene-Post **1954** with IM; or Kleene **1955b** §§ 2, 3.)

PROOF, from results in IM. First, φ and its representing predicate are general recursive each in the other, by ##14, C pp. 227–228 (with Theorem II p. 275), and Theorem III p. 279 with $\varphi(a_1, \ldots, a_n) = \mu w[\varphi(a_1, \ldots, a_n)=w]$. Now suppose φ is general recursive in arithmetical functions $\psi_1, \ldots, \psi_l$. By Theorem VII (d) p. 285, the representing predicates $Q_1, \ldots, Q_l$ of $\psi_1, \ldots, \psi_l$ are expressible each in one of the forms of Theorem V Part II p. 283, say with $k_1, \ldots, k_l$ quantifiers, respectively. Then, by introducing redundant quantifiers if necessary, all are expressible with $k = \max(k_1, \ldots, k_l)$ quantifiers. So by Post's theorem Theorem XI p. 293 (and Theorem VII (b)), the representing predicate of φ is arithmetical. Thus the arithmetical functions are closed under general recursiveness. As $T_1^\alpha(a, a, x)$ is primitive recursive in α (p. 292), hence by Theorem II general recursive in α, and by definition the arithmetical predicates are closed under number-quantification, the arithmetical functions are closed also under $'$. To show conversely that each arithmetical function is definable using only general recursive operations and $'$, it suffices by Theorem VII (d) to show it for predicates expressible in the forms of Theorem V. We do this by induction on the number k of the quantifiers. For $k = 0$, it is immediate. Say $P(a_1, \ldots, a_n)$ is expressible using $k+1$ quantifiers. If an existential quantifier (Ex) is outermost (otherwise we first consider $\overline{P}(a_1, \ldots, a_n)$), then (using $a_i = (\langle a_1, \ldots, a_n\rangle)_{i-1}$) $P(a_1, \ldots, a_n) \equiv (Ex)A(\langle a_1, \ldots, a_n\rangle, x)$, where $A(a, x)$ is expressible using k quantifiers and hence by hyp. ind. is definable by general recursive operations and $'$. Let $\beta = \lambda a\, \alpha((a)_0, (a)_1)$ where α is the representing function of A. Then A is recursive in β; so by p. 343 Example 2 for $l = 1$, $(Ex)A(a, x) \equiv (Ex)T_1^\beta(\psi(a), \psi(a), x)$ for a primitive recursive ψ. Thus $P(a_1, \ldots, a_n)$ is recursive in $(Ex)A(a, x)$, which is recursive in $(Ex)T_1^\beta(a, a, x)$, which comes by $'$ from β, which is recursive in $A(a, x)$.

REMARK 9.11^C. We now give informally a classical proof of the fan theorem via König's lemma **1926** (amplifying Kleene **1958** p. 139 lines 11–12). The fan theorem *27.9 in the informal symbolism is

(a) $\quad (\alpha)_{B(\alpha)}(Ex)R(\bar{\alpha}(x)) \rightarrow (Ez)(\alpha)_{B(\alpha)}(Ex)_{x\leq z}R(\bar{\alpha}(x))$

where $B(\alpha)$ is $(t)\alpha(t) \leq \beta(\bar{\alpha}(t))$. This is equivalent via contraposition etc. to

(b) $\quad (z)(E\alpha)_{B(\alpha)}(x)_{x\leq z}\overline{R}(\bar{\alpha}(x)) \rightarrow (E\alpha)_{B(\alpha)}(x)\overline{R}(\bar{\alpha}(x))$,

which upon being expressed geometrically will be König's lemma. For the geometrical version, after representing the fan by a tree as in 6.10 ¶ 3, we now print in bold face along each path α all the vertices up to the first one (if any) inclusive (underlined) at which $R(\bar{\alpha}(x))$, i.e. those occupied by sequence numbers not past secured 6.3. Again we suppress the part of the tree which is not in bold face. Now the hypothesis $(z)(E\alpha)_{B(\alpha)}(x)_{x\leq z}\overline{R}(\bar{\alpha}(x))$ of (b) says that in the tree remaining there are arbitrarily long finite paths (not illustrated by Figure 1 in 6.5). The conclusion $(E\alpha)_{B(\alpha)}(x)\overline{R}(\bar{\alpha}(x))$ says there is an infinite path. Indeed, we trace an infinite path by the following rule (in general, not effective). Under the rule, we start of course at the initial vertex marked [ ]. Suppose we have traced the path as far as any vertex a (either [ ] or a later vertex), and that a has the property of belonging to arbitrarily long finite paths (as [ ] does by hypothesis). Then one of the next vertices $a*2^{s+1}$ $(s = 0, \ldots, \beta(a))$ will have the same property; for, if all paths through $a*2^{s+1}$ were of length $\leq z_s+1$ $(s = 0, \ldots, \beta(a))$, all through a itself would be of length $\leq \max(z_0+1, \ldots, z_{\beta(a)}+1)$, contradicting our supposition that a has the property. The rule says to pick as the next vertex on the path one of the $a*2^{s+1}$ which has the property, say the one with least s. Thus, starting from

[ ], we are able successively to pick a next vertex, always with the property, ad infinitum.

LEMMA 9.12^{C}. *For any class* C *of functions closed under general recursiveness and the jump operation* $'$ *(e.g. by Lemma* 9.10, *the arithmetical functions): The fan theorem holds informally in the version* (a) *of Remark* 9.11, *when* β *and the representing function* ρ *of* R *belong to* C, *and the function variable* α *ranges over* C.

PROOF. By the proof in Remark 9.11, it will suffice to show that the α represented by the infinite path determined by the rule in that proof belongs to C. We analyze that rule by using a sequence number g, subject to appropriate conditions, to represent a finite path of length $z+1$ through a. Thus $g = \bar{\gamma}(z)$ for some γ such that $B(\gamma)$, i.e. $(t)\gamma(t) \leq \beta(\bar{\gamma}(t))$. It follows that $g \leq \gamma(\beta, z)$ where

$$\gamma(\beta, 0) = 1,$$
$$\gamma(\beta, t') = \gamma(\beta, t)\cdot p_t^{1+\max_{s\leq\gamma(\beta,t)}\beta(s)}$$

(Kleene **1956** Footnote 8). Thus the α represented by the path in question satisfies the equation

$$\text{(c)} \quad \alpha(x) = \mu s_{s\leq\beta(\bar{\alpha}(x))}(z)\{z>x \rightarrow (Eg)_{g\leq\gamma(\beta,z)}\{\mathrm{Seq}(g)\ \&\ \mathrm{lh}(g)=z\ \&$$
$$(t)_{t<z}[(g)_t\leq 1+\beta(\Pi_{i<t}p_i^{(g)_i})]\ \&\ \Pi_{i<x}p_i^{(g)_i}=\bar{\alpha}(x)\ \&$$
$$(g)_x=s+1\ \&\ (t)_{t\leq z}\bar{R}(\Pi_{i<t}p_i^{(g)_i})\}\},$$

which defines it by induction. Let $\sigma = \langle\beta, \rho\rangle$, so σ is primitive recursive in β, ρ, and β and ρ are each primitive recursive in σ. Now the right side of (c) is of the form $\mu s_{s\leq\beta(\bar{\alpha}(x))}(z)R^{\sigma}(\langle\bar{\alpha}(x), s\rangle, z)$ where R^{σ} is primitive recursive in σ. Applying the jump operation to σ, let $S(a) \equiv (Ez)T_1^{\sigma}(a, a, z)$. Since β, $\rho \in C$, so does σ, and hence so does (the representing function of) S. By IM p. 343 Example 2 for $l = 1$, $(Ez)\bar{R}^{\sigma}(a, z) \equiv S(\psi(a))$ for a primitive recursive ψ. Letting $\chi(a) = \mu s_{s\leq\beta(a)}\bar{S}(\psi(\langle a, s\rangle))$, (c) becomes

$$\text{(d)} \quad \alpha(x) = \chi(\bar{\alpha}(x)).$$

By IM #G p. 231 adapted to use $\bar{\alpha}(x)$ instead of $\tilde{\alpha}(x)$, α is primitive recursive in χ, which is primitive recursive in β, S, which $\in C$; so $\alpha \in C$.

THEOREM 9.13^{C}. *For any class* C *of functions closed under general recursiveness and the jump operation* $'$ *(e.g. the arithmetical functions): If* $\Gamma \vdash$ E *in the intuitionistic formal system of analysis with the fan theorem* *26.6 *(or via deducibility relationships,* *26.7, *27.7–*27.10) *replacing the bar theorem* ×26.3 *as axiom schema, and the formulas* Γ *are C/realizable [C/realizable/T, where* $T \subset C$*], then* E *is C/realizable [C/realizable/T].*

PROOF. To the proof of Theorem 9.7, we add a treatment of the fan theorem with the present C.

*26.6c. $\forall\alpha_{B(\alpha)}\exists!xR(\bar\alpha(x)) \supset \exists z\forall\alpha_{B(\alpha)}\exists x_{x\leq z}R(\bar\alpha(x))$, where $B(\alpha)$ is $\forall t\alpha(t)\leq\beta(\bar\alpha(t))$, and say $x\leq z$ and $\alpha(t)\leq\beta(\bar\alpha(t))$ are prime. In the following, the intuitive function variables shall range over C. Consider any β as interpretation of β, and any functions Ψ as interpretations of the other free function variables Ψ in $\forall\alpha_{B(\alpha)}\exists!xR(\bar\alpha(x))$. Assume that π C/realizes-β, Ψ $\forall\alpha_{B(\alpha)}\exists!xR(\bar\alpha(x))$. By Lemma C/8.4a: **(0)** If ρ C/realizes-β, α $B(\alpha)$, then $B(\alpha)$. Conversely, if $B(\alpha)$, then $\epsilon_{B(\alpha)}$ C/realizes-β, α $B(\alpha)$. So: **(1)** For each α such that $B(\alpha)$, $(\{\{\pi\}[\alpha]\}[\epsilon_{B(\alpha)}])_1$ C/realizes-Ψ, α, x $R(\bar\alpha(x))\ \&\ \forall y(R(\bar\alpha(y)) \supset x=y)$ for $x = (\{\{\pi\}[\alpha]\}[\epsilon_{B(\alpha)}](0))_0$. Paralleling some steps for Axiom Schema ×26.3c in the proof of Theorem 9.3: **(4)** For each σ, α, y such that $B(\alpha)$, if σ C/realizes-Ψ, α, y $R(\bar\alpha(y))$, then $y = x$ for the x of (1). Let

$$R(\pi, a) \equiv (E\alpha)_{B(\alpha)}[a = \bar\alpha(x) \text{ for } x = (\{\{\pi\}[\alpha]\}[\epsilon_{B(\alpha)}](0))_0].$$

Clearly **(5)** $(\alpha)_{B(\alpha)}(Ex)R(\pi, \bar\alpha(x))$. Define a partial recursive R_1 by

$$R_1(\pi, a) \simeq [a = \overline{\alpha_1}(x_1)\ \&\ (t)_{t<x_1}\alpha_1(t)\leq\beta(\overline{\alpha_1}(t)) \text{ for } \alpha_1 = \lambda t\ (a)_t \dot{-} 1 \text{ and } x_1 = (\{\{\pi\}[\alpha_1]\}[\epsilon_{B(\alpha)}](0))_0].$$

Using (1) and (4): **(6)** $R(\pi, a) \equiv R_1(\pi, a)$. Also, if $B(\alpha)\ \&\ R_1(\pi, \bar\alpha(x))$, then putting $a = \bar\alpha(x)$ and $\alpha_1 = \lambda t\ (a)_t\dot{-}1$, $B(\alpha_1)$ holds, and as in the basis for Axiom Schema ×26.3c $(\{\{\pi\}[\alpha_1]\}[\epsilon_{B(\alpha)}])_{1,0}$ C-realizes-Ψ, a $R(a)$, and so C/realizes-Ψ, α, x $R(\bar\alpha(x))$. Thus **(10)** $B(\alpha)\ \&\ R_1(\pi, \bar\alpha(x)) \rightarrow (\{\{\pi\}[\lambda t\ (\bar\alpha(x))_t\dot{-}1]\}[\epsilon_{B(\alpha)}])_{1,0}$ C/realizes-Ψ, α, x $R(\bar\alpha(x))$. But $\lambda a\ R_1(\pi, a)$ is general recursive in β, π, which $\in C$, so $\lambda a\ R_1(\pi, a) \in C$; and the range of α in (5) is C. So by Lemma 9.12 we can apply the fan theorem informally to (5) with (6) to obtain **(11)** $(Ez)(\alpha)_{B(\alpha)}(Ex)_{x\leq z}R_1(\pi, \bar\alpha(x))$. Using also (0) and (10), $\exists z\forall\alpha_{B(\alpha)}\exists x_{x\leq z}R(\bar\alpha(x))$ is C/realized-β, Ψ by $\langle\mu z(\alpha)_{B(\alpha)}(Ex)_{x\leq z}R_1(\pi, \bar\alpha(x)), \Lambda\alpha\,\Lambda\rho\,\langle\mu x R_1(\pi, \bar\alpha(x)), \langle\lambda t\,0, (\{\{\pi\}[\lambda t\ (\bar\alpha(\mu x R_1(\pi, \bar\alpha(x))))_t\dot{-}1]\}[\epsilon_{B(\alpha)}])_{1,0}\rangle\rangle\rangle$. To avoid the function quantifier $(\alpha)_{B(\alpha)}$ in this expression, we can introduce a number quantifier (g), subjected to conditions which effect that $g = \bar\alpha(z)$ for some α such that $B(\alpha)$, as in the proof of Lemma 9.12; thus **(12)** $(\alpha)_{B(\alpha)}(Ex)_{x\leq z}R_1(\pi, \bar\alpha(x)) \equiv (g)_{g\leq\gamma(\beta,z)}\{\mathrm{Seq}(g)\ \&\ \mathrm{lh}(g)=z\ \&\ (t)_{t<z}[(g)_t\leq 1+\beta(\Pi_{i<t}p_i^{(g)_i})\ \&\ (Ex)_{x\leq z}R_1(\pi, \Pi_{i<x}p_i^{(g)_i})\}$. Using (12), our

expression C/realizing-β, Ψ $\exists z \forall \alpha_{B(\alpha)} \exists x_{x \leq z} R(\bar{\alpha}(x))$ takes the form $\varphi[\beta, \pi]$ with a partial recursive φ. Now $\Lambda\pi\, \varphi[\beta, \pi]$ C/realizes-β, Ψ *26.6c.

9.5$^{\mathbf{C}}$**.** A predicate is *analytic* Kleene **1955** § 2, if it is expressible in terms of general recursive predicates of number and function variables by the operations of the predicate calculus. Two number-theoretic predicates are of the same *degree* Kleene-Post **1954**, if each is general recursive in the other.

THEOREM 9.14^{C}. *We can correlate to each arithmetical* [*analytic*] {*analytic*} *predicate* $P(a)$ *a system* S *between the intuitionistic formal system of number theory* T *and the classical one* [*a system* S *between the basic formal system* T *and the present classical formal system of analysis*] {*a consistent supersystem* S *of the intuitionistic formal system of analysis* T} *so that, when* $P_1(a)$ *and* $P_2(a)$ *are of distinct degrees, the correlated systems* S_1 *and* S_2 *are distinct,* i.e. have different classes of provable formulas. Each system S arises from T by adjoining an axiom of the form $\mathrm{P(a)} \vee \neg \mathrm{P(a)}$ (or equivalently by IM p. 120 Remark 1, $\neg\neg(\mathrm{P(a)} \vee \neg \mathrm{P(a)}) \supset \mathrm{P(a)} \vee \neg \mathrm{P(a)}$).

PROOF. A given expression for $P(a)$ is equivalent to a prenex expression (by the informal counterpart of IM Theorem 19 p. 167) with a general recursive scope (by IM #D p. 228), say e.g. $(x)(Ey)(z)R(a, x, y, z)$. By Lemma 8.5, we can express $R(a, x, y, z)$ by a formula $\mathrm{R(a, x, y, z)}$ not containing $\vee$ or $\exists$. Let $\mathrm{P(a)}$ be $\forall \mathrm{x} \neg \forall \mathrm{y} \neg \forall \mathrm{z} \mathrm{R(a, x, y, z)}$, and form S as stated. Using cases $(P(a) \vee \bar{P}(a))$, Lemma 8.4a (ii) and Clauses 3 and 5 in 8.5, for each a $\mathrm{P(a)} \vee \neg \mathrm{P(a)}$ is realized-a by $\langle \pi(a), \epsilon_{\mathrm{P(a)}} \rangle$ where $\pi(a)$ is the representing function of $P(a)$. Thus $\mathrm{P(a)} \vee \neg \mathrm{P(a)}$ is realizable/P. So by Theorem 9.3 (b) each provable formula of S is realizable/P, and by Corollary 9.4 S is (simply) consistent.

When S_1 and S_2 are thus constructed for $P_1(a)$ and $P_2(a)$ of different degrees, with say P_1 not recursive in P_2, the axiom $\mathrm{P_1(a)} \vee \neg \mathrm{P_1(a)}$ of S_1 is not provable in S_2. For if it were, it would be realizable/P_2, say with the realization function φ general recursive in P_2. Then $P_1(a) \equiv (\varphi[a](0))_0 = 0$, since by Lemma 8.4a (i) $(\varphi[a])_1$ realizes-a $\mathrm{P_1(a)}$ $(\neg \mathrm{P_1(a)})$ only if $P_1(a)$ $(\bar{P}_1(a))$. So P_1 would be recursive in $\lambda a t\, \varphi[a](t)$ and thence in P_2, contrary to hyp.

REMARK 9.15^{C}. In the system S formed similarly by adjoining as axiom $\mathrm{P}_k\mathrm{(a)} \vee \neg \mathrm{P}_k\mathrm{(a)}$ (or any formula realizable/P_k) for each of a list of predicates $P_0(a), P_1(a), P_2(a), \ldots,$ or $P_0(a), \ldots, P_n(a)$ (then let $P_{n+i'}(a) \equiv P_n(a)$), each provable formula is realizable/$\lambda ak\ P_k(a)$.

REMARK 9.16C. Were $\vee$ not excluded in Lemmas 8.4a and 8.4b, *27.17 [$\mathrm{P(a)} \vee \neg\mathrm{P(a)}$ for $P(a) \equiv (Ex)T_1(a, a, x)$] would be a classical counterexample to (i) [(ii)]. Counterexamples for unrestricted $\exists$ follow by $\vdash \mathrm{A} \vee \mathrm{B} \sim \exists\mathrm{x}((\mathrm{x}=0 \supset \mathrm{A}) \;\&\; (\mathrm{x}\neq 0 \supset \mathrm{B}))$ and *0.6.

§ 10. Special realizability. 10.1. Early in the investigations of 1945-realizability (since 1941), formulas were encountered whose realizability was only proved classically (e.g. Kleene 1945 p. 114 (h) and (l), G. F. Rose **1953** p. 11). In such a case, the realizability interpretation fails to exclude the formula's being provable intuitionistically, but on the other hand, we lack adequate grounds for affirming that it should hold intuitionistically, though we know classically that it can consistently be adjoined to the intuitionistic system as a postulate. The situation is the same with the present notion of realizability.

One such formula (by Theorem 11.7 (a) below with *25.3) is

$$\mathrm{M}_1: \qquad \forall\mathrm{z}\forall\mathrm{x}[\neg\forall\mathrm{y}\neg\mathrm{T}_1(\mathrm{z}, \mathrm{x}, \mathrm{y}) \supset \exists\mathrm{y}\mathrm{T}_1(\mathrm{z}, \mathrm{x}, \mathrm{y})]$$

where $\mathrm{T}_1(\mathrm{z, x, y})$ is a formula numeralwise expressing $T_1(z, x, y)$ (IM p. 281) chosen by (the method of proof of) Lemma 8.5. In view of IM §§ 62, 63, this formalizes Markov's principle **1954a** (introduced in lectures in 1952–53), known to us from the statement of it in Šanin **1958a**, "If the process of application of the algorithm $\mathfrak{Q}$ to an initially given P does not continue infinitely, then $\mathfrak{Q}$ shall apply to P." The like formulas M_n with $\mathrm{T}_n(\mathrm{z}, \mathrm{x}_1, \ldots, \mathrm{x}_n, \mathrm{y})$ for $n > 1$ in place of $\mathrm{T}_1(\mathrm{z, x, y})$ follow under the interpretation from M_1 by contraction of $\mathrm{x}_1, \ldots, \mathrm{x}_n$ (IM (18) p. 285) and IM Theorem IV p. 281.

Especially in connection with the author's investigations of **1957**-realizability (since 1951), it became apparent to him (before he learned of Markov's **1954a**) that considerable interest attaches to the result of adjoining this principle to intuitionistic systems, because of a variety of results which it then (or only then) becomes possible to obtain. An example recently come to light is Gödel's that, if "strong completeness" of the predicate calculus is provable intuitionistically, so is each instance (particularizing z, x to numerals) of Markov's principle. This result appears in Kreisel **1958a** Remark 2.1 and **1959a**, with sketch of proof in **1962b**, and as Theorem 2 of **1962**. Gödel seems not to have published it himself. Kreisel states the principle as "$\neg(x)A(x) \rightarrow (Ex)\neg A(x)$ for each primitive recursive A" (or similarly) without mentioning Markov. The bearing of the principle on Brouwer's theory of the continuum will be discussed in Chapter IV Remark 18.6.

Kreisel in **1959a** with **1959** para. 3.52 shows that this principle is not provable in intuitionistic number theory or in the formal system of Kleene **1957**. Kreisel obtains this result by applying an interpretation which he describes as "closely related to Kleene's realizability" (1945 or IM § 82), but which uses ideas from an interpretation of Gödel's (Kreisel **1959** paras. 3.1–3.3 and Gödel **1958**), from which it also differs. Kreisel deals further with these matters in **1962**. The formal system of Kleene **1957** is not as strong as the present formal system (letter from Kreisel, 22 November 1963).

We now give, in this and the next section, a proof of Kreisel's result for the present formal system. We arrived at this proof by attacking the problem directly with the help of some inspirations gained from Kreisel **1959a** and Gödel **1958**, and we have not determined the precise relationship of the interpretation used in it ('$_s$realizability') to the one Kreisel used (**1959** para. 3.52).

10.2. In realizability as treated above (§§ 8, 9), an implication $A \supset B$ is realized-Ψ by ϵ, if $\{\epsilon\}[\alpha]$ is any partial recursive function $\varphi[\alpha]$ such that, for each α which realizes-Ψ A, $\varphi[\alpha]$ is completely defined and realizes-Ψ B; for an α which does not realize-Ψ A, $\varphi[\alpha]$ need not be completely defined. In the 'special realizability' (briefly, '$_s$realizability') to be introduced in this section, we shall use instead of $\{\epsilon\}[\alpha]$ an analogous operation on ϵ which will produce a function $\varphi[\alpha]$ (termed 'special recursive') which is partial recursive but such that $\varphi[\alpha]$ is completely defined for every α of the appropriate sort or 'order' (determined by A) whether or not α '$_s$realizes-Ψ' A.

To carry this out, we first assign 'orders' to the formulas, and to the one-place (total) number-theoretic functions. A function to realize a formula must have the same order as the formula. We begin with an inductive definition of the 'orders' to be used.

(1) 1 is an *order*. (2) If a is an *order*, so is $a+1$. (3) If a_0 and a_1 are orders, so is (a_0, a_1). The only *orders* are those given by these three clauses. Orders differently generated by use of these three clauses are different.

An order given by Clause (2) is a *successor order* (s-*order*); by (3), a *pair order* (p-*order*). The orders 1+1, (1+1)+1, ((1+1)+1)+1, ... we write 2, 3, 4, ... simply. If a is an s-order, a−1 is the order b such that a = b+1. The variables ${}^{\mathrm{a}}\alpha$, ${}^{\mathrm{a}}\beta$, ..., ${}^{\mathrm{a}}\alpha_1$, ${}^{\mathrm{a}}\alpha_2$, ... will be used for functions of order a (as specified next).

For any one-place number-theoretic functions α and β, we define

$$(10.1)\qquad \{\alpha\}(\beta) \simeq \alpha(\beta) \simeq \alpha(\bar{\beta}(\mu y\alpha(\bar{\beta}(y))>0)) \dot{-} 1,$$

and say $\alpha(\beta)$ is *properly defined* when $(E!y)\alpha(\bar{\beta}(y))>0$.

Now we specify which one-place number-theoretic functions are 'of a given order'. (1) All such functions are *of order* 1. (2) To each order a, the functions ${}^{\mathrm{a}+1}\alpha$ *of order* a+1 are those such that, to each (function *of order* a) ${}^{\mathrm{a}}\alpha$, there exists a unique y such that ${}^{\mathrm{a}+1}\alpha({}^{\mathrm{a}}\bar{\alpha}(y)) > 0$ (so ${}^{\mathrm{a}+1}\alpha({}^{\mathrm{a}}\alpha)$ is properly defined, and, for that y, ${}^{\mathrm{a}+1}\alpha({}^{\mathrm{a}}\bar{\alpha}(y)) = {}^{\mathrm{a}+1}\alpha({}^{\mathrm{a}}\alpha)+1$). (3) To each pair of orders a_0 and a_1, the functions ${}^{\mathrm{a}}\alpha$ *of order* a = $(\mathrm{a}_0, \mathrm{a}_1)$ are those such that $({}^{\mathrm{a}}\alpha)_0$ is *of order* a_0, and $({}^{\mathrm{a}}\alpha)_1$ is *of order* a_1.

An object ${}^{\mathrm{a}+1}\alpha$ of an s-order has a dual role; it is a number-theoretic function $\lambda s\, {}^{\mathrm{a}+1}\alpha(s)$, and it serves via (10.1) as an operator $\lambda^{\mathrm{a}}\alpha\, {}^{\mathrm{a}+1}\alpha({}^{\mathrm{a}}\alpha)$. The operators of orders 3, 4, 5, ... are not simply countable functionals of the types 3, 4, 5, ... (Kleene **1959a**), since e.g. ${}^{3}\alpha({}^{2}\alpha)$ will depend in general on $\lambda s\, {}^{2}\alpha(s)$ and not simply on $\lambda^{1}\alpha\, {}^{2}\alpha({}^{1}\alpha)$. (An interpretation using functionals here seemed to work for all postulates except Brouwer's principle ×27.1.)

(We could extend our theory of orders to include the natural numbers as objects of order 0.)

Any function of an order a ≠ 1 is also of order 1, and there are other possibilities for functions to be of more than one order.

By ${}^{1}0$ we mean $\lambda s\, 0$. For any order a, ${}^{\mathrm{a}+1}0$ is defined by

$${}^{\mathrm{a}+1}0(1) = 1, \quad {}^{\mathrm{a}+1}0(s) = 0 \text{ for } s \neq 1.$$

For any orders a_0 and a_1, if a = $(\mathrm{a}_0, \mathrm{a}_1)$, ${}^{\mathrm{a}}0 = \langle {}^{\mathrm{a}_0}0, {}^{\mathrm{a}_1}0\rangle$. Now, for each order a, ${}^{\mathrm{a}}0$ is primitive recursive and of order a.

10.3. By a *special recursive function* (a *function special recursive in T*) we mean a function $\varphi(\mathfrak{a})$, where $\mathfrak{a}$ is a list of zero or more distinct number variables, and of zero or more distinct one-place number-theoretic function variables for each of which a respective order is specified, such that: (a) when the ranges of the function variables among $\mathfrak{a}$ are restricted to their specified orders, $\varphi(\mathfrak{a})$ is completely defined, and (b) when the ranges of the function variables among $\mathfrak{a}$ are unrestricted, $\varphi(\mathfrak{a})$ is partial recursive (partial recursive in T). The notions extend to function-valued functions $\varphi[\mathfrak{b}] = \lambda s\, \varphi(\mathfrak{b}, s)$.

A primitive or general recursive function $\varphi(\mathfrak{a})$ or $\varphi[\mathfrak{b}]$ is a fortiori special recursive.

By (10.1) with the definition of 'function of order a+1', ${}^{a+1}\alpha({}^{a}\alpha) = \varphi({}^{a+1}\alpha, {}^{a}\alpha)$ with φ special recursive.

LEMMA 10.1. *To each partial recursive function* $\varphi(\mathfrak{b}, {}^{a}\alpha)$, *there is a primitive recursive function* $\psi[\mathfrak{b}] = \Lambda^{a}\alpha\, \varphi(\mathfrak{b}, {}^{a}\alpha)$ *such that, for each* $\mathfrak{b}$ *for which* $\varphi(\mathfrak{b}, {}^{a}\alpha)$ *is defined for all* ${}^{a}\alpha$ *of order* a: $\Lambda^{a}\alpha\, \varphi(\mathfrak{b}, {}^{a}\alpha)$ *is of order* a+1, *and for each* ${}^{a}\alpha$ *of order* a

$$(10.2) \qquad \{\Lambda^{a}\alpha\, \varphi(\mathfrak{b}, {}^{a}\alpha)\}({}^{a}\alpha) = \varphi(\mathfrak{b}, {}^{a}\alpha).$$

Ordinarily our $\varphi(\mathfrak{b}, {}^{a}\alpha)$ will be special recursive, so the conclusions will apply whenever (the functions among) $\mathfrak{b}$ are of the specified orders. Here for brevity we have introduced the Λ-notation right in the lemma; cf. Lemma 8.1. If $\varphi(z, \mathfrak{b}, {}^{a}\alpha)$ is partial recursive and $\psi[z, \mathfrak{b}] = \Lambda^{a}\alpha\, \varphi(z, \mathfrak{b}, {}^{a}\alpha)$, and we put $\varphi(\mathfrak{b}, {}^{a}\alpha) = \varphi(e, \mathfrak{b}, {}^{a}\alpha)$ for a fixed e, then $\psi[e, \mathfrak{b}] = \Lambda^{a}\alpha\, \varphi(e, \mathfrak{b}, {}^{a}\alpha)$ is a $\Lambda^{a}\alpha\, \varphi(\mathfrak{b}, {}^{a}\alpha)$.

PROOF OF LEMMA 10.1. Say e.g. $\mathfrak{b}$ is $(a, {}^{b}\beta)$. By the normal form theorem IM pp. 292, 330, but using $\bar{\alpha}$, $\bar{\beta}$ instead of $\tilde{\alpha}$, $\tilde{\beta}$ (end 8.2), for any Gödel number e of φ, for each $a, {}^{b}\beta, {}^{a}\alpha$:

$$\text{(i)} \qquad \varphi(a, {}^{b}\beta, {}^{a}\alpha) \simeq U(\mu y T_1^{1,1}({}^{b}\bar{\beta}(y), {}^{a}\bar{\alpha}(y), e, a)),$$

$$\text{(ii)} \qquad T_1^{1,1}({}^{b}\bar{\beta}(y), {}^{a}\bar{\alpha}(y), e, a) \text{ for at most one } y.$$

So let $\psi[a, {}^{b}\beta] = \lambda s\, \psi(a, {}^{b}\beta, s)$ where

$$\psi(a, {}^{b}\beta, s) = \begin{cases} U(\mathrm{lh}(s))+1 & \text{if } T_1^{1,1}({}^{b}\bar{\beta}(\mathrm{lh}(s)), s, e, a), \\ 0 & \text{otherwise.} \end{cases}$$

10.4. Let ${}^{1}\alpha =_1 {}^{1}\beta \equiv {}^{1}\alpha = {}^{1}\beta \equiv (x)[{}^{1}\alpha(x) = {}^{1}\beta(x)]$. When a is an s-order, let ${}^{a}\alpha =_a {}^{a}\beta \equiv ({}^{a-1}\gamma)[{}^{a}\alpha({}^{a-1}\gamma) = {}^{a}\beta({}^{a-1}\gamma)]$. When a is a p-order (a_0, a_1), let ${}^{a}\alpha =_a {}^{a}\beta \equiv ({}^{a}\alpha)_0 =_{a_0} ({}^{a}\beta)_0 \,\&\, ({}^{a}\alpha)_1 =_{a_1} ({}^{a}\beta)_1$.

Now ${}^{a}\alpha = {}^{a}\beta \rightarrow {}^{a}\alpha =_a {}^{a}\beta$, but (for a $\neq$ 1) not in general conversely.

For any two choices $\Lambda_1{}^{a}\alpha\, \varphi(\mathfrak{b}, {}^{a}\alpha)$ and $\Lambda_2{}^{a}\alpha\, \varphi(\mathfrak{b}, {}^{a}\alpha)$ of $\Lambda^{a}\alpha\, \varphi(\mathfrak{b}, {}^{a}\alpha)$, by (10.2) $\Lambda_1{}^{a}\alpha\, \varphi(\mathfrak{b}, {}^{a}\alpha) =_{a+1} \Lambda_2{}^{a}\alpha\, \varphi(\mathfrak{b}, {}^{a}\alpha)$ (for each $\mathfrak{b}$ as supposed for (10.2)).

For each two orders a and b, we now define an order a∗b; and for e = a∗b we choose a special recursive function $\lambda^{e}\epsilon^{a}\alpha\; {}^{b}\{{}^{e}\epsilon\}[{}^{a}\alpha]$ so that, for each ${}^{e}\epsilon$, ${}^{a}\alpha$ of the specified orders, ${}^{b}\{{}^{e}\epsilon\}[{}^{a}\alpha]$ is of order b. The definitions are by recursion on b, corresponding to the inductive definition of 'order' (cf. IM p. 260). When $b = 1$, $e = a{*}b = (a, 1)+1$ and ${}^{b}\{{}^{e}\epsilon\}[{}^{a}\alpha] = \lambda s\, {}^{e}\epsilon(\langle {}^{a}\alpha, \lambda t\, s\rangle)$. When b is an s-order, $e = a{*}b = (a, b-1)+1$ and (using Lemma 10.1) ${}^{b}\{{}^{e}\epsilon\}[{}^{a}\alpha] = \Lambda^{b-1}\beta\, {}^{e}\epsilon(\langle {}^{a}\alpha, {}^{b-1}\beta\rangle)$, for some choice of the latter. When b is a p-order (b_0, b_1), $e = a{*}b = (a{*}b_0, a{*}b_1)$ and ${}^{b}\{{}^{e}\epsilon\}[{}^{a}\alpha] = \langle {}^{b_0}\{({}^{e}\epsilon)_0\}[{}^{a}\alpha], {}^{b_1}\{({}^{e}\epsilon)_1\}[{}^{a}\alpha]\rangle$ (where ${}^{b_0}\{({}^{e}\epsilon)_0\}[{}^{a}\alpha] = (\lambda^{a*b_0}\epsilon^{a}\alpha\, {}^{b_0}\{{}^{a*b_0}\epsilon\}[{}^{a}\alpha])[({}^{e}\epsilon)_0, {}^{a}\alpha]$, etc.). By ${}^{b}\{{}^{e}\epsilon\}[\alpha]$ we mean ${}^{b}\{{}^{e}\epsilon\}[{}^{1}\alpha]$ for ${}^{1}\alpha = \alpha$.

For different choices $\lambda^{e}\epsilon^{a}\alpha\, {}^{b}\{{}^{e}\epsilon\}[{}^{a}\alpha]_1$ and $\lambda^{e}\epsilon^{a}\alpha\, {}^{b}\{{}^{e}\epsilon\}[{}^{a}\alpha]_2$ of $\lambda^{e}\epsilon^{a}\alpha\, {}^{b}\{{}^{e}\epsilon\}[{}^{a}\alpha]$, by induction on b ${}^{b}\{{}^{e}\epsilon\}[{}^{a}\alpha]_1 =_b {}^{b}\{{}^{e}\epsilon\}[{}^{a}\alpha]_2$ (for each ${}^{e}\epsilon$, ${}^{a}\alpha$ of the specified orders). Moreover:

LEMMA 10.2. *If* ${}^{e}\epsilon_1 =_e {}^{e}\epsilon_2$, *then* ${}^{b}\{{}^{e}\epsilon_1\}[{}^{a}\alpha] =_b {}^{b}\{{}^{e}\epsilon_2\}[{}^{a}\alpha]$ (*where* $e = a{*}b$).

PROOF, by ind. on b. CASE 1: $b = 1$. Then, for all x, ${}^{b}\{{}^{e}\epsilon_1\}[{}^{a}\alpha](x) = {}^{e}\epsilon_1(\langle {}^{a}\alpha, \lambda t\, x\rangle)$ [def. etc.] $= {}^{e}\epsilon_2(\langle {}^{a}\alpha, \lambda t\, x\rangle)$ [hyp.] $= {}^{b}\{{}^{e}\epsilon_2\}[{}^{a}\alpha](x)$; i.e. ${}^{b}\{{}^{e}\epsilon_1\}[{}^{a}\alpha] =_b {}^{b}\{{}^{e}\epsilon_2\}[{}^{a}\alpha]$. CASE 2: b is an s-order. Then, for all ${}^{b-1}\beta$, ${}^{b}\{{}^{e}\epsilon_1\}[{}^{a}\alpha]({}^{b-1}\beta) = {}^{e}\epsilon_1(\langle {}^{a}\alpha, {}^{b-1}\beta\rangle)$ [def., (10.2)] $= {}^{e}\epsilon_2(\langle {}^{a}\alpha, {}^{b-1}\beta\rangle)$ [hyp.] $= {}^{b}\{{}^{e}\epsilon_2\}[{}^{a}\alpha]({}^{b-1}\beta)$; i.e. ${}^{b}\{{}^{e}\epsilon_1\}[{}^{a}\alpha] =_b {}^{b}\{{}^{e}\epsilon_2\}[{}^{a}\alpha]$. CASE 3: b is a p-order (b_0, b_1). Then ${}^{b}\{{}^{e}\epsilon_1\}[{}^{a}\alpha] = \langle {}^{b_0}\{({}^{e}\epsilon_1)_0\}[{}^{a}\alpha], {}^{b_1}\{({}^{e}\epsilon_1)_1\}[{}^{a}\alpha]\rangle$ [def.] $=_b \langle {}^{b_0}\{({}^{e}\epsilon_2)_0\}[{}^{a}\alpha], {}^{b_1}\{({}^{e}\epsilon_2)_1\}[{}^{a}\alpha]\rangle$ [hyp. ind.] $= {}^{b}\{{}^{e}\epsilon_2\}[{}^{a}\alpha]$.

LEMMA 10.3. *To each partial recursive function* $\varphi[\mathfrak{b}, {}^{a}\alpha]$, *there is a primitive recursive function* $\psi[\mathfrak{b}] = {}^{e}_{b}\Lambda^{a}\alpha\, \varphi[\mathfrak{b}, {}^{a}\alpha]$ *such that, for each* $\mathfrak{b}$ *for which* $\varphi[\mathfrak{b}, {}^{a}\alpha]$ *is (completely defined and) of order* b *for all* ${}^{a}\alpha$ *of order* a: ${}^{e}_{b}\Lambda^{a}\alpha\, \varphi[\mathfrak{b}, {}^{a}\alpha]$ *is of order* $e = a{*}b$, *and for each* ${}^{a}\alpha$ *of order* a

$$(10.3) \qquad {}^{b}\{{}^{e}_{b}\Lambda^{a}\alpha\, \varphi[\mathfrak{b}, {}^{a}\alpha]\}[{}^{a}\alpha] =_b \varphi[\mathfrak{b}, {}^{a}\alpha].$$

Ordinarily our $\varphi[\mathfrak{b}, {}^{a}\alpha]$ will be special recursive with values of order b for all $\mathfrak{b}$, ${}^{a}\alpha$ of the specified orders, so the conclusions will apply whenever $\mathfrak{b}$ are of the specified orders. We shall understand ${}^{e}_{b}\Lambda^{a}\alpha\, \varphi[\mathfrak{b}, {}^{a}\alpha]$ to be constructed by the method of the proof (below). Again, by the properties: If $\varphi[z, \mathfrak{b}, {}^{a}\alpha]$ is partial recursive and $\psi[z, \mathfrak{b}] = {}^{e}_{b}\Lambda^{a}\alpha\, \varphi[z, \mathfrak{b}, {}^{a}\alpha]$, and we put $\varphi[\mathfrak{b}, {}^{a}\alpha] = \varphi[e, \mathfrak{b}, {}^{a}\alpha]$ for a fixed e, then $\psi[e, \mathfrak{b}] = {}^{e}_{b}\Lambda^{a}\alpha\, \varphi[e, \mathfrak{b}, {}^{a}\alpha]$ is an ${}^{e}_{b}\Lambda^{a}\alpha\, \varphi[\mathfrak{b}, {}^{a}\alpha]$. The subscript b in ${}^{e}_{b}\Lambda^{a}\alpha\, \varphi[\mathfrak{b}, {}^{a}\alpha]$ may be omitted when the context makes it clear that the intended order of the operand $\varphi[\mathfrak{b}, {}^{a}\alpha]$ for the operation by ${}^{e}\Lambda^{a}\alpha$ is b. By ${}^{e}_{b}\Lambda\alpha\, \varphi[\mathfrak{b}, \alpha]$ we mean ${}^{e}_{b}\Lambda^{1}\alpha\, \varphi[\mathfrak{b}, {}^{1}\alpha]$.

PROOF OF LEMMA 10.3, by ind. on the order b. CASE 1: $b = 1$. Put $e = a{*}b = (a, 1)+1$. Let $\psi[\mathfrak{b}] = \Lambda^{e-1}\beta\, \varphi(\mathfrak{b}, ({}^{e-1}\beta)_0, ({}^{e-1}\beta(0))_1)$. Then by Lemma 10.1, for $\mathfrak{b}$, ${}^{a}\alpha$ of the specified orders, $\psi[\mathfrak{b}]$ is of order e, and, for all s, ${}^{\mathrm{b}}\{\psi[\mathfrak{b}]\}[{}^{\mathrm{a}}\alpha](s) = (\lambda s\, \psi[\mathfrak{b}](\langle {}^{\mathrm{a}}\alpha, \lambda t\, s\rangle))(s) = \psi[\mathfrak{b}](\langle {}^{\mathrm{a}}\alpha, \lambda t\, s\rangle) = \varphi(\mathfrak{b}, {}^{\mathrm{a}}\alpha, s)$ [(10.2) etc.] $= \varphi[\mathfrak{b}, {}^{\mathrm{a}}\alpha](s)$, i.e. ${}^{\mathrm{b}}\{\psi[\mathfrak{b}]\}[{}^{\mathrm{a}}\alpha] =_{\mathrm{b}} \varphi[\mathfrak{b}, {}^{\mathrm{a}}\alpha]$. Case 2: b is an s-order. Put $\mathrm{e} = \mathrm{a}*\mathrm{b} = (\mathrm{a}, \mathrm{b}-1)+1$. Let $\psi[\mathfrak{b}] = \Lambda^{\mathrm{e}-1}\beta\, \varphi[\mathfrak{b}, ({}^{\mathrm{e}-1}\beta)_0](({}^{\mathrm{e}-1}\beta)_1)$. Then by Lemma 10.1, $\psi[\mathfrak{b}]$ is of order e, and, for all ${}^{\mathrm{b}-1}\beta$, ${}^{\mathrm{b}}\{\psi[\mathfrak{b}]\}[{}^{\mathrm{a}}\alpha]({}^{\mathrm{b}-1}\beta) = \{\Lambda^{\mathrm{b}-1}\beta\, \psi[\mathfrak{b}](\langle {}^{\mathrm{a}}\alpha, {}^{\mathrm{b}-1}\beta\rangle)\}({}^{\mathrm{b}-1}\beta) = \psi[\mathfrak{b}](\langle {}^{\mathrm{a}}\alpha, {}^{\mathrm{b}-1}\beta\rangle) = \varphi[\mathfrak{b}, {}^{\mathrm{a}}\alpha]({}^{\mathrm{b}-1}\beta)$, i.e. ${}^{\mathrm{b}}\{\psi[\mathfrak{b}]\}[{}^{\mathrm{a}}\alpha] =_{\mathrm{b}} \varphi[\mathfrak{b}, {}^{\mathrm{a}}\alpha]$. Case 3: b is a p-order $(\mathrm{b}_0, \mathrm{b}_1)$. Put $\mathrm{e} = \mathrm{a}*\mathrm{b} = (\mathrm{a}*\mathrm{b}_0, \mathrm{a}*\mathrm{b}_1)$. By hyp. ind. we can find $\psi_0[\mathfrak{b}]$, $\psi_1[\mathfrak{b}]$ with values (for each $\mathfrak{b}$ as supposed) of orders $\mathrm{a}*\mathrm{b}_0$, $\mathrm{a}*\mathrm{b}_1$ such that for all ${}^{\mathrm{a}}\alpha$, ${}^{\mathrm{b}_0}\{\psi_0[\mathfrak{b}]\}[{}^{\mathrm{a}}\alpha] =_{\mathrm{b}_0} (\varphi[\mathfrak{b}, {}^{\mathrm{a}}\alpha])_0$, ${}^{\mathrm{b}_1}\{\psi_1[\mathfrak{b}]\}[{}^{\mathrm{a}}\alpha] =_{\mathrm{b}_1} (\varphi[\mathfrak{b}, {}^{\mathrm{a}}\alpha])_1$. Let $\psi[\mathfrak{b}] = \langle \psi_0[\mathfrak{b}], \psi_1[\mathfrak{b}]\rangle$, which has values of order $(\mathrm{a}*\mathrm{b}_0, \mathrm{a}*\mathrm{b}_1) = \mathrm{e}$. Now ${}^{\mathrm{b}}\{\psi[\mathfrak{b}]\}[{}^{\mathrm{a}}\alpha] = \langle {}^{\mathrm{b}_0}\{\psi_0[\mathfrak{b}]\}[{}^{\mathrm{a}}\alpha], {}^{\mathrm{b}_1}\{\psi_1[\mathfrak{b}]\}[{}^{\mathrm{a}}\alpha]\rangle$, whence $({}^{\mathrm{b}}\{\psi[\mathfrak{b}]\}[{}^{\mathrm{a}}\alpha])_0 =_{\mathrm{b}_0} (\varphi[\mathfrak{b}, {}^{\mathrm{a}}\alpha])_0$ & $({}^{\mathrm{b}}\{\psi[\mathfrak{b}]\}[{}^{\mathrm{a}}\alpha])_1 =_{\mathrm{b}_1} (\varphi[\mathfrak{b}, {}^{\mathrm{a}}\alpha])_1$, i.e. ${}^{\mathrm{b}}\{\psi[\mathfrak{b}]\}[{}^{\mathrm{a}}\alpha] =_{\mathrm{b}} \varphi[\mathfrak{b}, {}^{\mathrm{a}}\alpha]$. —

By ind. on b, any two choices of ${}^{\mathrm{e}}_{\mathrm{b}}\Lambda^{\mathrm{a}}\alpha\, \varphi[\mathfrak{b}, {}^{\mathrm{a}}\alpha]$ are $=_{\mathrm{e}}$ (for each $\mathfrak{b}$ as supposed).

For $\mathrm{e} = 1*\mathrm{b}$, we define ${}^{\mathrm{b}}\{{}^{\mathrm{e}}\epsilon\}[x] = {}^{\mathrm{b}}\{{}^{\mathrm{e}}\epsilon\}[\lambda t\, x]$ $(= \chi[{}^{\mathrm{e}}\epsilon, x]$ with a special recursive $\chi)$. For $\mathrm{e} = 1*\mathrm{b}$ and $\varphi[\mathfrak{b}, x]$ a partial recursive function, we write ${}^{\mathrm{e}}_{\mathrm{b}}\Lambda x\, \varphi[\mathfrak{b}, x] = {}^{\mathrm{e}}_{\mathrm{b}}\Lambda\alpha\, \varphi[\mathfrak{b}, \alpha(0)]$. Then, for $\mathfrak{b}$ such that $\varphi[\mathfrak{b}, x]$ is of order b for all x: ${}^{\mathrm{e}}_{\mathrm{b}}\Lambda x\, \varphi[\mathfrak{b}, x]$ is of order e, and (using (10.3)) for each x

(10.3a) $\quad {}^{\mathrm{b}}\{{}^{\mathrm{e}}_{\mathrm{b}}\Lambda x\, \varphi[\mathfrak{b}, x]\}[x] =_{\mathrm{b}} \varphi[\mathfrak{b}, x]$.

Ordinarily our $\varphi[\mathfrak{b}, x]$ will be special recursive with values of order b for all $\mathfrak{b}$, x ($\mathfrak{b}$ of the specified orders), so the conclusions will apply whenever $\mathfrak{b}$ are of the specified orders.

We define ${}^{1}\{{}^{2}\epsilon\} = \lambda s\, {}^{2}\epsilon(\lambda t\, s)$ $(= \chi[{}^{2}\epsilon]$ with a special recursive $\chi)$. For $\varphi[\mathfrak{b}]$ a partial recursive function, we write ${}^{2}\Lambda\, \varphi[\mathfrak{b}] = \Lambda^{1}\alpha\, \varphi(\mathfrak{b}, {}^{1}\alpha(0))$. Then, for $\mathfrak{b}$ such that $\varphi[\mathfrak{b}]$ is completely defined: ${}^{2}\Lambda\, \varphi[\mathfrak{b}]$ is of order 2, and (using (10.2))

(10.3b) $\quad {}^{1}\{{}^{2}\Lambda\, \varphi[\mathfrak{b}]\} = \varphi[\mathfrak{b}]$.

Ordinarily our $\varphi[\mathfrak{b}]$ will be special recursive, so the conclusions will apply whenever $\mathfrak{b}$ are of the specified orders.

Also (8.1c)–(8.3c) for $k+l = 2$ have analogs; but all we shall use is ${}^{\mathrm{b}}\{{}^{\mathrm{e}}\epsilon\}[{}^{\mathrm{a}_0}\rho_0, {}^{\mathrm{a}_1}\rho_1]$ as abbreviation for ${}^{\mathrm{b}}\{{}^{\mathrm{e}}\epsilon\}[\langle {}^{\mathrm{a}_0}\rho_0, {}^{\mathrm{a}_1}\rho_1\rangle]$.

10.5. Now we shall assign to each formula E an order e (= 'order E'). Simultaneously, for each list Ψ of variables including all which occur free in E, we shall define when a function ${}^{\mathrm{e}}\epsilon$ of order e 'specially realizes' E for a given assignment Ψ of numbers and functions as the values of Ψ, or briefly when $^{e}\epsilon$ '${}_{s}$realizes-Ψ' E; only a function $^{e}\epsilon$ of order e = order E can ${}_{s}$*realize*-Ψ E.

1. A prime formula P is *of order* 1. $^{1}\epsilon$ ${}_{s}$*realizes*-Ψ P, if P is true-Ψ.

2. e = *order* A & B = (a, b) where a = *order* A and b = *order* B. $^{e}\epsilon$ ${}_{s}$*realizes*-Ψ A & B, if $(^{e}\epsilon)_0$ ${}_{s}$*realizes*-Ψ A and $(^{e}\epsilon)_1$ ${}_{s}$*realizes*-Ψ B.

3. e = *order* A ∨ B = (1, (a, b)) where a = *order* A and b = *order* B. $^{e}\epsilon$ ${}_{s}$*realizes*-Ψ A ∨ B, if $(^{e}\epsilon(0))_0=0$ and $(^{e}\epsilon)_{1,0}$ ${}_{s}$*realizes*-Ψ A, or $(^{e}\epsilon(0))_0\neq 0$ and $(^{e}\epsilon)_{1,1}$ ${}_{s}$*realizes*-Ψ B.

4. e = *order* A ⊃ B = a∗b where a = *order* A and b = *order* B. $^{e}\epsilon$ ${}_{s}$*realizes*-Ψ A ⊃ B, if, for each $^{a}\alpha$, if $^{a}\alpha$ ${}_{s}$*realizes*-Ψ A then $^{b}\{^{e}\epsilon\}[^{a}\alpha]$ ${}_{s}$*realizes*-Ψ B.

5. e = *order* ¬A = a∗1 where a = *order* A. $^{e}\epsilon$ ${}_{s}$*realizes*-Ψ ¬A, if, for each $^{a}\alpha$, not $^{a}\alpha$ ${}_{s}$*realizes*-Ψ A. (Equivalently, $^{e}\epsilon$ ${}_{s}$*realizes*-Ψ ¬A, if $^{e}\epsilon$ ${}_{s}$*realizes*-Ψ A ⊃ 1=0 under Clause 4.)

6. e = *order* ∀xA = 1∗a where a = *order* A. $^{e}\epsilon$ ${}_{s}$*realizes*-Ψ ∀xA, if, for each x, $^{a}\{^{e}\epsilon\}[x]$ ${}_{s}$*realizes*-Ψ, x A.

7. e = *order* ∃xA = (1, a) where a = *order* A. $^{e}\epsilon$ ${}_{s}$*realizes*-Ψ ∃xA, if $(^{e}\epsilon)_1$ ${}_{s}$*realizes*-Ψ, $(^{e}\epsilon(0))_0$ A.

8. e = *order* ∀αA = 1∗a where a = *order* A. $^{e}\epsilon$ ${}_{s}$*realizes*-Ψ ∀αA, if, for each α, $^{a}\{^{e}\epsilon\}[\alpha]$ ${}_{s}$*realizes*-Ψ, α A.

9. e = *order* ∃αA = (2, a) where a = *order* A. $^{e}\epsilon$ ${}_{s}$*realizes*-Ψ ∃αA, if $(^{e}\epsilon)_1$ ${}_{s}$*realizes*-Ψ, $^{1}\{(^{e}\epsilon)_0\}$ A.

LEMMA 10.4. *Let* Ψ_1 *be a list of those of the variables* Ψ *which occur free in* E, *and let* Ψ_1 *be the list of the corresponding ones of the numbers and functions* Ψ. *Then* $^{e}\epsilon$ ${}_{s}$*realizes*-Ψ_1 E *if and only if* $^{e}\epsilon$ ${}_{s}$*realizes*-Ψ E.

LEMMA 10.5. *If* $^{e}\epsilon_1$ ${}_{s}$*realizes*-Ψ E, *and* $^{e}\epsilon_1 =_{e} {}^{e}\epsilon_2$, *then* $^{e}\epsilon_2$ ${}_{s}$*realizes*-Ψ E.

PROOF, by induction, using Lemma 10.2 in Cases 4, 6, 8. —

We say a closed formula E is ${}_{s}$*realizable*, if a general recursive function $^{e}\epsilon$ ${}_{s}$realizes E; an open formula, if its closure is (FIRST DEFINITION).

Equivalently (proof below), E is ${}_{s}$*realizable*, if there is a general recursive function φ (a general recursive ${}_{s}$*realization function for* E *in* Ψ) such that, for each Ψ, $\varphi[\Psi]$ ${}_{s}$realizes-Ψ E (SECOND DEFINITION). This notion of ${}_{s}$realizability is independent of the choice of the list Ψ, as before (cf. 8.7).

To illustrate the equivalence of the two definitions when E is open, say e.g. the closure of E is ∀α∀xE. Write e, f, g for the orders of E, ∀xE, ∀α∀xE (f = 1∗e, g = 1∗f).

First, suppose φ is general recursive, and that **(1)** for each α and x, $\varphi[\alpha, x]$ (is of order e and) ${}_{\mathrm{s}}$realizes-α, x E. Using Lemma 10.3, ${}^{\mathrm{g}}\Lambda\alpha\, {}^{\mathrm{f}}\Lambda x\, \varphi[\alpha, x]$ is a primitive recursive function ${}^{\mathrm{g}}\alpha$, which we shall show ${}_{\mathrm{s}}$realizes $\forall\alpha\forall x$E. By Clause 8, we need that, for each α, ${}^{\mathrm{f}}\{{}^{\mathrm{g}}\alpha\}[\alpha]$ ${}_{\mathrm{s}}$realizes-α $\forall x$E. Consider any α. By (10.3), ${}^{\mathrm{f}}\{{}^{\mathrm{g}}\alpha\}[\alpha] =_{\mathrm{f}} {}^{\mathrm{f}}\Lambda x\, \varphi[\alpha, x]$; so by Lemma 10.5 it will suffice to show that ${}^{\mathrm{f}}\Lambda x\, \varphi[\alpha, x]$ ${}_{\mathrm{s}}$realizes-α $\forall x$E, i.e. by Clause 6 that, for each x, ${}^{\mathrm{e}}\{{}^{\mathrm{f}}\Lambda x\, \varphi[\alpha, x]\}[x]$ ${}_{\mathrm{s}}$realizes-α, x E. This follows from (1) by Lemma 10.5, since by (10.3a) ${}^{\mathrm{e}}\{{}^{\mathrm{f}}\Lambda x\, \varphi[\alpha, x]\}[x] =_{\mathrm{e}} \varphi[\alpha, x]$.

Inversely, if ${}^{\mathrm{g}}\epsilon$ is a general recursive function ${}_{\mathrm{s}}$realizing $\forall\alpha\forall x$E, then, using Clauses 8 and 6, $\lambda\alpha x\, {}^{\mathrm{e}}\{{}^{\mathrm{f}}\{{}^{\mathrm{g}}\epsilon\}[\alpha]\}[x]$ is a general recursive ${}_{\mathrm{s}}$realization function for E in α, x. —

The $/T$, -Θ and $C/$ modifications of ${}_{\mathrm{s}}$realizability (e.g. when $\Theta, T \subset C$, $C/$${}_{\mathrm{s}}$*realizable*-$\Theta/T$), and $C/$${}_{\mathrm{s}}$*realizes*-$\Psi$, can be formulated as before (cf. 8.5, 8.7, 9.3).

10.6. LEMMA 10.6. *Lemma* 8.3 *holds reading "*${}_{\mathrm{s}}$*realizes-" for "realizes-".*

PROOF. Cases 2 and 7 read exactly as before.

CASES 1 and 5: prime or $\neg$A. ${}^{\mathrm{e}}0$ (end 10.2) where e = order E.

CASE 3: A $\vee$ B. $\langle(\chi[\Theta](0))_0, \langle(\chi[\Theta])_{1,0\mathrm{A}}, (\chi[\Theta])_{1,1\mathrm{B}}\rangle\rangle$.

CASE 4: A $\supset$ B. ${}^{\mathrm{e}}\Lambda{}^{\mathrm{a}}\alpha\, {}^{\mathrm{b}}\{\chi[\Theta]\}[{}^{\mathrm{a}}\alpha]$ where a, b, e are the orders of A, B, A $\supset$ B.

The modifications in CASES 6, 8 and 9 are similar.

10.7. LEMMA 10.7. *To each formula* E *of order* e *containing free only the variables* Ψ *and not containing* $\vee$ *or* $\exists$, *there is a primitive recursive function* ${}^{\mathrm{e}}\epsilon_{\mathrm{E}}$ *of order* e *such that, for each* Ψ:

(i) *If* $(E{}^{\mathrm{e}}\epsilon)[{}^{\mathrm{e}}\epsilon$ ${}_{\mathrm{s}}$*realizes*-Ψ E], *then* E *is true*-Ψ.

(ii) *If* E *is true*-Ψ, *then* ${}^{\mathrm{e}}\epsilon_{\mathrm{E}}$ ${}_{\mathrm{s}}$*realizes*-Ψ E.

PROOF, by induction. According as E is of the form P (a prime formula), A & B, A $\supset$ B, $\neg$A, $\forall x$A, or $\forall\alpha$A, let ${}^{\mathrm{e}}\epsilon_{\mathrm{E}}$ be as follows, where a = order A and b = order B: $\lambda t\, 0$, $\langle{}^{\mathrm{a}}\epsilon_{\mathrm{A}}, {}^{\mathrm{b}}\epsilon_{\mathrm{B}}\rangle$, ${}^{\mathrm{e}}\Lambda{}^{\mathrm{a}}\alpha\, {}^{\mathrm{b}}\epsilon_{\mathrm{B}}$, ${}^{\mathrm{e}}0$, ${}^{\mathrm{e}}\Lambda x\, {}^{\mathrm{a}}\epsilon_{\mathrm{A}}$, or ${}^{\mathrm{e}}\Lambda\alpha\, {}^{\mathrm{a}}\epsilon_{\mathrm{A}}$. — Cf. Remark 11.9.

§ 11. Special realizability under deduction in the intuitionistic formal system.

11.1. LEMMAS 11.1, 11.2. *Lemmas* 9.1, 9.2 *hold reading "*${}^{\mathrm{e}}\epsilon$ ${}_{\mathrm{s}}$*realizes" in place of "*ϵ *realizes".*

THEOREM 11.3. (a) *If* $\Gamma \vdash \mathrm{E}$ *in the intuitionistic formal system of analysis, and the formulas* Γ *are* ${}_{\mathrm{s}}$*realizable, then* E *is* ${}_{\mathrm{s}}$*realizable.* (b) *Similarly, reading "*${}_{\mathrm{s}}$*realizable/T" in place of "*${}_{\mathrm{s}}$*realizable".*

(c), (d) *Theorem* 9.3 (c), (d) *hold reading "*${}_{\mathrm{s}}$*realizable" in place of "realizable".*

PROOF. (Cf. the proof of Theorem 9.3.) (a) We shall understand that A (A(x), A(a), A(α, β)), B, C, R(a) (when present) have orders a, b, c, r; the prime subformulas have order 1. We show the orders of other subformulas as superior prefixes on the formal operators (IM p. 73).

1a. $\mathrm{A}\ {}^{\mathrm{d}}\!\supset (\mathrm{B}\ {}^{\mathrm{c}}\!\supset \mathrm{A})$ is ${}_{\mathrm{s}}$realized-Ψ by ${}^{\mathrm{d}}\Lambda^{\mathrm{a}}\alpha\ {}^{\mathrm{c}}\Lambda^{\mathrm{b}}\beta\ {}^{\mathrm{a}}\alpha$. For, this is of the required order $\mathrm{d} = \mathrm{a}{*}\mathrm{c}$ where $\mathrm{c} = \mathrm{b}{*}\mathrm{a}$, by Lemma 10.3. Suppose **(1)** ${}^{\mathrm{a}}\alpha$ ${}_{\mathrm{s}}$realizes-Ψ A; by Clause 4 in 10.5 we must infer from (1) that ${}^{\mathrm{c}}\{{}^{\mathrm{d}}\Lambda^{\mathrm{a}}\alpha\ {}^{\mathrm{c}}\Lambda^{\mathrm{b}}\beta\ {}^{\mathrm{a}}\alpha\}[{}^{\mathrm{a}}\alpha]$ ${}_{\mathrm{s}}$realizes-Ψ $\mathrm{B} \supset \mathrm{A}$. But by (10.3), ${}^{\mathrm{c}}\{{}^{\mathrm{d}}\Lambda^{\mathrm{a}}\alpha\ {}^{\mathrm{c}}\Lambda^{\mathrm{b}}\beta\ {}^{\mathrm{a}}\alpha\}[{}^{\mathrm{a}}\alpha] =_{\mathrm{c}} {}^{\mathrm{c}}\Lambda^{\mathrm{b}}\beta\ {}^{\mathrm{a}}\alpha$. So by Lemma 10.5 it will suffice to infer from (1) that ${}^{\mathrm{c}}\Lambda^{\mathrm{b}}\beta\ {}^{\mathrm{a}}\alpha$ ${}_{\mathrm{s}}$realizes-Ψ $\mathrm{B} \supset \mathrm{A}$. By Lemma 10.3 ${}^{\mathrm{c}}\Lambda^{\mathrm{b}}\beta\ {}^{\mathrm{a}}\alpha$ is of the required order $\mathrm{c} = \mathrm{b}{*}\mathrm{a}$. Suppose **(2)** ${}^{\mathrm{b}}\beta$ ${}_{\mathrm{s}}$realizes-Ψ B; we must infer from (1) and (2) that ${}^{\mathrm{a}}\{{}^{\mathrm{c}}\Lambda^{\mathrm{b}}\beta\ {}^{\mathrm{a}}\alpha\}[{}^{\mathrm{b}}\beta]$ ${}_{\mathrm{s}}$realizes-Ψ A. By (10.3), ${}^{\mathrm{a}}\{{}^{\mathrm{c}}\Lambda^{\mathrm{b}}\beta\ {}^{\mathrm{a}}\alpha\}[{}^{\mathrm{b}}\beta] =_{\mathrm{a}} {}^{\mathrm{a}}\alpha$; so by Lemma 10.5 what we need follows from (1).

1b, 7, 3, 4a, 4b, 10N, 10F, 11F, 11N are treated as before, supplying subscripts ${}_{\mathrm{s}}$ and order superscripts, and using Lemma 10.5.

5a. $\mathrm{A}\ {}^{\mathrm{d}}\!\supset \mathrm{A}\ {}^{\mathrm{c}}\!\vee \mathrm{B}$. ${}^{\mathrm{d}}\Lambda^{\mathrm{a}}\alpha\ \langle 0, \langle {}^{\mathrm{a}}\alpha, {}^{\mathrm{b}}0\rangle\rangle$ (end 10.2).

5b. $\mathrm{B}\ {}^{\mathrm{d}}\!\supset \mathrm{A}\ {}^{\mathrm{c}}\!\vee \mathrm{B}$. ${}^{\mathrm{d}}\Lambda^{\mathrm{b}}\beta\ \langle 1, \langle {}^{\mathrm{a}}0, {}^{\mathrm{b}}\beta\rangle\rangle$.

6. $(\mathrm{A}\ {}^{\mathrm{d}}\!\supset \mathrm{C})\ {}^{\mathrm{i}}\!\supset ((\mathrm{B}\ {}^{\mathrm{e}}\!\supset \mathrm{C})\ {}^{\mathrm{h}}\!\supset (\mathrm{A}\ {}^{\mathrm{f}}\!\vee \mathrm{B}\ {}^{\mathrm{g}}\!\supset \mathrm{C}))$. ${}^{\mathrm{i}}\Lambda^{\mathrm{d}}\pi\ {}^{\mathrm{h}}\Lambda^{\mathrm{e}}\rho\ {}^{\mathrm{g}}\Lambda^{\mathrm{f}}\sigma\ \lambda t\ \overline{\mathrm{sg}}(({}^{\mathrm{f}}\sigma(0))_0)\cdot({}^{\mathrm{c}}\{{}^{\mathrm{d}}\pi\}[({}^{\mathrm{f}}\sigma)_{1,0}])(t) + \mathrm{sg}(({}^{\mathrm{f}}\sigma(0))_0)\cdot({}^{\mathrm{c}}\{{}^{\mathrm{e}}\rho\}[({}^{\mathrm{f}}\sigma)_{1,1}])(t)$.

8^{I}. ${}^{\mathrm{c}}\neg\mathrm{A}\ {}^{\mathrm{e}}\!\supset (\mathrm{A}\ {}^{\mathrm{d}}\!\supset \mathrm{B})$. ${}^{\mathrm{e}}\Lambda^{\mathrm{c}}\pi\ {}^{\mathrm{d}}0$.

13. $\mathrm{A}(0)\ {}^{\mathrm{d}}\&\ {}^{\mathrm{c}}\forall \mathrm{x}(\mathrm{A}(\mathrm{x})\ {}^{\mathrm{b}}\!\supset \mathrm{A}(\mathrm{x}'))\ {}^{\mathrm{e}}\!\supset \mathrm{A}(\mathrm{x})$. ${}^{\mathrm{e}}_{\mathrm{a}}\Lambda^{\mathrm{d}}\alpha\ \rho[x, {}^{\mathrm{d}}\alpha]$, where ρ is defined by

$$\rho[0, {}^{\mathrm{d}}\alpha] = ({}^{\mathrm{d}}\alpha)_0,$$
$$\rho[x', {}^{\mathrm{d}}\alpha] = {}^{\mathrm{a}}\{{}^{\mathrm{b}}\{({}^{\mathrm{d}}\alpha)_1\}[x]\}[\rho[x, {}^{\mathrm{d}}\alpha]].$$

Writing $\rho[x, {}^{\mathrm{d}}\alpha] = \lambda t\, \rho(x, {}^{\mathrm{d}}\alpha, t)$, this takes the form

$$\rho(0, {}^{\mathrm{d}}\alpha, t) = \psi({}^{\mathrm{d}}\alpha, t),$$
$$\rho(x', {}^{\mathrm{d}}\alpha, t) \simeq \chi(x, {}^{\mathrm{d}}\alpha, \lambda t\, \rho(x, {}^{\mathrm{d}}\alpha, t), t)$$

where ψ is primitive recursive, χ is partial recursive, and, for ${}^{\mathrm{d}}\alpha$, ${}^{\mathrm{a}}\beta$ of the specified orders, $\lambda t\, \psi({}^{\mathrm{d}}\alpha, t)$ and $\lambda t\, \chi(x, {}^{\mathrm{d}}\alpha, {}^{\mathrm{a}}\beta, t)$ are (completely defined and) of order a. So, by induction on x, for ${}^{\mathrm{d}}\alpha$ of the specified order, $\lambda t\, \rho(x, {}^{\mathrm{d}}\alpha, t)$ is of order a. That this $\rho(x, {}^{\mathrm{d}}\alpha, t)$ is partial, and hence special, recursive is seen as before.

14, 17, $^{\times}1.1$, 16, prime axioms, $^{\times}2.1$ as before. 15: $^{1*}10$.

RULES OF INFERENCE. 2. A, A ⊃ B / B. Noting 10.5, we choose Ψ to include all variables free in A ⊃ B. By the hyp. ind., there are general recursive functions α and ψ such that, for each Ψ, $\alpha[\Psi]$ ${}_{\mathrm{s}}$realizes-Ψ A and $\psi[\Psi]$ ${}_{\mathrm{s}}$realizes-Ψ A ⊃ B. Let $\varphi[\Psi] = {}^{\mathrm{b}}\{\psi[\Psi]\}[\alpha[\Psi]]$. For each Ψ, $\alpha[\Psi]$ is of order a and $\psi[\Psi]$ is of order e = a∗b, so ${}^{\mathrm{b}}\{\psi[\Psi]\}[\alpha[\Psi]]$ is (defined and) of order b; so $\varphi[\Psi]$ $(= \lambda s\, \varphi[\Psi, s])$ is general recursive. For each Ψ, $\varphi[\Psi]$ ${}_{\mathrm{s}}$realizes-Ψ B.

9N, 9F, 12F, 12N as before.

AXIOM SCHEMA $^{\times}26.3\mathrm{c}$. $^{\mathrm{f}}\forall\alpha^{\mathrm{e}}\exists \mathrm{x}[\mathrm{R}(\bar{\alpha}(\mathrm{x}))\ {}^{\mathrm{d}}\&\ {}^{\mathrm{c}}\forall \mathrm{y}(\mathrm{R}(\bar{\alpha}(\mathrm{y}))\ {}^{\mathrm{b}}\supset \mathrm{x}=\mathrm{y})]\ {}^{\mathrm{j}}\&\ {}^{\mathrm{i}}\forall \mathrm{a}[\mathrm{Seq}(\mathrm{a})\ {}^{\mathrm{g}}\&\ \mathrm{R}(\mathrm{a})\ {}^{\mathrm{h}}\supset \mathrm{A}(\mathrm{a})]\ {}^{\mathrm{p}}\&\ {}^{\mathrm{n}}\forall \mathrm{a}[\mathrm{Seq}(\mathrm{a})\ {}^{\mathrm{l}}\&\ {}^{\mathrm{k}}\forall \mathrm{s}\mathrm{A}(\mathrm{a}*2^{\mathrm{s}+1})\ {}^{\mathrm{m}}\supset \mathrm{A}(\mathrm{a})]\ {}^{\mathrm{q}}\supset \mathrm{A}(1)$.

We proceed as before down through (8), now supplying subscripts ${}_{\mathrm{s}}$ and order superscripts. Moreover, we now observe that, for ${}^{\mathrm{p}}\pi$ of order p, $R({}^{\mathrm{p}}\pi, a)$ and $R_1({}^{\mathrm{p}}\pi, a)$ are *always* defined, i.e. whether or not ${}^{\mathrm{p}}\pi$ ${}_{\mathrm{s}}$realizes-Ψ the antecedent. Furthermore, the definition of $\eta({}^{\mathrm{p}}\pi, a, u)$ coincides for $a \in S_1^{{}^{\mathrm{p}}\pi}$ with a recursion of form corresponding to the inductive definition of $S_1^{{}^{\mathrm{p}}\pi}$ (cf. IM p. 260), call it a *bar recursion.* By the corresponding form of proof (IM p. 259), always: **(9a)** $a \in S_1^{{}^{\mathrm{p}}\pi} \rightarrow \{\eta[{}^{\mathrm{p}}\pi, a]$ is (completely defined and) of order a$\}$. But we don't know that $\eta[{}^{\mathrm{p}}\pi, 1]$ is always of order a, which we now would require to complete the proof as before, since the proof of (8) uses the assumption that ${}^{\mathrm{p}}\pi$ ${}_{\mathrm{s}}$realizes the antecedent.

In the following this assumption is not made, except as indicated.

We define a second special recursive predicate $R_2({}^{\mathrm{p}}\pi, a)$ thus.

$$R_2({}^{\mathrm{p}}\pi, a) \simeq [a = \overline{\alpha_1}(x) \text{ for } \alpha_1 = \lambda t\, (a)_t \dot{-} 1,\ x \geq ({}^{\mathrm{e}}\{({}^{\mathrm{p}}\pi)_{0,0}\}[\alpha_1](0))_0].$$

Clearly: **(10)** $R_1({}^{\mathrm{p}}\pi, a) \rightarrow R_2({}^{\mathrm{p}}\pi, a)$.

We now show that: **(11)** $(\alpha)(Ex)R_2({}^{\mathrm{p}}\pi, \bar{\alpha}(x))$. Consider any α. Let $x_0 = ({}^{\mathrm{e}}\{({}^{\mathrm{p}}\pi)_{0,0}\}[\alpha](0))_0$. For a fixed ${}^{\mathrm{p}}\pi$ of order p, since $\lambda\alpha\, x_0$ is partial recursive in ${}^{\mathrm{p}}\pi$, the computation of x_0 uses only the first y_0 values of α, for some y_0 (cf. IM pp. 330, 292). Put $x = \max(x_0, y_0)$. Since $x \geq y_0$, $x_0 = ({}^{\mathrm{e}}\{({}^{\mathrm{p}}\pi)_{0,0}\}[\alpha_1](0))_0$ for $\alpha_1 = \lambda t\, (\bar{\alpha}(x))_t \dot{-} 1$. Now $\bar{\alpha}(x) = \overline{\alpha_1}(x)$ with $x \geq x_0$. So $R_2({}^{\mathrm{p}}\pi, \bar{\alpha}(x))$.

Also: **(12)** $R_2({}^{\mathrm{p}}\pi, \bar{\alpha}(y))\ \&\ \bar{R}_1({}^{\mathrm{p}}\pi, \bar{\alpha}(y)) \rightarrow (Ex)_{x<y}R({}^{\mathrm{p}}\pi, \bar{\alpha}(x))$. For, put $\bar{\alpha}(y) = \overline{\alpha_1}(x)$ for $\alpha_1 = \lambda t\, (\bar{\alpha}(y))_t \dot{-} 1$, $x = y > ({}^{\mathrm{e}}\{({}^{\mathrm{p}}\pi)_{0,0}\}[\alpha_1](0))_0 = x_1$. Then $\bar{\alpha}(x_1) = \overline{\alpha_1}(x_1)$, whence $R({}^{\mathrm{p}}\pi, \bar{\alpha}(x_1))$.

Let $S_2^{{}^{\mathrm{p}}\pi}$ be the set of the sequence numbers barred with respect to $\lambda a\, R_2({}^{\mathrm{p}}\pi, a)$.

Similarly to $\eta[{}^{\mathrm{p}}\pi, a]$, we find a partial recursive $\eta_2[{}^{\mathrm{p}}\pi, a] = \lambda u\, \eta_2({}^{\mathrm{p}}\pi, a, u)$ satisfying

$$\eta_2({}^{\mathrm{p}}\pi, a, u) \simeq \begin{cases} {}^{\mathrm{a}}0(u) \text{ if } R_2({}^{\mathrm{p}}\pi, a) \;\&\; \bar{R}_1({}^{\mathrm{p}}\pi, a), \\ \text{as before if } R_1({}^{\mathrm{p}}\pi, a), \\ \text{as before otherwise.} \end{cases}$$

Using (10), similarly to (9a): **(13)** $a \in S_2^{{}^{\mathrm{p}}\pi} \rightarrow \{\eta_2[{}^{\mathrm{p}}\pi, a] \text{ is of order a}\}$. By (11): **(14)** $1 \in S_2^{{}^{\mathrm{p}}\pi}$. Hence $\eta_2[{}^{\mathrm{p}}\pi, 1]$ is of order a.

To show that ${}^{\mathrm{q}}\Lambda\, {}^{\mathrm{p}}\pi\, \eta_2[{}^{\mathrm{p}}\pi, 1]$ ${}_{\mathrm{s}}$realizes-Ψ the axiom, it remains for us to show that $\eta_2[{}^{\mathrm{p}}\pi, 1]$ ${}_{\mathrm{s}}$realizes-Ψ A(1) when ${}^{\mathrm{p}}\pi$ ${}_{\mathrm{s}}$realizes-Ψ the antecedent. So, with that assumption in force again, consider the set $S^{{}^{\mathrm{p}}\pi}$ of the sequence numbers securable but not past secured with respect to $\lambda a\, R_1({}^{\mathrm{p}}\pi, a)$. (Using (10), $S^{{}^{\mathrm{p}}\pi} \subset S_1^{{}^{\mathrm{p}}\pi} \subset S_2^{{}^{\mathrm{p}}\pi}$.) By (12) with (6), $a \in S^{{}^{\mathrm{p}}\pi} \rightarrow \overline{R_2({}^{\mathrm{p}}\pi, a) \;\&\; \bar{R}_1({}^{\mathrm{p}}\pi, a)}$. Therefore on $S^{{}^{\mathrm{p}}\pi}$, $\eta_2({}^{\mathrm{p}}\pi, a, u)$ satisfies the same bar recursion as $\eta({}^{\mathrm{p}}\pi, a, u)$; hence it is the same function to within possible differences in the results of the operations by ${}^{\mathrm{k}}\Lambda s$ in the "otherwise" case. Hence for the same reasons as (7): **(15)** $a \in S^{{}^{\mathrm{p}}\pi} \rightarrow \{\eta_2[{}^{\mathrm{p}}\pi, a]$ ${}_{\mathrm{s}}$realizes-Ψ, a A(a)$\}$. But using (8): **(16)** $1 \in S^{{}^{\mathrm{p}}\pi}$. By (15), (16) and Lemma 11.1 (a), $\eta_2[{}^{\mathrm{p}}\pi, 1]$ ${}_{\mathrm{s}}$realizes-Ψ A(1).

Axiom Schema ${}^{\mathrm{x}}$27.1. ${}^{\mathrm{c}}\forall\alpha\, {}^{\mathrm{b}}\exists\beta \mathrm{A}(\alpha, \beta)\ {}^{\mathrm{q}}\supset$
${}^{\mathrm{p}}\exists\tau\, {}^{\mathrm{n}}\forall\alpha\{{}^{\mathrm{h}}\forall \mathrm{t}\, {}^{\mathrm{g}}\exists \mathrm{y}[\tau(2^{\mathrm{t}+1} * \bar{\alpha}(\mathrm{y})) > 0\ {}^{\mathrm{f}}\&\ {}^{\mathrm{e}}\forall \mathrm{z}(\tau(2^{\mathrm{t}+1} * \bar{\alpha}(\mathrm{z})) > 0\ {}^{\mathrm{d}}\supset \mathrm{y} = \mathrm{z})]$
${}^{\mathrm{m}}\&\ {}^{\text{ł}}\forall\beta[{}^{\mathrm{j}}\forall \mathrm{t}\, {}^{\mathrm{i}}\exists \mathrm{y}\tau(2^{\mathrm{t}+1} * \bar{\alpha}(\mathrm{y})) = \beta(\mathrm{t}) + 1\ {}^{\mathrm{k}}\supset \mathrm{A}(\alpha, \beta)]\}$.

Assume that ${}^{\mathrm{c}}\pi$ ${}_{\mathrm{s}}$realizes-Ψ $\forall\alpha\exists\beta\mathrm{A}(\alpha, \beta)$. Then **(1)** for each α, $({}^{\mathrm{b}}\{{}^{\mathrm{c}}\pi\}[\alpha])_1$ ${}_{\mathrm{s}}$realizes-Ψ, α, β_1 $\mathrm{A}(\alpha, \beta)$ for $\beta_1 = {}^{1}\{({}^{\mathrm{b}}\{{}^{\mathrm{c}}\pi\}[\alpha])_0\}$. Let $\tau_1[{}^{\mathrm{c}}\pi, t] = \lambda s\, \tau_1({}^{\mathrm{c}}\pi, t, s) = \Lambda^1\alpha\, \beta_1(t) = \Lambda^1\alpha\, ({}^{\mathrm{b}}\{{}^{\mathrm{c}}\pi\}[{}^{1}\alpha])_0(\lambda x\, t)$. By Lemma 10.1, for each t, $\tau_1[{}^{\mathrm{c}}\pi, t]$ is of order 2, so, for each α, **(2a)** $(t)(E!y)\tau_1({}^{\mathrm{c}}\pi, t, \bar{\alpha}(y)) > 0$, and (by (10.2)) $\{\tau_1[{}^{\mathrm{c}}\pi, t]\}(\alpha) = \beta_1(t)$, whence by (10.1): **(3a)** $(t)\tau_1({}^{\mathrm{c}}\pi, t, \bar{\alpha}(y_t)) = \beta_1(t) + 1$ where $y_t = \mu y \tau_1({}^{\mathrm{c}}\pi, t, \bar{\alpha}(y)) > 0$. Let $\tau = \lambda s\, \tau_1({}^{\mathrm{c}}\pi, (s)_0 \dot{-} 1, \Pi_{i < \mathrm{lh}(s) \dot{-} 1}\, p_i^{(s)_{i+1}})$. Consider any α. By (2a), (3a): **(2)** $(t)(E!y)\tau(2^{t+1} * \bar{\alpha}(y)) > 0$, **(3)** $(t)\tau(2^{t+1} * \bar{\alpha}(y_t)) = \beta_1(t) + 1$ where $y_t = \mu y \tau(2^{t+1} * \bar{\alpha}(y)) > 0$.

We continue as before. So putting ${}^{\mathrm{h}}\rho_0 = {}^{\mathrm{h}}\Lambda t\, \langle \mu y \tau(2^{t+1} * \bar{\alpha}(y)) > 0, \langle \lambda s\, 0, {}^{\mathrm{e}}\Lambda z\, {}^{\mathrm{d}}\Lambda\sigma\, \lambda s\, 0\rangle\rangle$, ${}^{\text{ł}}\rho_1 = {}^{\text{ł}}\Lambda\beta\, {}^{\mathrm{k}}\Lambda^{\mathrm{j}}\sigma\, ({}^{\mathrm{b}}\{{}^{\mathrm{c}}\pi\}[\alpha])_1$, the axiom is ${}_{\mathrm{s}}$realized-Ψ by ${}^{\mathrm{q}}\Lambda\, {}^{\mathrm{c}}\pi\, \langle {}^{2}\Lambda\, \tau, {}^{\mathrm{n}}\Lambda\alpha\, \langle {}^{\mathrm{h}}\rho_0, {}^{\text{ł}}\rho_1\rangle\rangle$.

11.2. Corollaries 11.4–11.6. *Corollaries* 9.4, 9.5 (*with different proof*) *and* 9.6 *hold reading "${}_{\mathrm{s}}$realizable" in place of "realizable".*

11.3. Theorem 11.7. (a)^{C} *Let* A(x, y) *be picked by Lemma* 8.5 *to numeralwise express a general recursive predicate* $A(x, y)$. *Classically,*

$\forall \mathrm{x}[\neg\forall \mathrm{y}\neg \mathrm{A}(\mathrm{x}, \mathrm{y}) \supset \exists \mathrm{y}\mathrm{A}(\mathrm{x}, \mathrm{y})]$ *is realizable.* (b) *Let* $\mathrm{T}(\mathrm{x}, \mathrm{y})$ *be* $\mathrm{T}_1((\mathrm{x})_0, (\mathrm{x})_1, \mathrm{y})$ *where* $\mathrm{T}_1(\mathrm{z}, \mathrm{x}, \mathrm{y})$ *is picked by (the method of proof of) Lemma* 8.5 *to numeralwise express* $T_1(z, x, y)$ *(with* x *free for* z, *and* z *free for* y). *Then* $\forall \mathrm{x}[\neg\forall \mathrm{y}\neg \mathrm{T}(\mathrm{x}, \mathrm{y}) \supset \exists \mathrm{y}\mathrm{T}(\mathrm{x}, \mathrm{y})]$ *is un*$_{\mathrm{s}}$*realizable.* By #19 and Lemmas 5.2 and 4.2, $\mathrm{T}(\mathrm{x}, \mathrm{y})$ numeralwise expresses $T_1((x)_0, (x)_1, y)$.

(c)[C] *Let* $\mathrm{A}_0(\mathrm{x}, \mathrm{y})$ *and* $\mathrm{A}_1(\mathrm{x}, \mathrm{y})$ *by picked by Lemma* 8.5 *to numeralwise express general recursive predicates* $A_0(x, y)$ *and* $A_1(x, y)$, *respectively. Classically,* $\forall \mathrm{x}[\neg(\forall \mathrm{y}\mathrm{A}_0(\mathrm{x}, \mathrm{y})\ \&\ \forall \mathrm{y}\mathrm{A}_1(\mathrm{x}, \mathrm{y})) \supset \neg\forall \mathrm{y}\mathrm{A}_0(\mathrm{x}, \mathrm{y}) \vee \neg\forall \mathrm{y}\mathrm{A}_1(\mathrm{x}, \mathrm{y})]$ *is realizable.* (d) *Let* $\mathrm{W}_0(\mathrm{x}, \mathrm{y})$ *be* $\mathrm{T}_1((\mathrm{x})_1, \mathrm{x}, \mathrm{y})\ \&\ \forall \mathrm{z}_{\mathrm{z}\leq \mathrm{y}}\neg \mathrm{T}_1((\mathrm{x})_0, \mathrm{x}, \mathrm{z})$, *and* $\mathrm{W}_1(\mathrm{x}, \mathrm{y})$ *be* $\mathrm{T}_1((\mathrm{x})_0, \mathrm{x}, \mathrm{y})\ \&\ \forall \mathrm{z}_{\mathrm{z}\leq \mathrm{y}}\neg \mathrm{T}_1((\mathrm{x})_1, \mathrm{x}, \mathrm{z})$, *with* $\mathrm{z}\leq \mathrm{y}$ *prime. Then* $\forall \mathrm{x}[\neg(\forall \mathrm{y}\neg \mathrm{W}_0(\mathrm{x}, \mathrm{y})\ \&\ \forall \mathrm{y}\neg \mathrm{W}_1(\mathrm{x}, \mathrm{y})) \supset \neg\forall \mathrm{y}\neg \mathrm{W}_0(\mathrm{x}, \mathrm{y}) \vee \neg\forall \mathrm{y}\neg \mathrm{W}_1(\mathrm{x}, \mathrm{y})]$ *is un*$_{\mathrm{s}}$*realizable.* By IM p. 202 (C) and (E), $\mathrm{W}_0(\mathrm{x}, \mathrm{y})$ and $\mathrm{W}_1(\mathrm{x}, \mathrm{y})$ numeralwise express $W_0(x, y)$ and $W_1(x, y)$, respectively, IM p. 308.

PROOF. (a) It is realized by $\Lambda x\, \Lambda \alpha\, \langle \mu y A(x, y), \epsilon_{\mathrm{A(x,y)}}\rangle$, using Lemma 8.4a. For, consider any x. Suppose α realizes-x $\neg\forall \mathrm{y}\neg \mathrm{A}(\mathrm{x}, \mathrm{y})$ Then by Lemma 8.4a (i), $\overline{(y)}\bar{A}(x, y)$, whence classically $(Ey)A(x, y)$, so $\mathrm{A}(\mathrm{x}, \mathrm{y})$ is true-x, $\mu y A(x, y)$, so by Lemma 8.4a (ii) $\epsilon_{\mathrm{A(x,y)}}$ realizes-x, $\mu y A(x, y)$ $\mathrm{A}(\mathrm{x}, \mathrm{y})$, so $\langle \mu y A(x, y), \epsilon_{\mathrm{A(x,y)}}\rangle$ realizes-x $\exists \mathrm{y}\mathrm{A}(\mathrm{x}, \mathrm{y})$.

(b) Suppose a general recursive function ${}^{\mathrm{f}}\epsilon$ $_{\mathrm{s}}$realizes ${}^{\mathrm{f}}\forall \mathrm{x}[{}^{\mathrm{c}}\neg{}^{\mathrm{b}}\forall \mathrm{y}{}^{\mathrm{a}}\neg{}^{\mathrm{t}}\mathrm{T}(\mathrm{x}, \mathrm{y})\ {}^{\mathrm{e}}\supset {}^{\mathrm{d}}\exists \mathrm{y}{}^{\mathrm{t}}\mathrm{T}(\mathrm{x}, \mathrm{y})]$. Then, for all x: **(1)** ${}^{\mathrm{e}}\{{}^{\mathrm{f}}\epsilon\}[x]$ $_{\mathrm{s}}$realizes-x $\neg\forall \mathrm{y}\neg \mathrm{T}(\mathrm{x}, \mathrm{y}) \supset \exists \mathrm{y}\mathrm{T}(\mathrm{x}, \mathrm{y})$. By Lemma 10.7 (ii) (and proof): **(2)** $\overline{(y)}\bar{T}_1((x)_0, (x)_1, y) \rightarrow \{{}^{\mathrm{c}}0$ $_{\mathrm{s}}$realizes-x $\neg\forall \mathrm{y}\neg \mathrm{T}(\mathrm{x}, \mathrm{y})\}$. Now if $(Ey)T_1((x)_0, (x)_1, y)$, then $\overline{(y)}\bar{T}_1((x)_0, (x)_1, y)$, whence by (2) and (1) ${}^{\mathrm{d}}\{{}^{\mathrm{e}}\{{}^{\mathrm{f}}\epsilon\}[x]\}[{}^{\mathrm{c}}0]$ $_{\mathrm{s}}$realizes-x $\exists \mathrm{y}\mathrm{T}(\mathrm{x}, \mathrm{y})$, whence $({}^{\mathrm{d}}\{{}^{\mathrm{e}}\{{}^{\mathrm{f}}\epsilon\}[x]\}[{}^{\mathrm{c}}0])_1$ $_{\mathrm{s}}$realizes-x, $v(x)$ $\mathrm{T}(\mathrm{x}, \mathrm{y})$ for $v(x) = ({}^{\mathrm{d}}\{{}^{\mathrm{e}}\{{}^{\mathrm{f}}\epsilon\}[x]\}[{}^{\mathrm{c}}0](0))_0$, whence by Lemma 10.7 (i) $T_1((x)_0, (x)_1, v(x))$. The converse is immediate. Thus: **(3)** $(Ey)T_1((x)_0, (x)_1, y) \equiv T_1((x)_0, (x)_1, v(x))$ for the general recursive function $v(x)$ just introduced. Substituting $\langle x, x\rangle$ for x: **(4)** $(Ey)T_1(x, x, y) \equiv T_1(x, x, v(\langle x, x\rangle))$, contradicting that $(Ey)T_1(x, x, y)$ is not general recursive (IM p. 283).

(c) $\Lambda x\, \Lambda \alpha\, \langle \rho(x), \lambda t\, 0\rangle$ where

$$\rho(x) \simeq \begin{cases} 0 \text{ if } \bar{A}_0(x, \mu y[\bar{A}_0(x, y) \vee \bar{A}_1(x, y)]), \\ 1 \text{ if } A_0(x, \mu y[\bar{A}_0(x, y) \vee \bar{A}_1(x, y)]). \end{cases}$$

(d) Suppose a general recursive function ${}^{\mathrm{h}}\epsilon$ $_{\mathrm{s}}$realizes ${}^{\mathrm{h}}\forall \mathrm{x}[{}^{\mathrm{d}}\neg({}^{\mathrm{b}}\forall \mathrm{y}{}^{\mathrm{a}}\neg{}^{\mathrm{w}}\mathrm{W}_0(\mathrm{x}, \mathrm{y})\ {}^{\mathrm{c}}\&\ {}^{\mathrm{b}}\forall \mathrm{y}{}^{\mathrm{a}}\neg{}^{\mathrm{w}}\mathrm{W}_1(\mathrm{x}, \mathrm{y}))\ {}^{\mathrm{g}}\supset {}^{\mathrm{e}}\neg{}^{\mathrm{b}}\forall \mathrm{y}{}^{\mathrm{a}}\neg{}^{\mathrm{w}}\mathrm{W}_0(\mathrm{x}, \mathrm{y})\ {}^{\mathrm{f}}\vee\ {}^{\mathrm{e}}\neg{}^{\mathrm{b}}\forall \mathrm{y}{}^{\mathrm{a}}\neg{}^{\mathrm{w}}\mathrm{W}_1(\mathrm{x}, \mathrm{y})]$. Then, for all x: **(1)** ${}^{\mathrm{g}}\{{}^{\mathrm{h}}\epsilon\}[x]$ $_{\mathrm{s}}$realizes-x $\neg(\forall \mathrm{y}\neg \mathrm{W}_0(\mathrm{x}, \mathrm{y})$

& $\forall y \neg W_1(x, y)) \supset \neg \forall y \neg W_0(x, y) \vee \neg \forall y \neg W_1(x, y)$. By Lemma 10.7 (ii): **(2)** $\overline{(y)\overline{W}_0(x, y) \,\&\, (y)\overline{W}_1(x, y)} \rightarrow \{{}^{d}0 \ {}_{s}\text{realizes}\ \neg(\forall y \neg W_0(x, y) \,\&\, \forall y \neg W_1(x, y))\}$. Suppose **(A)** $(Ey)W_0(x, y)$. Then $\overline{(y)}\overline{W}_0(x, y)$, whence **(B)** $\overline{(y)\overline{W}_0(x, y) \,\&\, (y)\overline{W}_1(x, y)}$. Also by IM p. 308 (51) $\overline{(Ey)}W_1(x, y)$, whence **(C)** $(y)\overline{W}_1(x, y)$. By (B), (2) and (1): **(D)** ${}^{f}\{{}^{g}\{{}^{h}\epsilon\}[x]\}[{}^{d}0]$ ${}_{s}$realizes-x $\neg \forall y \neg W_0(x, y) \vee \neg \forall y \neg W_1(x, y)$. Put $v(x) = ({}^{f}\{{}^{g}\{{}^{h}\epsilon\}[x]\}[{}^{d}0](0))_0$. Now **(E)** $v(x) = 0$, as otherwise $({}^{f}\{{}^{g}\{{}^{h}\epsilon\}[x]\}[{}^{d}0])_{1,1}$ would ${}_{s}$realize-x $\neg \forall y \neg W_1(x, y)$, so by Lemma 10.7 (i) $\overline{(y)}\overline{W}_1(x, y)$, contradicting (C). Summarizing (from (A)): **(3)** $(Ey)W_0(x, y) \rightarrow v(x) = 0$. Similarly: **(4)** $(Ey)W_1(x, y) \rightarrow v(x) \neq 0$. Thus $D_2 = \hat{x}[v(x) = 0]$ and $D_3 = \hat{x}[v(x) \neq 0] = \overline{D}_2$ are disjoint general recursive (a fortiori, recursively enumerable) classes, containing $C_0 = \hat{x}(Ey)W_0(x, y)$ and $C_1 = \hat{x}(Ey)W_1(x, y)$, respectively, whose union is all natural numbers. This contradicts IM pp. 311–312.

Remark 11.8. The reasoning under (a) fails with ${}_{s}$realizability, because $\langle \mu y A(x, y), \epsilon_{A(x,y)} \rangle$ is not in general completely defined; e.g. it is not when $A(x, y) \equiv T_1((x)_0, (x)_1, y)$. So ${}^{f}\Lambda x\, {}^{e}\Lambda {}^{c}\alpha \langle \mu y A(x, y), {}^{a}\epsilon_{A(x,y)} \rangle$ is not in general a function of order f. The reasoning under (b) fails for realizability, because $v(x) = (\{\{\epsilon\}[x]\}[\lambda x\, 0](0))_0$ is not a general recursive function, but only partial recursive.

Remark 11.9^{C}. Theorem 11.7 (a) for a primitive recursive $A(x, y)$ is immediate by a classical application of Lemma 8.4b (ii), if the symbolism and postulates of Group D are extended (within the framework laid down in 5.1) to give a prime formula A(x, y) expressing $A(x, y)$. Therefore Lemma 8.4b cannot hold for ${}_{s}$realizability independently of how far we extend our Postulate Group D; in particular, it fails if we extend that to reach the representing function of $T_1(z, x, y)$, as the application just described would then contradict Theorem 11.7 (b).

Corollary 11.10^{C}. *The following classically provable formulas are formally undecidable in the intuitionistic formal system of analysis*:

(a) $\forall z \forall x[\neg \forall y \neg T_1(z, x, y) \supset \exists y T_1(z, x, y)]$ (Markov's principle M_1).

(b) $\forall x[\neg(\forall y \neg W_0(x, y) \,\&\, \forall y \neg W_1(x, y)) \supset \neg \forall y \neg W_0(x, y) \vee \neg \forall y \neg W_1(x, y)]$ (an instance of De Morgan's classical law).

(c) $\forall x[\exists y T_1((x)_0, (x)_1, y) \vee \neg \exists y T_1((x)_0, (x)_1, y)]$ (Law of excluded middle with one quantifier).

(We show intuitionistically that each of the three formulas is unprovable, classically that its negation is unprovable.)

PROOF. (a) Using *25.3, M_1 is equivalent to $\forall x[\neg\forall y \neg T(x, y) \supset \exists y T(x, y)]$, which by (b) of the theorem with Theorem 11.3 (a) is unprovable. By (a) of the theorem with Corollary 9.4 and Theorem 9.3 (a), $\neg\forall x[\neg\forall y \neg T(x, y) \supset \exists y T(x, y)]$ is unprovable.

(c) Similarly (using only §§ 8, 9) from the results of 8.6 ¶ 6, adapted inessentially to this example. —

Note that (c) ⊢ (a). Also it is not hard to see that (c) ⊢ (b). It seems to be an open question whether the double negations of these formulas are provable. By the results with 1945-realizability (8.6 ¶ 6, IM p. 511), $\neg\neg$(c) is unprovable in intuitionistic number theory, even including the postulates of Group D (however extended) with only number variables.

CHAPTER III

THE INTUITIONISTIC CONTINUUM

by RICHARD E. VESLEY [7]

§ 12. **Introduction.** We shall develop the intuitionistic theory of the continuum in the formal system of Chapter I. Our aims are, first, to investigate the adequacy of the system for this development, and, second, to provide an exposition of the theory. As Beth has observed (**1959** p. 422), "the central place in intuitionistic mathematics is occupied by the *theory of the continuum*". Any formal system for intuitionistic analysis should provide the means for a development of this theory at least through the well-known uniform continuity theorem (§ 15 below). Such a development may clarify for some readers the intuitionistic sources, which (except for Heyting 1930, 1930a) are deliberately non-formal.

Of the many sources (Brouwer **1918–9**, **1924**, **1927**, **1928a**, etc., Heyting **1952–3**, **1956**), we rely on two primarily, but not entirely. In §§ 14, 15 we follow rather closely the exposition of Heyting **1956**. In § 16 we prove formally some less well-known theorems from Brouwer **1928a** (not appearing in Heyting **1952–3** or **1956**).

Although the main objectives here are foundational analysis and clarification of the existing intuitionistic theory of the continuum, some additions are made to that theory. As one detail which seems to be developed here for the first time, we prove (*R14.11 below) the equivalence of the notion of sharp difference ($\neq_s$ in our symbolism), which Brouwer introduced in **1928a**, to his notion of apartness $\#$, introduced in **1918–9** II and frequently used in intuitionistic writings (e.g. in his **1954**, and Heyting's **1956**), of real number generators. Our proof of the equivalence uses Brouwer's principle (§ 7 above).

This equivalence of $\neq_s$ to $\#$ we also use in simplifying the hy-

[7] This chapter is a revised version of a Ph. D. thesis, written under the direction of Professor S. C. Kleene, and accepted by the University of Wisconsin May 28, 1962.

pothesis in Brouwer's formulation **1928a** of the free-connectedness property (our *R14.13–*R14.14), and (in Remark 16.1) we indicate a reason why Brouwer could not have established the property quite as he formulated it.

§ 13. Real number generators and real numbers. Brouwer studies various formulations of the continuum in different papers. In general, he takes the following two steps. First, he describes a particular species of "points" (**1927** p. 60) or "real number generators" (Heyting's term in **1956** p. 16). (Other terms are also used.) These are infinite sequences and may be, for example, infinite sequences of rational numbers or dual fractions with some convergence condition (Brouwer **1928a** p. 5, Heyting **1956** p. 16), sequences of nested dual intervals with a convergence condition (Brouwer **1927** p. 60, **1949** p. 122), or choice sequences of the rational numbers producing objects analogous to Dedekind cuts (Brouwer **1924–7** II p. 467). Second, he defines an "equality" (or "coincidence") predicate over these infinite sequences and a "point core" (Brouwer **1927** p. 60) or "real number" (Heyting **1956** p. 37) as an equivalence class with respect to this equality predicate.

In this form, as a "species of second order" (cf. Heyting **1956** p. 38), the continuum appears not to be an object for study in the formal system. But the elements of the underlying species of points or real number generators can be studied and the theory developed on this basis. Properties of real numbers will then appear as just those properties of real number generators which depend only on the equivalence classes with respect to the equality (or coincidence) predicate to which the generators belong.

We use from now on the abbreviation "r.n.g." for real number generator(s).

Of the possible species of r.n.g. mentioned above we choose to consider the species of convergent sequences of dual fractions. As a preliminary we could without difficulty develop the theory of dual fractions (or of rational numbers in general) in the formal system. But we find it simpler to proceed directly to the convergent sequences. We consider only sequences of the form $\alpha(0), \alpha(1)/2, \alpha(2)/2^2, \alpha(3)/2^3, \ldots$, which we study in the formal system through the sequences $\alpha(0), \alpha(1), \alpha(2), \alpha(3), \ldots$ of the numerators of successive fractions. We may write the convergence condition for these sequences as $(k)(Ex)(p)$ $|\alpha(x)/2^x - \alpha(x+p)/2^{x+p}| < 1/2^k$, which formalized is the right side of *R0.1. Taking "$\alpha \in R$" as an abbreviation for this formula, *R0.1 holds by IM *19. We abbreviate $\alpha \in R$ & $\beta \in R$ as "$\alpha, \beta \in R$", etc., and similarly with other uses of "$\in$" (also $\neg\alpha\in$ as "$\alpha\notin$").

*R0.1. $\vdash \alpha \in R \sim \forall k \exists x \forall p 2^k |2^p \alpha(x) - \alpha(x+p)| < 2^{x+p}$.

This formulation of the convergence criterion corresponds to Heyting's definition **1956** p. 16 of a "Cauchy sequence", except that we are using sequences of dual fractions instead of sequences of rational numbers in general, and are confining our attention to non-negative dual fractions, for reasons of convenience.

Cauchy convergence is usually written a little differently, namely, in our context, $(k)(Ex)(p)(q)|\alpha(x+p)/2^{x+p} - \alpha(x+q)/2^{x+q}| < 1/2^k$. This formalizes as the right side of *R0.2, which we abbreviate "$\alpha \in R_1$".

*R0.2. $\vdash \alpha \in R_1 \sim \forall k \exists x \forall p \forall q 2^k |2^q \alpha(x+p) - 2^p \alpha(x+q)| < 2^{x+p+q}$.

But these two formulations of Cauchy convergence are equivalent.

*R0.3. $\vdash \alpha \in R \sim \alpha \in R_1$.

Proof. I. Assume $\alpha \in R$, and after $\forall$-elim. and prior to $\exists$-elim., **(i)** $\forall p 2^{k+1} |2^p \alpha(x) - \alpha(x+p)| < 2^{x+p}$. Thence by $\forall$-elims., **(ii)** $2^{k+1} |2^p \alpha(x) - \alpha(x+p)| < 2^{x+p}$ and **(iii)** $2^{k+1} |2^q \alpha(x) - \alpha(x+q)| < 2^{x+q}$. Now $2^{k+1} |2^q \alpha(x+p) - 2^p \alpha(x+q)| \leq 2^{k+1} |2^q \alpha(x+p) - 2^{p+q} \alpha(x)| + 2^{k+1} |2^{p+q} \alpha(x) - 2^p \alpha(x+q)|$ [*11.5; *145b with *3.9] $= 2^q 2^{k+1} |2^p \alpha(x) - \alpha(x+p)| + 2^p 2^{k+1} |2^q \alpha(x) - \alpha(x+q)|$ [*11.9 with *3.3, *11.4] $< 2^q 2^{x+p} + 2^p 2^{x+q}$ [(ii), (iii), *145a, *144a, *134a] $= 2 \cdot 2^{x+p+q}$, whence by *145a, **(iv)** $2^k |2^q \alpha(x+p) - 2^p \alpha(x+q)| < 2^{x+p+q}$, and by $\forall$-, $\exists$- and $\forall$-introd., $\alpha \in R_1$. II. Use $p=0$ in $\alpha \in R_1$, and change the bound variable q to p.

We follow Heyting **1956** p. 41 in considering also the "canonical" r.n.g. (abbreviated "c.r.n.g."). These are sequences of the already-described kind with a prescribed rate of convergence given by $(x)|\alpha(x)/2^x - \alpha(x')/2^{x'}| \leq 1/2^{x'}$. This leads to the formula on the right side of *R0.4, which formula we abbreviate "$\alpha \in R'$".

*R0.4. $\vdash \alpha \in R' \sim \forall x |2\alpha(x) - \alpha(x')| \leq 1$.

Equivalent formulations are given by:

*R0.5a–c. $\vdash |2\alpha(x) - \alpha(x')| \leq 1$
$\sim [2\alpha(x) \leq \alpha(x') + 1 \;\&\; \alpha(x') \leq 2\alpha(x) + 1]$
$\sim [\alpha(x') + 1 = 2\alpha(x) \vee \alpha(x') = 2\alpha(x) \vee \alpha(x') = 2\alpha(x) + 1]$
$\sim [2\alpha(x) \dot{-} 1 \leq \alpha(x') \leq 2\alpha(x) + 1]$.

Proof. Call the formula "A $\sim$ B $\sim$ C $\sim$ D". Using *2 and *16, it will suffice to prove **(a)** $A \sim B$, **(b)** $B \supset C$, **(c)** $C \supset D$, **(d)** $D \supset B$. (a) By *11.15a. (b) B gives four cases: $(=, <)$, $(<, <)$, $(<, =)$, $(=, =)$. But $(=, =)$ is impossible, while $(=, <)$, $(<, <)$, $(<, =)$ give the three cases of C, respectively (using *138a). (c) Using *6.15, $2\alpha(x) \dot{-} 1 \leq 2\alpha(x) < 2\alpha(x)+1$. Using this, D follows in each of the three cases of C. (d) Assume D. Then $2\alpha(x) \leq (2\alpha(x) \dot{-} 1)+1$ [$^{\times}$8.1, *8.4] $\leq \alpha(x')+1$ [D, *144b]. Thus B.

The reason for our considering two species of r.n.g. simultaneously is that the approach to the theory through R has some advantage for real number arithmetic (cf. Remark 14.1), while the approach through R′ permits us to bring in the spread concept in 14.1. These two approaches give rise to species of real numbers which are identical (in the sense of Brouwer **1924–7** I pp. 245–6), as will be established formally by showing that to each element of one species there is an equal element of the other. A mapping from R′ to R is provided by *R0.7; one from R to R′ will be established in *R1.11 of 14.3 after introduction of the equality predicate for r.n.g.

*R0.6. $\vdash \alpha \in R' \supset \forall p \forall x |2^p \alpha(x) - \alpha(x+p)| < 2^p$.

*R0.7. $\vdash \alpha \in R' \supset \alpha \in R$.

Proofs. *R0.6. Assume $\alpha \in R'$. We shall deduce $\forall x |2^p \alpha(x) - \alpha(x+p)| < 2^p$ by ind. on p. Ind. step. $|2^{p'}\alpha(x) - \alpha(x+p')| \leq |2^{p'}\alpha(x) - 2^p\alpha(x')| + |2^p\alpha(x') - \alpha(x+p')| = 2^p|2\alpha(x) - \alpha(x')| + |2^p\alpha(x') - \alpha(x'+p)| < 2^p + 2^p$ [$\alpha \in R'$, hyp. ind.] $= 2^{p'}$.

*R0.7. Assume $\alpha \in R'$. Then $2^k|2^p\alpha(k) - \alpha(k+p)| < 2^{k+p}$ [*R0.6]. By $\forall$-, $\exists$- and $\forall$-introd., $\forall k \exists x \forall p 2^k |2^p \alpha(x) - \alpha(x+p)| < 2^{x+p}$, i.e. $\alpha \in R$.

§ 14. The spread representation; basic properties of the continuum.

14.1. In *R0.8 we show that the r.n.g. of the second kind constitute a spread. For the meanings of "Spr(σ)" and "α∈σ" see 6.9.

*R0.8. $\vdash \exists \sigma [\mathrm{Spr}(\sigma) \;\&\; \sigma(1)=0 \;\&\; \forall \alpha (\alpha \in \sigma \sim \alpha \in R')]$.

Proof. We introduce σ by Lemma 5.5 (c) (cf. the proof of *26.4a), using *23.5 [*23.2 with *6.3] to express a [σ(b) for b<a] in terms of $\bar{\sigma}(a)$, thus:

$$\text{(A)} \qquad \forall a \sigma(a) = \begin{cases} 0 \text{ if } a=1 \vee [\mathrm{Seq}(a) \;\&\; \mathrm{lh}(a)=1] \vee \\ \quad [\mathrm{Seq}(a) \;\&\; \mathrm{lh}(a)>1 \;\&\; \\ \quad |2((a)_{\mathrm{lh}(a) \dot{-} 2} \dot{-} 1) - ((a)_{\mathrm{lh}(a) \dot{-} 1} \dot{-} 1)| \leq 1 \;\&\; \\ \quad \sigma(\Pi_{i<\mathrm{lh}(a) \dot{-} 1} p_i^{(a)_i}) = 0], \\ 1 \text{ otherwise.} \end{cases}$$

By *22.3, *B4 (with *6.7), *22.1, *B6 and *143b (with *3.10, *18.5), $\mathrm{Seq}(a) \,\&\, \mathrm{lh}(a)>1 \supset \Pi_{i<\mathrm{lh}(a)\dot{-}1} p_i^{(a)_i} < a$.

I. $\sigma(1)=0$ is immediate.

IIa. Toward $\mathrm{Spr}(\sigma)$, $\forall a[\sigma(a)=0 \supset \mathrm{Seq}(a)]$ is immediate.

IIb. To deduce $\forall a[\sigma(a)=0 \supset \exists s\sigma(a*2^{s+1})=0]$, assume (prior to $\supset$- and $\forall$-introd.), $\sigma(a)=0$. CASE 1: $a=1$. Then $\sigma(a*2^{0+1}) = \sigma(1*2^{0+1}) = \sigma(2^1) = 0$, using *22.5, *22.7, *20.3, (A). By $\exists$-introd., $\exists s\sigma(a*2^{s+1})=0$. CASE 2: $a\neq 1$. Then by (A), $\mathrm{Seq}(a) \,\&\, \mathrm{lh}(a)\geq 1$. Let t abbreviate $a*2^{2((a)_{\mathrm{lh}(a)\dot{-}1}\dot{-}1)+1}$. Using *22.8, etc., $\mathrm{Seq}(t) \,\&\, \mathrm{lh}(t)=\mathrm{lh}(a)+1>1$. Also $|2((t)_{\mathrm{lh}(t)\dot{-}2}\dot{-}1)-((t)_{\mathrm{lh}(t)\dot{-}1}\dot{-}1)| = |2((a)_{\mathrm{lh}(a)\dot{-}1}\dot{-}1)-2((a)_{\mathrm{lh}(a)\dot{-}1}\dot{-}1)| = 0 \leq 1$, and $\sigma(\Pi_{i<\mathrm{lh}(t)\dot{-}1} p_i^{(t)_i}) = \sigma(a) = 0$. So by (A), $\sigma(t) = 0$, whence by $\exists$-introd., $\exists s\sigma(a*2^{s+1})=0$.

IIc. To deduce $\forall a[\mathrm{Seq}(a) \,\&\, \sigma(a)>0 \supset \forall s\sigma(a*2^{s+1})>0]$, assume $\mathrm{Seq}(a) \,\&\, \sigma(a)>0$. By (A), $a\neq 1$, whence $\mathrm{lh}(a*2^{s+1})>1$. So, since $\sigma(\Pi_{i<\mathrm{lh}(a*2^{s+1})\dot{-}1} p_i^{(a*2^{s+1})_i}) = \sigma(a) > 0$, (A) gives $\sigma(a*2^{s+1}) = 1 > 0$.

IIIa. Toward $\forall\alpha[\alpha\in\sigma \sim \alpha\in R']$, first assume $\alpha\in\sigma$. By $\forall$-elim., $\sigma(\bar{\alpha}(x''))=0$. Then by (A), $|2((\bar{\alpha}(x''))_x\dot{-}1)-((\bar{\alpha}(x''))_{x'}\dot{-}1)| \leq 1$, whence $|2\alpha(x)-\alpha(x')|\leq 1$, and by $\forall$-introd., $\alpha\in R'$.

IIIb. Conversely, assume $\alpha\in R'$. By ind. on x with a double basis (IM p. 193), using (A), $\sigma(\bar{\alpha}(x))=0$. By $\forall$-introd., $\alpha\in\sigma$.

14.2. In *R0.9, p, x_1, x are distinct variables, and A(p) is a formula in which x_1 and x are free for p.

*R0.9. $\vdash \forall p A(x_1+p) \,\&\, x_1\leq x \supset \forall p A(x+p)$.

*R0.10. $\vdash \forall p\forall q 2^k|2^q\alpha(x_1+p)-2^p\alpha(x_1+q)|<2^{x_1+p+q} \,\&\, x_1\leq x \supset$
$\forall p\forall q 2^k|2^q\alpha(x+p)-2^p\alpha(x+q)|<2^{x+p+q}$.

*R0.11. $\vdash \forall p 2^{k+1}|2^p\alpha(x_1)-\alpha(x_1+p)|<2^{x_1+p} \,\&\, x_1\leq x \supset$
$\forall p 2^k|2^p\alpha(x)-\alpha(x+p)|<2^{x+p}$.

PROOFS. *R0.9. Assume $\forall p A(x_1+p) \,\&\, x_1\leq x$. Assume $x=x_1+c$ (cf. 5.5 ¶ 4). By $\forall$-elim., $A(x_1+c+p)$, whence $A(x+p)$. By $\forall$-introd., $\forall p A(x+p)$.

*R0.10. Assume the antecedent and $x=x_1+c$. Using $\forall$-elim. with c+p and c+q for p and q, $2^k|2^{c+q}\alpha(x_1+c+p)-2^{c+p}\alpha(x_1+c+q)| < 2^{x_1+c+p+c+q}$. Dividing by 2^c (i.e. using *145a with *11.9 and *3.9), and replacing x_1+c by x, $2^k|2^q\alpha(x+p)-2^p\alpha(x+q)| < 2^{x+p+q}$.

*R0.11. Assume the antecedent. The steps from (i) to (iv) in the proof of *R0.3 and $\forall$-introd. give $\forall p\forall q 2^k|2^q\alpha(x_1+p)-2^p\alpha(x_1+q)| < 2^{x_1+p+q}$. So by *R0.10, $\forall p\forall q 2^k|2^q\alpha(x+p)-2^p\alpha(x+q)| < 2^{x+p+q}$, whence putting p=0, $\forall p 2^k|2^p\alpha(x)-\alpha(x+p)| < 2^{x+p}$.

14.3. The coincidence predicate for r.n.g. (Heyting **1956** p. 16) is expressed by the formula on the right in *R1.1, which formula we abbreviate as "$\alpha \doteq \beta$". In *R1.3 $\alpha=\beta$ is $\forall x(\alpha(x)=\beta(x))$ (cf. 4.5).

*R1.1. $\vdash \alpha \doteq \beta \sim \forall k \exists x \forall p 2^k|\alpha(x+p)-\beta(x+p)|<2^{x+p}$.

*R1.2. $\vdash \forall p \alpha(x+p)=\beta(x+p) \supset \alpha \doteq \beta$;
$\vdash \forall t_{t \geq x} \alpha(t)=\beta(t) \supset \alpha \doteq \beta$.

*R1.3. $\vdash \alpha=\beta \supset \alpha \doteq \beta$. *R1.4. $\vdash \alpha \doteq \alpha$.

*R1.5. $\vdash \alpha \doteq \beta \supset \beta \doteq \alpha$. *R1.6. $\vdash \alpha \doteq \beta \ \& \ \beta \doteq \gamma \supset \alpha \doteq \gamma$.

*R1.7. $\vdash \alpha \in R \ \& \ \alpha \doteq \beta \supset \beta \in R$.

PROOFS. *R1.2. Assume $\forall t_{t \geq x} \alpha(t)=\beta(t)$. Then $\alpha(x+p)=\beta(x+p)$, so $2^k|\alpha(x+p)-\beta(x+p)| = 0$ [*11.2] $< 2^{x+p}$, whence by $\forall$-, $\exists$- and $\forall$-introd., $\alpha \doteq \beta$.

*R1.4. From $\alpha=\alpha$ (4.5 ¶ 5) by *R1.3.

*R1.5. Use *11.4.

*R1.6. Assume $\alpha \doteq \beta$, $\beta \doteq \gamma$. By $\forall$-elims., and omitting $\exists x_1$, $\exists x_2$ preceding $\exists$-elims.: $\forall p 2^{k+1}|\alpha(x_1+p)-\beta(x_1+p)|<2^{x_1+p}$, $\forall p 2^{k+1}|\beta(x_2+p)-\gamma(x_2+p)|<2^{x_2+p}$. Letting $x=\max(x_1, x_2)$ and using *R0.9, *8.4: $\forall p 2^{k+1}|\alpha(x+p)-\beta(x+p)|<2^{x+p}$, $\forall p 2^{k+1}|\beta(x+p)-\gamma(x+p)|<2^{x+p}$. Now $2^{k+1}|\alpha(x+p)-\gamma(x+p)| \leq 2^{k+1}|\alpha(x+p)-\beta(x+p)|+2^{k+1}|\beta(x+p)-\gamma(x+p)|$ [*11.5, etc.] $< 2^{x+p}+2^{x+p} = 2 \cdot 2^{x+p}$. So $2^k|\alpha(x+p)-\gamma(x+p)|<2^{x+p}$. By $\forall$-, $\exists$- and $\forall$-introd., $\alpha \doteq \gamma$.

*R1.7. Assume $\alpha \in R$, $\alpha \doteq \beta$. By $\forall$-elims., and preceding $\exists$-elims.: $\forall p 2^{k+3}|2^p\alpha(x_1)-\alpha(x_1+p)|<2^{x_1+p}$, $\forall p 2^{k+2}|\alpha(x_2+p)-\beta(x_2+p)|<2^{x_2+p}$. Letting $x=\max(x_1, x_2)$, and using *R0.11 and *R0.9, with *8.4: **(i)** $\forall p 2^{k+2}|2^p\alpha(x)-\alpha(x+p)|<2^{x+p}$, **(ii)** $\forall p 2^{k+2}|\alpha(x+p)-\beta(x+p)|<2^{x+p}$, whence ($\forall$-elim. with 0 for p, *145a, etc.) **(iii)** $2^{k+2}|2^p\beta(x)-2^p\alpha(x)| < 2^{x+p}$. Now $2^{k+2}|2^p\beta(x)-\beta(x+p)| \leq 2^{k+2}|2^p\beta(x)-2^p\alpha(x)|+ 2^{k+2}|2^p\alpha(x)-\alpha(x+p)|+2^{k+2}|\alpha(x+p)-\beta(x+p)| < 2^{x+p}+2^{x+p}+2^{x+p}$ [(iii), (i), (ii)] $< 2^{x+p+2}$. So $2^k|2^p\beta(x)-\beta(x+p)| < 2^{x+p}$. By $\forall$-, $\exists$- and $\forall$-introd., $\beta \in R$.

*R1.8. $\vdash \exists b 2^{y+1}|2^y a-2^x b| \leq 2^{y+x}$.

*R1.9. $\vdash 2^{y+3}|2^p a-d|<2^{x+p} \ \& \ 2^{y+1}|2^y a-2^x b| \leq 2^{y+x} \supset$
$2^{y+3}|2^y d-2^{x+p}b|<2^{x+p+y}5$.

*R1.10. $\vdash \alpha, \gamma \in R \ \& \ \forall p 2^{z+4}|\alpha(w+p)-\gamma(w+p)|<2^{w+p} \supset$
$\exists \alpha'_{\alpha' \in R} \exists \gamma'_{\gamma' \in R}(\alpha' \doteq \alpha \ \& \ \gamma' \doteq \gamma \ \& \ \forall x_{x \leq z} \alpha'(x)=\gamma'(x))$.

*R1.11. $\vdash \alpha \in R \sim \exists \alpha'_{\alpha' \in R} \alpha' \doteq \alpha$.

PROOFS. *R1.8. CASE 1: $y \leq x$. Assume $y+c=x$, $a=2^c q+r$ & $r<2^c$ (using *146a). SUBCASE 1.1: $2^{y+1}|2^y a-2^x q| \leq 2^{y+x}$. Use $\exists$-introd. SUBCASE 1.2: $2^{y+x} < 2^{y+1}|2^y a-2^x q| = 2^{y+1}|2^y(2^c q+r)-2^{y+c}q| = 2^{y+1}(2^y r)$ [*11.8]. So assume $2^{y+x}+d' = 2^{y+1}(2^y r)$. Then $2^{y+1}|2^y a- 2^x(q+1)| = 2^{y+1}|2^y(2^c q+r)-2^{y+c}(q+1)| = 2^{y+1}|2^y r-2^{y+c}|$ [*11.7] $= 2^{y+1}(2^x \dot{-} 2^y r)$ [$r<2^c$, *11.1, *6.11] $= 2^{y+1+x} \dot{-} 2^{y+1}(2^y r)$ [*6.14] $= (2^{y+x}+2^{y+x}) \dot{-} (2^{y+x}+d') = 2^{y+x} \dot{-} d'$ [*6.8] $\leq 2^{y+x}$ [*6.15]. Use $\exists$-introd. CASE 2: $y>x$. Then $2^{y+1}|2^y a-2^x 2^{y \dot{-} x}a| = 0$ [*6.7, *11.2] $\leq 2^{y+x}$.

*R1.9. Assume **(i)** $2^{y+3}|2^p a-d|<2^{x+p}$ and **(ii)** $2^{y+1}|2^y a-2^x b| \leq 2^{y+x}$. Then $2^{y+3}|2^{y+x}d-2^{2x+p}b| \leq 2^{y+3}|2^{y+x}d-2^{y+x+p}a| + 2^{y+3}|2^{y+x+p}a-2^{2x+p}b| = 2^{y+x}2^{y+3}|2^p a-d|+2^{x+p+2}2^{y+1}|2^y a-2^x b| < 2^{y+x}2^{x+p}+2^{x+p+2}2^{y+x}$ [(i), (ii)] $= 2^{2x+p+y}5$. Thus $2^{y+3}|2^y d-2^{x+p}b| < 2^{x+p+y}5$.

*R1.10. Assume $\alpha, \gamma \in R$ and **(i)** $\forall p 2^{z+4}|\alpha(w+p)-\gamma(w+p)|<2^{w+p}$. First we shall deduce **(a)** $\exists b \exists c\{\exists x \forall p 2^{y+3}|2^y\alpha(x+p)-2^{x+p}b|<2^{x+p+y}5$ & $\exists x \forall p 2^{y+3}|2^y\gamma(x+p)-2^{x+p}c|<2^{x+p+y}5$ & $(y \leq z \supset b=c)\}$. CASE 1: $y \leq z$. Using $\alpha \in R$, assume prior to $\exists$-elim.: **(ii)** $\forall p 2^{z+5}|2^p\alpha(x_1)-\alpha(x_1+p)|< 2^{x_1+p}$. Letting $x=\max(w, x_1)$, and using (i) and (ii) with *R0.9 and *R0.11: **(iii)** $\forall p 2^{z+4}|\alpha(x+p)-\gamma(x+p)|<2^{x+p}$, **(iv)** $\forall p 2^{z+4}|2^p\alpha(x)- \alpha(x+p)|<2^{x+p}$. Using *11.5 (and *145a), $\forall p 2^{z+3}|2^p\alpha(x)-\gamma(x+p)|< 2^{x+p}$. Thence using the case hyp., *144b, *3.12, and $a \leq b \supset ac \leq bc$ [5.5 ¶ 4]: **(v)** $\forall p 2^{y+3}|2^p\alpha(x)-\gamma(x+p)|<2^{x+p}$. Similarly from (iv): **(vi)** $\forall p 2^{y+3}|2^p\alpha(x)-\alpha(x+p)|<2^{x+p}$. Using *R1.8, assume prior to $\exists$-elim.: **(vii)** $2^{y+1}|2^y\alpha(x)-2^x b| \leq 2^{y+x}$. By *R1.9, with (v) (or (vi)) and (vii): **(viii)** $2^{y+3}|2^y\gamma(x+p)-2^{x+p}b|<2^{x+p+y}5$, **(ix)** $2^{y+3}|2^y\alpha(x+p) -2^{x+p}b|<2^{x+p+y}5$. By *11, *100: **(x)** $y \leq z \supset b=b$. From (ix), (viii), (x) by $\forall$-, $\exists$- and &-introds., (a). CASE 2: $y>z$. Using $\alpha, \gamma \in R$, assume (prior to $\exists$-elim.) formulas from which by *R0.11: **(xi)** $\forall p 2^{y+3}|2^p\alpha(x)- \alpha(x+p)|<2^{x+p}$, **(xii)** $\forall p 2^{y+3}|2^p\gamma(x)-\gamma(x+p)|<2^{x+p}$. Using *R1.8, assume (prior to $\exists$-elims.): **(xiii)** $2^{y+1}|2^y\alpha(x)-2^x b| \leq 2^{y+x}$, **(xiv)** $2^{y+1}|2^y\gamma(x)-2^x c| \leq 2^{y+x}$. By (xi), (xiii) (or (xii), (xiv)) and *R1.9: $2^{y+3}|2^y\alpha(x+p)-2^{x+p}b|<2^{x+p+y}5$, $2^{y+3}|2^y\gamma(x+p)-2^{x+p}c|<2^{x+p+y}5$. Also, using case hyp., $y \leq z \supset b=c$. By $\forall$-, $\exists$- and &-introds., (a). Now from (a) by *2.2 (twice), $\exists\alpha'\exists\gamma'\{\forall y \exists x \forall p 2^{y+3}|2^y\alpha(x+p)-2^{x+p}\alpha'(y)|< 2^{x+p+y}5$ & $\forall y \exists x \forall p 2^{y+3}|2^y\gamma(x+p)-2^{x+p}\gamma'(y)|<2^{x+p+y}5$ & $\forall y_{y \leq z}\alpha'(y)= \gamma'(y)\}$. Prior to $\exists$-elims. from this, we assume a formula from which by &-elims: **(xv)** $\forall y \exists x \forall p 2^{y+3}|2^y\alpha(x+p)-2^{x+p}\alpha'(y)|<2^{x+p+y}5$, **(xvi)** $\forall y \exists x \forall p 2^{y+3}|2^y\gamma(x+p)-2^{x+p}\gamma'(y)|<2^{x+p+y}5$, **(xvii)** $\forall y_{y \leq z}\alpha'(y)=\gamma'(y)$.

From (xv) we deduce $\alpha'\in R'$, thus. After ∀-elims. (with y and y+1 for y) and preceding ∃-elims. we assume formulas from which by *R0.9 (and ∀-elim.): **(xviii)** $2^{y+3}|2^y\alpha(x+p)-2^{x+p}\alpha'(y)|<2^{x+p+y}5$ and **(xix)** $2^{y+4}|2^{y+1}\alpha(x+p)-2^{x+p}\alpha'(y+1)|<2^{x+p+y+1}5$. Now $2^{2y+x+p+4}|2\alpha'(y)-\alpha'(y+1)| = 2^{2y+4}|2^{x+p+1}\alpha'(y)-2^{x+p}\alpha'(y+1)| \le 2^{2y+4}|2^{x+p+1}\alpha'(y)-2^{y+1}\alpha(x+p)|+2^{2y+4}|2^{y+1}\alpha(x+p)-2^{x+p}\alpha'(y+1)| = 2^{y+2}2^{y+3}|2^y\alpha(x+p)-2^{x+p}\alpha'(y)|+2^y2^{y+4}|2^{y+1}\alpha(x+p)-2^{x+p}\alpha'(y+1)| < 2^{y+2}2^{x+p+y}5+2^y2^{x+p+y+1}5$ [(xviii), (xix)] $= 2^{2y+x+p+1}15$. Thence $8|2\alpha'(y)-\alpha'(y+1)|<15$, whence (by contradicting $|2\alpha'(y)-\alpha'(y+1)|>1$), $|2\alpha'(y)-\alpha'(y+1)|\le 1$. Thence **(xx)** $\alpha'\in R'$. Similarly from (xvi): **(xxi)** $\gamma'\in R'$. Toward deducing $\alpha'\doteq\alpha$ from (xv) and (xx), assume from (xv), after ∀-elim. (with $t=13\cdot 2^k$ for y) and prior to ∃-elim., $\forall p 2^{t+3}|2^t\alpha(x_1+p)-2^{x_1+p}\alpha'(t)|<2^{x_1+p+t}5$, whence by *R0.9, letting $x=\max(x_1, t)$, **(xxii)** $2^{t+3}|2^t\alpha(x+p)-2^{x+p}\alpha'(t)|<2^{x+p+t}5$. Assume (prior to ∃-elim.): **(xxiii)** $x=t+c$. Now $t2^t|\alpha(x+p)-\alpha'(x+p)| \le 2^t2^{t+3}|\alpha(x+p)-\alpha'(x+p)|$ [*3.10; $a\le b \supset ac\le bc$] $= 2^{t+3}|2^t\alpha(x+p)-2^t\alpha'(x+p)| \le 2^{t+3}|2^t\alpha(x+p)-2^{x+p}\alpha'(t)|+2^{t+3}|2^{x+p}\alpha'(t)-2^t\alpha'(x+p)| < 2^{x+p+t}5+2^{t+3}2^t|2^{c+p}\alpha'(t)-\alpha'(t+c+p)|$ [(xxii), (xxiii)] $< 2^{x+p+t}5+2^{t+3}2^t2^{c+p}$ [*R0.6, (xx)] $= 2^{x+p}2^t13$. Thus $2^k|\alpha(x+p)-\alpha'(x+p)|<2^{x+p}$. By ∀-, ∃- and ∀-introd.: **(xxiv)** $\alpha'\doteq\alpha$. Similarly, from (xvi) and (xxi): **(xxv)** $\gamma'\doteq\gamma$. Combining (xx), (xxi), (xxiv), (xxv), (xvii): $\exists\alpha'_{\alpha'\in R'}\exists\gamma'_{\gamma'\in R'}(\alpha'\doteq\alpha \;\&\; \gamma'\doteq\gamma \;\&\; \forall x_{x\le z}\alpha'(x)=\gamma'(x))$.

*R1.11. I. Assume $\alpha\in R$. Use *R1.10 with α, α for α, γ. II. Use *R0.7, *R1.7.

The law of the excluded middle does not hold for equality $\doteq$ of r.n.g. α and β; indeed, when it is generalized on one of the variables, say α, we can refute it, using Brouwer's principle *27.6. (Cf. Brouwer **1928a** p. 7 item 1, with **1929** p. 161; **1954** p. 5 lines 7–9.) This we shall do in *R9.23.

However, the law of double negation does hold for $\alpha \doteq \beta$ (cf. **1928a** p. 8 lines 23–35, **1954** p. 5 lines 6–7, Heyting **1956** p. 17); this we shall include in *R2.8.

14.4. We represent the apartness relation between r.n.g. (Brouwer **1954** p. 4 lines 13–14, Heyting **1956** p. 19) by the formula on the right in *R2.1, abbreviated "$\alpha\#\beta$". *R2.3, *R2.4, *R2.6, *R2.7 are from Heyting **1956** p. 20. Kleene shows in 18.2 of Chapter IV below that the converse of *R2.5 is not provable and (by classical reasoning) not refutable.

*R2.1. $\vdash \alpha \# \beta \sim \exists k \exists x \forall p 2^k|\alpha(x+p)-\beta(x+p)| \geq 2^{x+p}$.

*R2.2. $\vdash \neg \alpha \# \alpha$. *R2.3. $\vdash \alpha \# \beta \supset \beta \# \alpha$.

*R2.4. $\vdash \alpha \doteq \beta \ \& \ \alpha \# \gamma \supset \beta \# \gamma$. *R2.5. $\vdash \alpha \# \beta \supset \neg \alpha \doteq \beta$.

*R2.6. $\vdash \alpha, \beta, \gamma \in R \ \& \ \alpha \# \beta \supset \alpha \# \gamma \vee \beta \# \gamma$.

*R2.7. $\vdash \alpha, \beta \in R \ \& \ \neg \alpha \# \beta \supset \alpha \doteq \beta$.

*R2.8. $\vdash \alpha, \beta \in R \supset (\neg\neg \alpha \doteq \beta \sim \neg \alpha \# \beta \sim \alpha \doteq \beta)$.

Proofs. *R2.4. Assume $\alpha \doteq \beta$, $\alpha \# \gamma$. Prior to $\exists$-elims., assume $\forall p 2^k|\alpha(x_1+p)-\gamma(x_1+p)| \geq 2^{x_1+p}$; and after $\forall$-elim. and prior to $\exists$-elim., assume $\forall p 2^{k+1}|\alpha(x_2+p)-\beta(x_2+p)| < 2^{x_2+p}$. Letting $x = \max(x_1, x_2)$ and using *R0.9, $\forall$-elim. and *145b: **(i)** $2^{k+1}|\alpha(x+p)-\gamma(x+p)| \geq 2 \cdot 2^{x+p}$ and **(ii)** $2^{k+1}|\alpha(x+p)-\beta(x+p)| < 2^{x+p}$. Now $2^{k+1}|\beta(x+p)-\gamma(x+p)| \geq 2^{x+p}$, since otherwise $2^{k+1}|\alpha(x+p)-\gamma(x+p)| \leq 2^{k+1}|\alpha(x+p)-\beta(x+p)|+2^{k+1}|\beta(x+p)-\gamma(x+p)| < 2^{x+p}+2^{x+p}$ [using (ii)] $= 2 \cdot 2^{x+p}$, contradicting (i). By $\forall$- and $\exists$-introds., $\beta \# \gamma$.

*R2.5. By *R2.2 and *R2.4.

*R2.6. Assume $\alpha, \beta, \gamma \in R$ and $\alpha \# \beta$. Prior to $\exists$-elims. from $\alpha \# \beta$, and after &- and $\forall$-elims. and prior to $\exists$-elims. from $\alpha, \beta, \gamma \in R$, assume formulas from which, using *R0.9 (and *145b) and *R0.11: **(i)** $2^{k+3}|2^p\alpha(x)-2^p\beta(x)| \geq 8 \cdot 2^{x+p}$, **(ii)** $2^{k+3}|2^p\alpha(x)-\alpha(x+p)| < 2^{x+p}$, **(iii)** $2^{k+3}|2^p\beta(x)-\beta(x+p)| < 2^{x+p}$, **(iv)** $2^{k+3}|2^p\gamma(x)-\gamma(x+p)| < 2^{x+p}$. Now $2^{k+3}|2^p\alpha(x)-2^p\gamma(x)| \geq 4 \cdot 2^{x+p} \vee 2^{k+3}|2^p\beta(x)-2^p\gamma(x)| \geq 4 \cdot 2^{x+p}$, since otherwise $2^{k+3}|2^p\alpha(x)-2^p\beta(x)| \leq 2^{k+3}|2^p\alpha(x)-2^p\gamma(x)|+2^{k+3}|2^p\gamma(x)-2^p\beta(x)| < 4 \cdot 2^{x+p}+4 \cdot 2^{x+p} = 8 \cdot 2^{x+p}$, contradicting (i). Case 1: $2^{k+3}|2^p\alpha(x)-2^p\gamma(x)| \geq 4 \cdot 2^{x+p}$. Then **(a)** $2^{k+3}|\alpha(x+p)-\gamma(x+p)| \geq 2 \cdot 2^{x+p}$, since otherwise $2^{k+3}|2^p\alpha(x)-2^p\gamma(x)| \leq 2^{k+3}|2^p\alpha(x)-\alpha(x+p)|+2^{k+3}|\alpha(x+p)-\gamma(x+p)|+2^{k+3}|\gamma(x+p)-2^p\gamma(x)| < 2^{x+p}+2 \cdot 2^{x+p}+2^{x+p}$ [using (ii), (iv)] $= 4 \cdot 2^{x+p}$, contradicting case hyp. From (a) by *145b, $2^{k+2}|\alpha(x+p)-\gamma(x+p)| \geq 2^{x+p}$, whence by $\forall$- and $\exists$-introds., $\alpha \# \gamma$, and by $\vee$-introd., $\alpha \# \gamma \vee \beta \# \gamma$. Case 2: $2^{k+3}|2^p\beta(x)-2^p\gamma(x)| \geq 4 \cdot 2^{x+p}$. Similarly, using (iii) and (iv), $\beta \# \gamma$, and by $\vee$-introd., $\alpha \# \gamma \vee \beta \# \gamma$.

*R2.7. Assume $\alpha, \beta \in R$ and $\neg \alpha \# \beta$. After $\forall$-elims., and prior to $\exists$-elims., from $\alpha \in R$ and $\beta \in R$, we assume formulas from which by *R0.11: **(i)** $\forall p 2^{k+3}|2^p\alpha(x)-\alpha(x+p)| < 2^{x+p}$, **(ii)** $\forall p 2^{k+3}|2^p\beta(x)-\beta(x+p)| < 2^{x+p}$. For reductio ad absurdum, assume $2^{k+1}|\alpha(x)-\beta(x)| \geq 2^x$. Now $4 \cdot 2^{x+p} \leq 2^{k+3}|2^p\alpha(x)-2^p\beta(x)| \leq 2^{k+3}|2^p\alpha(x)-\alpha(x+p)|+2^{k+3}|\alpha(x+p)-\beta(x+p)|+2^{k+3}|\beta(x+p)-2^p\beta(x)| \leq 2^{x+p}+2^{k+3}|\alpha(x+p)-\beta(x+p)|+2^{x+p}$ [(i), (ii)] $\leq 2 \cdot 2^{x+p}+2^{k+3}|\alpha(x+p)-\beta(x+p)|$. Hence $2^{x+p} \leq 2^{k+2}|\alpha(x+p)-\beta(x+p)|$. By $\forall$- and $\exists$-introds., $\alpha \# \beta$, contradicting $\neg\alpha\#\beta$. Thus **(iii)** $2^{k+1}|\alpha(x)-\beta(x)|<2^{x}$. So $2^{k+3}|\alpha(x+p)-\beta(x+p)| \leq 2^{k+3}|\alpha(x+p)-2^{p}\alpha(x)|+2^{k+3}|2^{p}\alpha(x)-2^{p}\beta(x)|+2^{k+3}|2^{p}\beta(x)-\beta(x+p)| < 2^{x+p}+4\cdot 2^{x+p}+2^{x+p}$ [(i), (iii), (ii)] $= 6\cdot 2^{x+p} < 8\cdot 2^{x+p}$. So $2^{k}|\alpha(x+p)-\beta(x+p)|<2^{x+p}$. By $\forall$-, $\exists$- and $\forall$-introd., $\alpha \doteq \beta$.

*R2.8. Write it "$\alpha, \beta \epsilon R \supset (A \sim B \sim C)$". Assume $\alpha, \beta \epsilon R$. (a) $A \supset B$ by *R2.5, *12. (b) $B \supset C$ by *R2.7. (c) $C \supset A$ by *49a.

14.5. Of the operations of arithmetic we introduce only addition $\alpha+\beta$ and subtraction $\alpha \dot{-} \beta$, $|\alpha-\beta|$. There would be no difficulty in treating multiplication and division if required (cf. Heyting **1956** p. 21).

Our new axioms are of the first form provided in 5.1. In *R3.3, *R3.4, *R4.3–*R4.5, *R5.3–*R5.6 we have analogues to *117, *119, *6.3, *6.5, *6.8, ×11.1, *11.4, *11.7, *11.8, respectively (thus including all equalities from IM Theorem 25 and above ×6.1–*6.21, ×11.1–*11.15b not involving 0, ′ or multiplication). Each of these follows by $\forall$-introd. from (a substitution instance of) the corresponding theorem of IM or 5.5 above, and by *R1.3 each has a version with $=$ replaced by $\doteq$.

×R3.1. $(\alpha+\beta)(x)=\alpha(x)+\beta(x)$.
*R3.2. $\vdash \alpha, \beta \epsilon R \supset \alpha+\beta \epsilon R$.
*R3.3. $\vdash \alpha+(\beta+\gamma)=(\alpha+\beta)+\gamma$. *R3.4. $\vdash \alpha+\beta=\beta+\alpha$.
*R3.5. $\vdash \alpha \doteq \beta \sim \alpha+\gamma \doteq \beta+\gamma$. *R3.6. $\vdash \alpha \# \beta \sim \alpha+\gamma \# \beta+\gamma$.
×R4.1. $(\alpha \dot{-} \beta)(x)=\alpha(x) \dot{-} \beta(x)$.
*R4.2. $\vdash \alpha, \beta \epsilon R \supset \alpha \dot{-} \beta \epsilon R$.
*R4.3. $\vdash (\alpha+\beta) \dot{-} \beta=\alpha$. *R4.4. $\vdash \alpha \dot{-} (\beta+\gamma)=(\alpha \dot{-} \beta) \dot{-} \gamma$.
*R4.5. $\vdash (\alpha+\gamma) \dot{-} (\beta+\gamma)=\alpha \dot{-} \beta$.
*R4.6. $\vdash \alpha \doteq \beta \supset \alpha \dot{-} \gamma \doteq \beta \dot{-} \gamma$. *R4.7. $\vdash \alpha \doteq \beta \supset \gamma \dot{-} \alpha \doteq \gamma \dot{-} \beta$.
×R5.1. $(|\alpha-\beta|)(x)=|\alpha(x)-\beta(x)|$.
*R5.2. $\vdash \alpha, \beta \epsilon R \supset |\alpha-\beta| \epsilon R$.
*R5.3. $\vdash |\alpha-\beta|=(\alpha \dot{-} \beta)+(\beta \dot{-} \alpha)$. *R5.4. $\vdash |\alpha-\beta|=|\beta-\alpha|$.
*R5.5. $\vdash |(\alpha+\gamma)-(\beta+\gamma)|=|\alpha-\beta|$. *R5.6. $\vdash |(\alpha+\beta)-\beta|=\alpha$.
*R5.7. $\vdash \alpha \doteq \beta \supset |\alpha-\gamma| \doteq |\beta-\gamma|$.

Proofs. *R3.2. Use *R0.11, *11.5, *11.7.

*R3.5–*R3.6. Use *11.7.

Remark 14.1. As counterexample to *R3.2 with R replaced by R′, $\vdash \lambda x 1 \epsilon R'$ & $\neg \lambda x 1+\lambda x 1 \epsilon R'$. Similarly, counterexamples to *R4.2 and *R5.2 with R′ replacing R are obtained by taking α to be $\lambda x 2^{x'} \dot{-} 1$ and β to be $\lambda x 1$.

14.6. We shall study two ordering predicates for the continuum which are important intuitionistically: the "natural ordering" or "measurable natural ordering" $<\circ$ (Brouwer **1928a** p. 8, **1951**; called the "pseudo-ordering" and written "$<$" in Heyting **1956** pp. 107, 25), and the "virtual ordering" $\lessdot$ (written "$<$" in Brouwer **1928a** p. 9, **1951**, and "$\lessdot$" in Heyting **1956** p. 107).

Informally the natural ordering predicate $\alpha <\circ \beta$ is $(Ex)(Ek)(p)\beta(x+p)/2^{x+p} \dot{-} \alpha(x+p)/2^{x+p} \geq 1/2^k$. We express this formally by the right side of *R6.1, abbreviated "$\alpha <\circ \beta$" (read "α measurably less than β"; cf. Brouwer **1951**) or "$\beta \circ> \alpha$". We write "$\alpha \leq\circ \beta$" or "$\beta \circ\geq \alpha$" for $\alpha <\circ \beta \vee \alpha \doteq \beta$; "$\alpha \circ> \beta, \gamma$" for $\alpha \circ> \beta \;\&\; \alpha \circ> \gamma$, etc.

Analogously to "$\alpha = \beta$" for $\forall x(\alpha(x) = \beta(x))$, we adopt the abbreviations "$\alpha \leq \beta$" and "$\beta \geq \alpha$" for $\forall x(\alpha(x) \leq \beta(x))$; since we shall avoid employing the corresponding abbreviations "$\alpha < \beta$" and "$\beta > \alpha$" for $\forall x(\alpha(x) < \beta(x))$, which are of no separate use here, there will be no occasion to confuse $\alpha \leq \beta$ and $\beta \geq \alpha$ with the disjunction $\alpha < \beta \vee \alpha = \beta$.

We furthermore abbreviate $\neg \alpha <\circ \beta$ by "$\alpha \not<\circ \beta$" and "$\beta \circ\not> \alpha$". In number-theory there was no gain from such an abbreviation since $\neg a < b$ is equivalent to $a \geq b$ and to $b \leq a$.

Using *R1.3–*R1.6 we are justified in constructing chains of functors linked by $=$, $\doteq$. Now using in addition *R6.4, *R6.6–*R6.8, *R6.10–*R6.15 below we may construct chains of r.n.g. linked by $=$, $\doteq$, $<\circ$, $\leq$, $\leq\circ$, $\circ\not>$ (or by $=$, $\doteq$, $\circ>$, $\geq$, $\circ\geq$, $\not<\circ$). Cf. IM end § 26.

Thus constructing an ascending chain from α to β establishes in any case that $\alpha \circ\not> \beta$ (if $\leq$ and $\circ\not>$ are not used, that $\alpha \leq\circ \beta$; if $<\circ$ is used, that $\alpha <\circ \beta$). Such a chain appears in the proof of *R9.19; others will be used in 15.3 ff.

Various inequality formulas in arithmetic have analogues in the natural ordering of the continuum, though we cannot derive these analogues in quite as uniform a manner as we did the analogues of equalities in 14.5. The proofs of *R6.16–*R6.20 provide examples.

Most of the (next) theorems are from Heyting **1956** pp. 25–26.

*R6.1. $\vdash \alpha <\circ \beta \sim \exists k \exists x \forall p 2^k(\beta(x+p) \dot{-} \alpha(x+p)) \geq 2^{x+p}$.

*R6.2. $\vdash \alpha, \beta \in R \;\&\; \alpha \# \beta \supset \alpha <\circ \beta \vee \beta <\circ \alpha$.

*R6.3. $\vdash \alpha <\circ \beta \supset \alpha \# \beta$. *R6.4. $\vdash \alpha <\circ \beta \supset \neg \alpha \doteq \beta$.

*R6.5. $\vdash \alpha, \beta \in R \;\&\; \alpha \not<\circ \beta \;\&\; \beta \not<\circ \alpha \supset \alpha \doteq \beta$.

*R6.6. $\vdash \alpha <\circ \beta \;\&\; \beta <\circ \gamma \supset \alpha <\circ \gamma$.

*R6.7. $\vdash \alpha <\circ \beta \supset \beta \not<\circ \alpha$. *R6.8. $\vdash \alpha \not<\circ \alpha$.

*R6.9. $\vdash \alpha, \beta, \gamma \in R \;\&\; \alpha <\circ \beta \supset \alpha <\circ \gamma \vee \gamma <\circ \beta$.

*R6.10. $\vdash \alpha, \beta, \gamma \in R \ \& \ \alpha \circ\!\not> \beta \ \& \ \beta <\!\circ \gamma \supset \alpha <\!\circ \gamma$.
*R6.11. $\vdash \alpha, \beta, \gamma \in R \ \& \ \alpha <\!\circ \beta \ \& \ \beta \circ\!\not> \gamma \supset \alpha <\!\circ \gamma$.
*R6.12. $\vdash \alpha, \beta, \gamma \in R \ \& \ \alpha \circ\!\not> \beta \ \& \ \beta \circ\!\not> \gamma \supset \alpha \circ\!\not> \gamma$.
*R6.13. $\vdash \alpha \doteq \beta \ \& \ \alpha <\!\circ \gamma \supset \beta <\!\circ \gamma$.
*R6.14. $\vdash \alpha \doteq \beta \ \& \ \gamma <\!\circ \alpha \supset \gamma <\!\circ \beta$.
*R6.15. $\vdash \alpha \leq \beta \supset \alpha \circ\!\not> \beta$. *R6.16. $\vdash \alpha <\!\circ \beta \sim \alpha + \gamma <\!\circ \beta + \gamma$.
*R6.17. $\vdash |\alpha - \gamma| \circ\!\not> |\alpha - \beta| + |\beta - \gamma|$.
*R6.18. $\vdash \beta \circ\!\geq \gamma \supset (\alpha + \beta) \dot{-} \gamma \doteq \alpha + (\beta \dot{-} \gamma)$.
*R6.19. $\vdash \alpha \circ\!> \beta, \gamma \sim \alpha \dot{-} \gamma \circ\!> \beta \dot{-} \gamma$.
*R6.20. $\vdash \alpha <\!\circ \beta, \gamma \sim \gamma \dot{-} \alpha \circ\!> \gamma \dot{-} \beta$.

Proofs. *R6.2. Assume $\alpha, \beta \in R$ and $\alpha \# \beta$. Prior to $\exists$-elims. and after $\forall$-elims. assume formulas from which, using *R0.9 and *R0.11: **(i)** $\forall p 2^k |\alpha(x+p) - \beta(x+p)| \geq 2^{x+p}$, **(ii)** $\forall p 2^{k+2} |2^p \alpha(x) - \alpha(x+p)| < 2^{x+p}$, **(iii)** $\forall p 2^{k+2} |2^p \beta(x) - \beta(x+p)| < 2^{x+p}$. By $\forall$-elim. from (i) with 0 for p (and *145b), $|2^{k+2+p}\alpha(x) - 2^{k+2+p}\beta(x)| \geq 4 \cdot 2^{x+p}$. Case 1: $\alpha(x) \leq \beta(x)$. Then by *11.14, **(i′)** $2^{k+2+p}\beta(x) \geq 2^{k+2+p}\alpha(x) + 4 \cdot 2^{x+p}$. But by (ii) and (iii) with *11.15: **(ii′)** $2^{k+2}\alpha(x+p) < 2^{k+2+p}\alpha(x) + 2^{x+p}$, **(iii′)** $2^{k+2+p}\beta(x) < 2^{k+2}\beta(x+p) + 2^{x+p}$. Now $2^{k+2}\beta(x+p) \geq 2^{k+2}\alpha(x+p) + 2 \cdot 2^{x+p}$, or using (ii′) and (iii′) we could contradict (i′). Hence $2^{k+1}\beta(x+p) \dot{-} 2^{k+1}\alpha(x+p) \geq 2^{x+p}$. By $\forall$- and $\exists$-introds., $\alpha <\!\circ \beta$, and by $\vee$-introd., $\alpha <\!\circ \beta \vee \beta <\!\circ \alpha$.

*R6.4. By *R6.3, *R2.5.

*R6.5. By *R6.2, *R2.7 (with *63, *12).

*R6.7. Use *R0.9.

*R6.9. Assume $\alpha, \beta, \gamma \in R$ and $\alpha <\!\circ \beta$. By *R6.3, *R2.6 and *R6.2, $(\alpha <\!\circ \gamma \vee \gamma <\!\circ \alpha) \vee (\beta <\!\circ \gamma \vee \gamma <\!\circ \beta)$. But $\gamma <\!\circ \alpha$ with $\alpha <\!\circ \beta$ leads by *R6.6 to $\gamma <\!\circ \beta$, and $\beta <\!\circ \gamma$ to $\alpha <\!\circ \gamma$. Hence $\alpha <\!\circ \gamma \vee \gamma <\!\circ \beta$.

*R6.10–*R6.12. By *R6.9.

*R6.13. Assume $\alpha \doteq \beta$, $\alpha <\!\circ \gamma$. Prior to $\exists$-elims. and after $\forall$-elims. assume formulas from which by *R0.9: **(i)** $\forall p 2^{k+1} |\alpha(x+p) - \beta(x+p)| < 2^{x+p}$, **(ii)** $\forall p 2^k (\gamma(x+p) \dot{-} \alpha(x+p)) \geq 2^{x+p}$. Using (i) and *11.15: **(iii)** $2^{k+1}\beta(x+p) < 2^{k+1}\alpha(x+p) + 2^{x+p}$. Using (ii), $2^{k+1}\gamma(x+p) \dot{-} 2^{k+1}\alpha(x+p) \geq 2^{x+p+1} = 2^{x+p} + 2^{x+p}$, whence by *6.17, etc., $2^{k+1}\gamma(x+p) \dot{-} (2^{k+1}\alpha(x+p) + 2^{x+p}) \geq 2^{x+p}$. So $2^{k+1}\gamma(x+p) \dot{-} 2^{k+1}\beta(x+p) \geq 2^{k+1}\gamma(x+p) \dot{-} (2^{k+1}\alpha(x+p) + 2^{x+p})$ [*6.18, (iii)] $\geq 2^{x+p}$, whence $\beta <\!\circ \gamma$.

*R6.15. Assume $\alpha \leq \beta$, i.e. $\forall x \alpha(x) \leq \beta(x)$ and (prior to $\exists$-elims.) $\forall p 2^k (\alpha(x+p) \dot{-} \beta(x+p)) \geq 2^{x+p}$. Thence by $\forall$-elim. with 0 for p, and *6.11, $0 \geq 2^x$, contradicting *3.9.

*R6.16. Use *6.8.

*R6.17. By $\forall$-introd. from *11.5, and *R6.15.

*R6.18. Assume $\beta\circ\geq\gamma$. CASE 1: $\beta\circ>\gamma$. Assume $\forall p 2^k(\beta(x+p)\dot{-}\gamma(x+p))\geq 2^{x+p}$. Thence, after $\forall$-elim., $\beta(x+p)>\gamma(x+p)$. So by *6.6 with ^{x}R3.1 and ^{x}R4.1, $((\alpha+\beta)\dot{-}\gamma)(x+p)=(\alpha+(\beta\dot{-}\gamma))(x+p)$. Hence $2^h|((\alpha+\beta)\dot{-}\gamma)(x+p)-(\alpha+(\beta\dot{-}\gamma))(x+p)| = 0<2^{x+p}$. By $\forall$- and $\exists$-introds., $(\alpha+\beta)\dot{-}\gamma\doteq\alpha+(\beta\dot{-}\gamma)$. CASE 2: $\beta\doteq\gamma$. Then by *R3.5 (with *R3.4) and *R4.6: **(i)** $(\alpha+\beta)\dot{-}\gamma\doteq(\alpha+\gamma)\dot{-}\gamma$, **(ii)** $\alpha+(\beta\dot{-}\gamma)\doteq\alpha+(\gamma\dot{-}\gamma)$. By *R4.3, **(iii)** $(\alpha+\gamma)\dot{-}\gamma=\alpha$; and by $\forall$-introd. from (a substitution instance of) $a+(c\dot{-}c)=a$ [*6.3a], **(iv)** $\alpha+(\gamma\dot{-}\gamma)=\alpha$. Now, using the chain method with $=$, $\doteq$: $(\alpha+\beta)\dot{-}\gamma \doteq (\alpha+\gamma)\dot{-}\gamma$ [(i)] $=\alpha$ [(iii)] $=$ $\alpha+(\gamma\dot{-}\gamma)$ [(iv)] $\doteq \alpha+(\beta\dot{-}\gamma)$ [(ii)].

*R6.19. First we refine *6.19 to the following (proved by cases $b\leq c$, $b>c$): *6.19′. $d>0 \supset (a\geq b+d, c+d \sim a\dot{-}c\geq(b\dot{-}c)+d)$. I. Assume $\alpha\circ>\beta, \gamma$. Via *R0.9, $\forall p 2^k(\alpha(x+p)\dot{-}\beta(x+p))\geq 2^{x+p}$, $\forall p 2^k(\alpha(x+p)\dot{-}\gamma(x+p))\geq 2^{x+p}$. Thence $2^k\alpha(x+p)\geq 2^k\beta(x+p)+2^{x+p}$, $2^k\gamma(x+p)+2^{x+p}$. Thence by *6.19′ and ^{x}R4.1, $2^k(\alpha\dot{-}\gamma)(x+p)\geq 2^k(\beta\dot{-}\gamma)(x+p)+2^{x+p}$, whence $\alpha\dot{-}\gamma\circ>\beta\dot{-}\gamma$.

*R6.20. First prove by cases $b\geq c$, $b<c$: *6.20′. $d>0 \supset (a+d\leq b, c \sim c\dot{-}a\geq(c\dot{-}b)+d)$.

In *R6.21 $A(\alpha, \alpha'', \beta)$ is $\alpha''\doteq\alpha$ & $\alpha''\leq\beta$ & $\forall x(\alpha(x)\leq\beta(x) \sim \alpha''(x)=\alpha(x))$ & $\forall x(\alpha(x)\geq\beta(x) \sim \alpha''(x)=\beta(x))$. In *R6.23 $B(\alpha, \alpha'', \beta)$ is $\alpha''\doteq\alpha$ & $\alpha''\geq\beta$ & $\forall x(\alpha(x)\geq\beta(x) \sim \alpha''(x)=\alpha(x))$ & $\forall x(\alpha(x)\leq\beta(x) \sim \alpha''(x)=\beta(x))$.

*R6.21. $\vdash \alpha, \gamma, \beta\in R'$ & $\alpha, \gamma\circ\not>\beta$ & $\forall x_{x<z}\alpha(x)=\gamma(x)$
$\supset \exists\alpha''_{\alpha''\in R'}\exists\gamma''_{\gamma''\in R'}\{A(\alpha, \alpha'', \beta)$ &
$A(\gamma, \gamma'', \beta)$ & $\forall x_{x<z}\alpha''(x)=\gamma''(x)\}$.

*R6.22. $\vdash \alpha, \beta\in R' \supset [\alpha\circ\not>\beta \sim \exists\alpha''_{\alpha''\in R'}(\alpha''\doteq\alpha$ & $\alpha''\leq\beta)]$.

*R6.23. $\vdash \alpha, \gamma, \beta\in R'$ & $\alpha, \gamma\not<\circ\beta$ & $\forall x_{x<z}\alpha(x)=\gamma(x)$
$\supset \exists\alpha''_{\alpha''\in R'}\exists\gamma''_{\gamma''\in R'}\{B(\alpha, \alpha'', \beta)$ &
$B(\gamma, \gamma'', \beta)$ & $\forall x_{x<z}\alpha''(x)=\gamma''(x)\}$.

*R6.24. $\vdash \alpha, \beta\in R' \supset [\alpha\not<\circ\beta \sim \exists\alpha''_{\alpha''\in R'}(\alpha''\doteq\alpha$ & $\alpha''\geq\beta)]$.

PROOFS. *R6.21. Assume $\alpha, \gamma, \beta\in R'$, **(i)** $\alpha\circ\not>\beta$, **(ii)** $\gamma\circ\not>\beta$, **(iii)** $\forall x_{x<z}\alpha(x)=\gamma(x)$. Introduce α'' and γ'', using Lemma 5.5 (a):

(iv) $$\forall x\alpha''(x)=\begin{cases}\alpha(x) & \text{if } \alpha(x)\leq\beta(x),\\ \beta(x) & \text{if } \alpha(x)>\beta(x).\end{cases}$$

(v) $$\forall x\gamma''(x)=\begin{cases}\gamma(x) & \text{if } \gamma(x)\leq\beta(x),\\ \beta(x) & \text{if } \gamma(x)>\beta(x).\end{cases}$$

Assuming $x<z$, (iii) gives $\alpha(x)=\gamma(x)$, whence by (iv) and (v), $\alpha''(x)=\gamma''(x)$. By $\supset$- and $\forall$-introd.: **(vi)** $\forall x_{x<z}\alpha''(x)=\gamma''(x)$. Toward $\alpha''\in R'$, we establish $|2\alpha''(x)-\alpha''(x')|\leq 1$ thus. Case 1: $\alpha(x)\leq\beta(x)$, $\alpha(x')\leq\beta(x')$. By (iv), $\alpha''(x)=\alpha(x)$ and $\alpha''(x')=\alpha(x')$. Use $\alpha\in R'$. Case 2: $\alpha(x)\leq\beta(x)$, $\alpha(x')>\beta(x')$. Then $2\alpha''(x) = 2\alpha(x)$ [(iv)] $\leq 2\beta(x)$ [case hyp.] $\leq \beta(x')+1$ [$\beta\in R'$, *R0.5a] $= \alpha''(x')+1$ [(iv), case hyp.]. Also $\alpha''(x') = \beta(x') < \alpha(x') \leq 2\alpha(x)+1$ [$\alpha\in R'$, *R0.5a] $= 2\alpha''(x)+1$. So by *11.15a, $|2\alpha''(x)-\alpha''(x')|\leq 1$. Cases 3 and 4 are similar. By $\forall$-introd.: **(vii)** $\alpha''\in R'$. Easily from (iv): **(viii)** $\alpha''\leq\beta$, **(ix)** $\forall x(\alpha(x)\leq\beta(x) \sim \alpha''(x)=\alpha(x))$, **(x)** $\forall x(\alpha(x)\geq\beta(x) \sim \alpha''(x)=\beta(x))$. Toward $\alpha''\doteq\alpha$, assume $\alpha''\#\alpha$, whence: **(a)** $2^k|\alpha''(x+p)-\alpha(x+p)|\geq 2^{x+p}$. Then $\alpha''(x+p)\neq\alpha(x+p)$. By (ix), $\alpha(x+p)>\beta(x+p)$, whence by (iv), $\alpha''(x+p)=\beta(x+p)$. So (a) becomes $2^k(\alpha(x+p)\dot{-}\beta(x+p))\geq 2^{x+p}$. By $\forall$- and $\exists$-introds., $\alpha\circ>\beta$, contradicting (i). Hence $\neg\alpha''\#\alpha$. By *R2.7: **(xi)** $\alpha''\doteq\alpha$. Combining (vii), (xi), (viii), (ix) and (x): **(xii)** $\alpha''\in R'$ & $A(\alpha, \alpha'', \beta)$. Similarly: **(xiii)** $\gamma''\in R'$ & $A(\gamma, \gamma'', \beta)$. Combining (xii), (xiii) and (vi): $\exists\alpha''_{\alpha''\in R'}\exists\gamma''_{\gamma''\in R'}\{A(\alpha, \alpha'', \beta)$ & $A(\gamma, \gamma'', \beta)$ & $\forall x_{x<z}\alpha''(x)=\gamma''(x)\}$.

*R6.22. Assume $\alpha, \beta\in R'$. I. Apply *R6.21 with α, α, β for α, γ, β. II. Use *R6.15, *R6.14.

The virtual ordering predicate is expressed by the formula $\alpha\circ\not>\beta$ & $\neg\alpha\doteq\beta$, which we abbreviate "$\alpha\dot{<}\beta$" or "$\beta\dot{>}\alpha$". *R7.2–*R7.4, *R7.10 and *R7.11 are a rearrangement of the axioms for virtual order given in Brouwer **1924–7** II p. 453, Brouwer **1928a** p. 8, Heyting **1956** pp. 106–107. The first equivalence in *R7.8, *R7.9 is simply an unabbreviation of "$\alpha\not<\circ\beta$".

*R7.1. $\vdash \alpha\dot{<}\beta \sim \alpha\circ\not>\beta$ & $\neg\alpha\doteq\beta$.

*R7.2. $\vdash \alpha\doteq\beta$ & $\alpha\dot{<}\gamma \supset \beta\dot{<}\gamma$.

*R7.3. $\vdash \alpha\doteq\beta$ & $\gamma\dot{<}\alpha \supset \gamma\dot{<}\beta$.

*R7.4. $\vdash \alpha, \beta, \gamma\in R$ & $\alpha\dot{<}\beta$ & $\beta\dot{<}\gamma \supset \alpha\dot{<}\gamma$.

*R7.5. $\vdash \alpha, \beta\in R$ & $\alpha\dot{<}\beta \supset \neg\alpha\dot{>}\beta$. *R7.6. $\vdash \neg\alpha\dot{<}\alpha$.

*R7.7. $\vdash \alpha<\circ\beta \supset \alpha\dot{<}\beta$.

*R7.8. $\vdash \alpha, \beta\in R \supset (\alpha\not<\circ\beta \sim \neg\alpha<\circ\beta \sim \neg\alpha\dot{<}\beta)$.

*R7.9. $\vdash \alpha, \beta\in R \supset (\neg\alpha\not<\circ\beta \sim \neg\neg\alpha<\circ\beta \sim \neg\neg\alpha\dot{<}\beta \sim \alpha\dot{<}\beta)$.

*R7.10. $\vdash \alpha, \beta\in R \supset (\alpha\doteq\beta \sim \neg\alpha\dot{<}\beta$ & $\neg\alpha\dot{>}\beta)$.

*R7.11. $\vdash \alpha, \beta\in R \supset (\alpha\dot{<}\beta \sim \neg\alpha\dot{>}\beta$ & $\neg\alpha\doteq\beta)$.

Proofs. *R7.4. Assume $\alpha, \beta, \gamma\in R$, $\alpha\dot{<}\beta$ whence **(i)** $\alpha\circ\not>\beta$ and **(ii)** $\neg\alpha\doteq\beta$, and $\beta\dot{<}\gamma$ or simply **(iii)** $\beta\circ\not>\gamma$. Using (i) and (iii) (besides $\alpha, \beta, \gamma\in R$) in *R6.12, $\alpha\circ\not>\gamma$. Toward $\neg\alpha\doteq\gamma$, assume $\alpha\doteq\gamma$. Using this and (iii) in *R6.13, $\alpha \not<\circ \beta$; and using the latter and (i) in *R6.5, $\alpha \doteq \beta$, contradicting (ii).

*R7.9. By *R7.8, *25, *49b.

The formula on the right in *R8.1, abbreviated "$\alpha \in [\delta_1, \delta_2]$", expresses that α, δ_1, δ_2 are r.n.g. such that α belongs to the closed interval $[\delta_1, \delta_2]$ (cf. Brouwer **1924–7** II p. 454, **1928a** p. 9 lines 17–13 from below, Heyting **1956** pp. 40–41).

*R8.1. $\vdash \alpha \in [\delta_1, \delta_2] \sim \alpha, \delta_1, \delta_2 \in R \;\&$
$\neg(\alpha \dot{<} \delta_1 \;\&\; \alpha \dot{<} \delta_2) \;\&\; \neg(\alpha \dot{>} \delta_1 \;\&\; \alpha \dot{>} \delta_2)$.

*R8.2. $\vdash \alpha \doteq \beta \;\&\; \alpha \in [\delta_1, \delta_2] \supset \beta \in [\delta_1, \delta_2]$.

*R8.3. $\vdash \delta_1 \doteq \delta_1' \;\&\; \alpha \in [\delta_1, \delta_2] \supset \alpha \in [\delta_1', \delta_2]$.

*R8.4. $\vdash \delta_2 \doteq \delta_2' \;\&\; \alpha \in [\delta_1, \delta_2] \supset \alpha \in [\delta_1, \delta_2']$.

*R8.5. $\vdash \alpha, \beta \in R \supset \alpha, \beta \in [\alpha, \beta]$.

*R8.6. $\vdash \alpha, \delta_1, \delta_2 \in R \;\&\; \delta_1 \circ\!\not> \delta_2 \supset$
$\{\alpha \in [\delta_1, \delta_2] \sim (\alpha \not<\circ \delta_1 \;\&\; \alpha \circ\!\not> \delta_2)\}$.

PROOFS. *R8.5. (Cf. preceding *R0.1.) Use *R7.6.

*R8.6. Assume $\alpha, \delta_1, \delta_2 \in R$ and **(i)** $\delta_1 \circ\!\not> \delta_2$. I. Assume **(ii)** $\neg(\alpha \dot{<} \delta_1 \;\&\; \alpha \dot{<} \delta_2)$ and **(iii)** $\neg(\alpha \dot{>} \delta_1 \;\&\; \alpha \dot{>} \delta_2)$. If $\alpha <\circ \delta_1$, then by (i) and *R6.11, $\alpha <\circ \delta_2$; so by *R7.7 (twice) and &-introd., $\alpha \dot{<} \delta_1 \;\&\; \alpha \dot{<} \delta_2$, contradicting (ii). Thus $\alpha \not<\circ \delta_1$. Similarly, using *R6.10 to contradict (iii), $\alpha \circ\!\not> \delta_2$. II. Assume $\alpha \not<\circ \delta_1 \;\&\; \alpha \circ\!\not> \delta_2$. By *R7.8, $\neg \alpha \dot{<} \delta_1 \;\&\; \neg \alpha \dot{>} \delta_2$, whence $\neg(\alpha \dot{<} \delta_1 \;\&\; \alpha \dot{<} \delta_2) \;\&\; \neg(\alpha \dot{>} \delta_1 \;\&\; \alpha \dot{>} \delta_2)$; so $\alpha \in [\delta_1, \delta_2]$.

14.7. It will be convenient to have available formally some of the theory of the species of finite dual fractions, or, more precisely, the theory of certain r.n.g. which correspond to finite dual fractions. It can be verified easily that the functor $a \cdot 2^{-m}$ (usually abbreviated "$a2^{-m}$") of Axiom ˣR9.1 gives under the interpretation a r.n.g. corresponding to the dual fraction $a \cdot 2^{-m}$. The notation is unambiguous since there is no other way in which negative exponents have been used.

We shall adopt the abbreviation "a" for $a \cdot 2^{-0}$ in contexts making it clear that a is a functor (not a term). Thus in *R9.18 "0" abbreviates $0 \cdot 2^{-0}$.

ˣR9.1. $(a \cdot 2^{-m})(x) = [a/2^{m \dot{-} x}] \cdot 2^{x \dot{-} m}$.

*R9.2. $\vdash a2^{-m} \in R'$.

*R9.3. $\vdash 2^k a 2^{-(m+k)} = a2^{-m}$.

*R9.4. $\vdash a2^{-m} \overset{\circ}{=} b2^{-n} \sim 2^n a = 2^m b \sim a2^{-m} = b2^{-n}$.
*R9.5. $\vdash a2^{-m} \overset{\circ}{=} b2^{-m} \sim a = b \sim a2^{-m} = b2^{-m}$.
*R9.6. $\vdash m = n \sim a2^{-m} \overset{\circ}{=} a2^{-n}$.
*R9.7. $\vdash a2^{-m} <_{\circ} b2^{-n} \sim 2^n a < 2^m b$.
*R9.8. $\vdash a2^{-m} <_{\circ} b2^{-m} \sim a < b$.
*R9.9. $\vdash a2^{-m} + b2^{-n} \overset{\circ}{=} (2^n a + 2^m b)2^{-(m+n)}$.
*R9.10. $\vdash a2^{-m} + b2^{-m} \overset{\circ}{=} (a+b)2^{-m}$.
*R9.11. $\vdash a2^{-m} \dot{-} b2^{-n} \overset{\circ}{=} (2^n a \dot{-} 2^m b)2^{-(m+n)}$.
*R9.12. $\vdash a2^{-m} \dot{-} b2^{-m} \overset{\circ}{=} (a \dot{-} b)2^{-m}$.
*R9.13. $\vdash |a2^{-m} - b2^{-n}| \overset{\circ}{=} |2^n a - 2^m b| 2^{-(m+n)}$.
*R9.14. $\vdash |a2^{-m} - b2^{-m}| \overset{\circ}{=} |a-b| 2^{-m}$.

PROOFS. *R9.2. By *R0.4, ${}^{\mathrm{x}}$R9.1 we need to prove (prior to $\forall$-introd.): **(a)** $|2[a/2^{m \dot{-} x}]2^{x \dot{-} m} - [a/2^{m \dot{-} x'}]2^{x' \dot{-} m}| \leq 1$. CASE 1: $x < m$. Then **(i)** $x \dot{-} m = x' \dot{-} m = 0$, **(ii)** $m \dot{-} x' < m \dot{-} x$ [*6.20], **(iii)** $m \dot{-} x = (m \dot{-} x') + 1$. Using (i), (a) reduces to: **(b)** $|2[a/2^{m \dot{-} x}] - [a/2^{m \dot{-} x'}]| \leq 1$. By *13.4: **(iv)** $a = 2^{m \dot{-} x}[a/2^{m \dot{-} x}] + \mathrm{rm}(a, 2^{m \dot{-} x}) = 2^{m \dot{-} x'}[a/2^{m \dot{-} x'}] + \mathrm{rm}(a, 2^{m \dot{-} x'})$. By *12.3 with *3.9: **(v)** $\mathrm{rm}(a, 2^{m \dot{-} x}) < 2^{m \dot{-} x}$, $\mathrm{rm}(a, 2^{m \dot{-} x'}) < 2^{m \dot{-} x'} < 2^{m \dot{-} x}$ [using also (ii) and *3.12]. So $2^{m \dot{-} x'}|2[a/2^{m \dot{-} x}] - [a/2^{m \dot{-} x'}]| = |2^{m \dot{-} x}[a/2^{m \dot{-} x}] - 2^{m \dot{-} x'}[a/2^{m \dot{-} x'}]|$ [using (iii)] $= |(a \dot{-} \mathrm{rm}(a, 2^{m \dot{-} x})) - (a \dot{-} \mathrm{rm}(a, 2^{m \dot{-} x'}))|$ [(iv)] $= |\mathrm{rm}(a, 2^{m \dot{-} x}) - \mathrm{rm}(a, 2^{m \dot{-} x'})|$ [*11.11 with (iv)] $< 2^{m \dot{-} x}$ [*11.12; *8.6 with (v)]. Hence using (iii), $|2[a/2^{m \dot{-} x}] - [a/2^{m \dot{-} x'}]| < 2$, whence (b). CASE 2: $x \geq m$. Now (a) reduces to $|2[a/2^0]2^{x \dot{-} m} - [a/2^0]2^{(x \dot{-} m)+1}| \leq 1$. Use *13.6, *11.2.

*R9.3. We need to prove **(a)** $[2^k a/2^{(m+k) \dot{-} x}]2^{x \dot{-} (m+k)} = [a/2^{m \dot{-} x}]2^{x \dot{-} m}$. CASE 1: $x < m$. Then $x \dot{-} m = x \dot{-} (m+k) = 0$ and $k + (m \dot{-} x) = (m+k) \dot{-} x$; and (a) reduces to $[2^k a/2^{(m+k) \dot{-} x}] = [a/2^{m \dot{-} x}]$. By *13.5, this will follow if we establish $2^k a = 2^{(m+k) \dot{-} x}[a/2^{m \dot{-} x}] + 2^k \mathrm{rm}(a, 2^{m \dot{-} x})$ & $2^k \mathrm{rm}(a, 2^{m \dot{-} x}) < 2^{(m+k) \dot{-} x}$. But by *13.4 and *12.3 (with *107 and *145a), $2^k a = 2^{k+(m \dot{-} x)}[a/2^{m \dot{-} x}] + 2^k \mathrm{rm}(a, 2^{m \dot{-} x})$ & $2^k \mathrm{rm}(a, 2^{m \dot{-} x}) < 2^{k+(m \dot{-} x)}$. CASE 2: $m \leq x < m+k$. Now (a) reduces to $[2^k a/2^{(m+k) \dot{-} x}] = [a/1]2^{x \dot{-} m}$. But $k = (m+k) \dot{-} m \geq (m+k) \dot{-} x$ [*6.18]. So $[2^k a/2^{(m+k) \dot{-} x}] = [2^{k \dot{-} ((m+k) \dot{-} x)} a 2^{(m+k) \dot{-} x} + 0/2^{(m+k) \dot{-} x}] = 2^{k \dot{-} ((m+k) \dot{-} x)} a$ [*13.8, ${}^{\mathrm{x}}$13.1] $= 2^{x \dot{-} m} a$ [*6.9, etc.] $= [a/1]2^{x \dot{-} m}$ [*13.6]. CASE 3: $m+k \leq x$. Using *13.6, (a) reduces to $2^k a 2^{x \dot{-} (m+k)} = a2^{x \dot{-} m}$.

*R9.4. Call this $A \sim B \sim C$. It will suffice to prove **(a)** $A \supset B$, **(b)** $B \supset C$, **(c)** $C \supset A$. (a) Assume $a2^{-m} \overset{\circ}{=} b2^{-n}$. Using *R1.1 and ${}^{\mathrm{x}}$R9.1, assume $\forall p 2^{m+n}|[a/2^{m \dot{-} (x+p)}]2^{(x+p) \dot{-} m} - [b/2^{n \dot{-} (x+p)}]2^{(x+p) \dot{-} n}| < 2^{x+p}$. Thence by $\forall$-elim., $2^{m+n}|[a/2^{m \dot{-} (x+m+n)}]2^{(x+m+n) \dot{-} m} -$

$[b/2^{n \dot{-} (x+m+n)}]2^{(x+m+n) \dot{-} n}| < 2^{x+m+n}$. This reduces to $2^{m+n+x}|a2^n - b2^m| < 2^{m+n+x}$, whence $|a2^n - b2^m| < 1$, and thus $|a2^n - b2^m| = 0$. So by *11.2, $2^n a = 2^m b$. (b) Assume $2^n a = 2^m b$. CASE 1: $m \leq n$. Let $n = m+k$. Now $2^{m+k}a = 2^m b$, whence $2^k a = b$. So $b2^{-n} = 2^k a 2^{-n} = 2^k a 2^{-(m+k)} = a2^{-m}$ [*R9.3]. CASE 2: $m > n$. Similarly. (c) By *R1.3.

*R9.7. I. As for (a) under *R9.4, using now *R6.1 instead of *R1.1, $2^k(b2^m \dot{-} a2^n) \geq 2^{m+n}$, whence $b2^m \dot{-} a2^n > 0$, whence by *6.12, $b2^m > a2^n$. II. First we prove **(i)** $a < b \supset a2^{-m} <_\circ b2^{-m}$ thus. Assume $a < b$. Then $2^m((b2^{-m})(m+p) \dot{-} (a2^{-m})(m+p)) = 2^{m+p}(b \dot{-} a) \geq 2^{m+p}$, whence $a2^{-m} <_\circ b2^{-m}$. Now we adapt (b) for *R9.4. Thus in Case 1, now $2^k a < b$. So using (i), $b2^{-n} \circ\!> 2^k a 2^{-n} =$ etc.

*R9.9. By *R1.2, it will follow from $(a2^{-m} + b2^{-n})(m+n+p) = ((2^n a + 2^m b)2^{-(m+n)})(m+n+p)$, which is easily deduced using ^{x}R3.1, ^{x}R9.1, etc.

*R9.10. From *R9.9 by *R9.3.

*R9.15. $\vdash \alpha \in R' \supset |\alpha(m)2^{-m} - \alpha| \circ\!\not> 1 \cdot 2^{-m}$.
*R9.16. $\vdash \alpha \in R' \supset \alpha \not<_\circ (\alpha(m) \dot{-} 1)2^{-m}$.
*R9.17. $\vdash \alpha \in R' \supset \alpha \circ\!\not> (\alpha(m)+1)2^{-m}$.
*R9.18. $\vdash \alpha \not<_\circ 0$.
*R9.19. $\vdash \alpha, \beta \in R \;\&\; \alpha <_\circ \beta \supset \exists a \exists m(\alpha <_\circ a2^{-m} <_\circ \beta)$.
*R9.20. $\vdash \beta \in R' \supset \exists \alpha_{\alpha \in R'}(\bar{\alpha}(y) = \bar{\beta}(y) \;\&\; \alpha \doteqdot (\beta(y) \dot{-} 1)2^{-y})$.
*R9.21. $\vdash \beta \in R' \supset \exists \alpha_{\alpha \in R'}(\bar{\alpha}(y) = \bar{\beta}(y) \;\&\; \alpha \doteqdot (\beta(y)+1)2^{-y})$.

PROOFS. *R9.15. Assume $\alpha \in R'$ and $|\alpha(m)2^{-m} - \alpha| \circ\!> 1 \cdot 2^{-m}$. Assume, prior to $\exists$-elims., $\forall p 2^k(|[\alpha(m)/2^{m \dot{-} (x+p)}]2^{(x+p) \dot{-} m} - \alpha(x+p)| \dot{-} [1/2^{m \dot{-} (x+p)}]2^{(x+p) \dot{-} m}) \geq 2^{x+p}$. Using $\forall$-elim. with $m+p$ for p, this reduces to $2^k(|\alpha(m)2^{x+p} - \alpha(m+x+p)| \dot{-} 2^{x+p}) \geq 2^{m+x+p}$, whence by *3.9 and *6.12, $|2^{x+p}\alpha(m) - \alpha(m+x+p)| > 2^{x+p}$, contradicting $\alpha \in R'$ by *R0.6.

*R9.16. Assume $\alpha \in R'$ and $\alpha <_\circ (\alpha(m) \dot{-} 1)2^{-m}$ whence, prior to $\exists$-elims., after $\forall$-elim., $2^k(((\alpha(m) \dot{-} 1)2^{-m})(x+m+p) \dot{-} \alpha(x+m+p)) \geq 2^{x+m+p}$. Then $((\alpha(m) \dot{-} 1)2^{-m})(x+m+p) \dot{-} \alpha(x+m+p) > 0$, whence $[\alpha(m) \dot{-} 1/2^{m \dot{-} (x+m+p)}]2^{(m+x+p) \dot{-} m} \dot{-} \alpha(x+m+p) > 0$, which reduces to $(\alpha(m)2^{x+p} \dot{-} 2^{x+p}) \dot{-} \alpha(x+m+p) > 0$. Then $2^{x+p}\alpha(m) \dot{-} \alpha(x+m+p) > 2^{x+p}$, whence $|2^{x+p}\alpha(m) - \alpha(m+x+p)| > 2^{x+p}$, contradicting $\alpha \in R'$ by *R0.6.

*R9.19. Assume $\alpha, \beta \in R$ and **(i)** $\alpha <_\circ \beta$. Using *R1.11, assume **(ii)** $\alpha' \in R' \;\&\; \alpha' \doteqdot \alpha$, **(iii)** $\beta' \in R' \;\&\; \beta' \doteqdot \beta$. By (i) and *R6.13–*R6.14, **(iv)** $\alpha' <_\circ \beta'$. Assume $\forall p 2^k(\beta'(x+p) \dot{-} \alpha'(x+p)) \geq 2^{x+p}$. Thence $\beta'(x+2+k)$

$\dot{-}\alpha'(x+2+k) \geq 2^{x+2} > 3$, whence **(v)** $\beta'(x+2+k) \dot{-} 1 > \alpha'(x+2+k) +2$. So $\alpha \doteq \alpha'$ [(ii)] $\circ\!\not> (\alpha'(x+2+k)+1)2^{-(x+2+k)}$ [*R9.17] $<\circ (\alpha'(x+2+k)+2)2^{-(x+2+k)}$ [*R9.8] $<\circ (\beta'(x+2+k) \dot{-} 1)2^{-(x+2+k)}$ [(v), *R9.8] $\circ\!\not> \beta'$ [*R9.16] $\doteq \beta$ [(iii)]. By $\exists$-introds., $\exists a \exists m(\alpha <\circ a2^{-m} <\circ \beta)$.

*R9.20. Assume $\beta \epsilon R'$. Using Lemma 5.5 (a), let

$$\forall x \alpha(x) = \begin{cases} \beta(x) \text{ if } x \leq y, \\ 2^{x \dot{-} y}\beta(y) \dot{-} (2^{x \dot{-} y} \dot{-} 1) \text{ if } x > y. \end{cases}$$

Then $\bar{\alpha}(y) = \bar{\beta}(y)$. We shall deduce **(a)** $|2\alpha(x) - \alpha(x')| \leq 1$. CASE 1: $x < y$. Use $\beta \epsilon R'$. CASE 2: $x = y$. Then $|2\alpha(x) - \alpha(x')| = |2\beta(y) - (2\beta(y) \dot{-} (2 \dot{-} 1))| = |(2\beta(y) \dot{-} 0) - (2\beta(y) \dot{-} 1)|$ [$^{\times}$6.1, Lemma 5.2] $\leq |0-1|$ [*11.13] $= 1$. CASE 3: $x > y$. Then $|2\alpha(x) - \alpha(x')| = |2(2^{x \dot{-} y}\beta(y) \dot{-} (2^{x \dot{-} y} \dot{-} 1)) - (2^{x' \dot{-} y}\beta(y) \dot{-} (2^{x' \dot{-} y} \dot{-} 1))| = |(2^{x' \dot{-} y}\beta(y) \dot{-} (2^{x' \dot{-} y} \dot{-} 2)) - (2^{x' \dot{-} y}\beta(y) \dot{-} (2^{x' \dot{-} y} \dot{-} 1))|$ [*6.14, *6.6 with case hyp.] $\leq |2-1|$ [*11.13] $= 1$. From (a) by $\forall$-introd., $\alpha \epsilon R'$. Finally, $2^k|\alpha(y+k'+p) - ((\beta(y) \dot{-} 1)2^{-y})(y+k'+p)| = 2^k|(2^{k'+p}\beta(y) \dot{-} (2^{k'+p} \dot{-} 1)) - (\beta(y) \dot{-} 1)2^{k'+p}|$ [$^{\times}$R9.1, etc.] $\leq 2^k|(2^{k'+p} \dot{-} 1) - 2^{k'+p}|$ [*6.14, *11.13] $\leq 2^k|1-0|$ [*11.13] $= 2^k < 2^{y+k'+p}$. By $\forall$-, $\exists$- and $\forall$-introd., $\alpha \doteq (\beta(y) \dot{-} 1)2^{-y}$.

*R9.21. Similarly, letting

$$\forall x \alpha(x) = \begin{cases} \beta(x) \text{ if } x \leq y, \\ 2^{x \dot{-} y}\beta(y) + (2^{x \dot{-} y} \dot{-} 1) \text{ if } x > y. \end{cases}$$

Using *R9.20 and *R9.21 with Brouwer's principle (for numbers) we next refute the (generalized) law of the excluded middle for $\doteq$. This proof of *R9.22 via *R9.20 and *R9.21 is essentially due to Kleene, who in March 1963 gave a simpler proof of *R9.22 than that in the author's thesis (where the result appeared right after the present *R1.11). We have adapted Kleene's proof to obtain successive proofs of *R9.20, *R9.21 and *R9.22.

*R9.22. $\vdash \beta \epsilon R' \supset \neg \forall \alpha_{\alpha \epsilon R'}(\alpha \doteq \beta \vee \neg \alpha \doteq \beta)$.

*R9.23. $\vdash \beta \epsilon R \supset \neg \forall \alpha_{\alpha \epsilon R}(\alpha \doteq \beta \vee \neg \alpha \doteq \beta)$.

*R9.24. $\vdash \beta \epsilon R \supset \neg \forall \alpha_{\alpha \epsilon R}(\alpha <\circ \beta \vee \alpha \doteq \beta \vee \alpha \circ> \beta)$.

*R9.25. $\vdash \beta \epsilon R \supset \neg \forall \alpha_{\alpha \epsilon R}(\alpha \dot{<} \beta \vee \alpha \doteq \beta \vee \alpha \dot{>} \beta)$.

PROOFS. *R9.22. Assume **(i)** $\beta \epsilon R'$ and $\forall \alpha_{\alpha \epsilon R'}(\alpha \doteq \beta \vee \neg \alpha \doteq \beta)$. Using *27.6 with *R0.8, and omitting $\exists \tau$ prior to $\exists$-elim., assume **(ii)** $\forall \alpha_{\alpha \epsilon R'} \exists y \{\forall x[\tau(\bar{\alpha}(x)) > 0 \supset y = x] \ \& \ \{(\alpha \doteq \beta \ \& \ \tau(\bar{\alpha}(y)) = 1) \vee (\neg \alpha \doteq \beta \ \& \ \tau(\bar{\alpha}(y)) = 2)\}\}$. Thence, using (i) and omitting $\exists y$: **(iii)** $\forall x[\tau(\bar{\beta}(x)) > 0$

$\supset y=x]$ & $\{(\beta \doteq \beta$ & $\tau(\bar{\beta}(y))=1) \vee (\neg\beta \doteq \beta$ & $\tau(\bar{\beta}(y))=2)\}$. By *R1.4, $\beta \doteq \beta$. So by (iii): **(iv)** $\tau(\bar{\beta}(y))=1$. Using *R9.20 and *R9.21 with (i), assume **(v)** $\alpha_1 \in R'$ & $\bar{\alpha}_1(y)=\bar{\beta}(y)$ & $\alpha_1 \doteq (\beta(y) \dot{-} 1)2^{-y}$ and **(vi)** $\alpha_2 \in R'$ & $\bar{\alpha}_2(y)=\bar{\beta}(y)$ & $\alpha_2 \doteq (\beta(y)+1)2^{-y}$. Now $(\beta(y) \dot{-} 1)2^{-y} \doteq \alpha_1$ [(v)] $\doteq \beta$ [(v), (iv), (ii)] $\doteq \alpha_2$ [(vi), (iv), (ii)] $\doteq (\beta(y)+1)2^{-y}$ [(vi)]. So by *R9.5, $\beta(y) \dot{-} 1 = \beta(y)+1$, which by cases ($\beta(y) \leq 1$, $\beta(y) > 1$) is absurd.

*R9.23. Assume $\beta \in R$. Using *R1.11, assume prior to $\exists$-elim.: $\beta' \in R'$ & $\beta' \doteq \beta$. By *R9.22, **(i)** $\neg\forall\alpha_{\alpha \in R'}(\alpha \doteq \beta' \vee \neg\alpha \doteq \beta')$. By *R0.7 and $\beta' \doteq \beta$: $[\alpha \in R \supset \alpha \doteq \beta \vee \neg\alpha \doteq \beta] \supset [\alpha \in R' \supset \alpha \doteq \beta' \vee \neg\alpha \doteq \beta']$. Thence by *69 and *12: **(ii)** $\neg\forall\alpha_{\alpha \in R'}(\alpha \doteq \beta' \vee \neg\alpha \doteq \beta') \supset \neg\forall\alpha_{\alpha \in R}(\alpha \doteq \beta \vee \neg\alpha \doteq \beta)$. Use (i) with (ii).

*R9.24. Assume $\beta \in R$ and $\forall\alpha_{\alpha \in R}(\alpha <\circ \beta \vee \alpha \doteq \beta \vee \alpha \circ> \beta)$. Using *R6.4, $\forall\alpha_{\alpha \in R}(\alpha \doteq \beta \vee \neg\alpha \doteq \beta)$. But by $\supset$-elim. from *R9.23, $\neg\forall\alpha_{\alpha \in R}(\alpha \doteq \beta \vee \neg\alpha \doteq \beta)$.

§ 15. The uniform continuity theorem. 15.1. We shall establish formally the theorem on the uniform continuity of a function defined for every real number represented by r.n.g. in the closed interval $[\delta_1, \delta_2]$ where $\delta_1 \circ\!\not> \delta_2$. (For the theorem without the condition $\delta_1 \circ\!\not> \delta_2$, cf. Heyting **1956** p. 46; the original versions in Brouwer **1923a** p. 5, **1924** p. 193, **1927** p. 67 are for [0,1].)

A preliminary result *R10.1 states that for each pair of r.n.g. α and γ sufficiently close together in the sense that $|\alpha - \gamma|$ is small, there are c.r.n.g. α' and γ' (with $\alpha' \doteq \alpha$ and $\gamma' \doteq \gamma$) "close together" in the sense that initial segments coincide.

*R10.1. $\vdash \alpha,\gamma \in R$ & $|\alpha - \gamma| <\circ 1 \cdot 2^{-(z+4)} \supset$
$$\exists\alpha'_{\alpha' \in R'}\exists\gamma'_{\gamma' \in R'}(\alpha' \doteq \alpha \;\&\; \gamma' \doteq \gamma \;\&\; \forall x_{x \leq z}\alpha'(x)=\gamma'(x)).$$

PROOF. Assume $\alpha,\gamma \in R$, $|\alpha - \gamma| <\circ 1 \cdot 2^{-(z+4)}$. Assume $\forall p 2^k((1 \cdot 2^{-(z+4)})(x+p) \dot{-} (|\alpha - \gamma|)(x+p)) \geq 2^{x+p}$. Using $\forall$-elim. with $z+4+p$ for p, writing $w=x+z+4$, and reducing, $2^k(2^{(w+p) \dot{-} (z+4)} \dot{-} |\alpha(w+p) - \gamma(w+p)|) \geq 2^{w+p}$. So $2^{(w+p) \dot{-} (z+4)} \dot{-} |\alpha(w+p) - \gamma(w+p)| > 0$, whence by *6.12, etc., $2^{z+4}|\alpha(w+p) - \gamma(w+p)| < 2^{w+p}$, and by $\forall$-introd., $\forall p 2^{z+4}|\alpha(w+p) - \gamma(w+p)| < 2^{w+p}$. Now use *R1.10.

Let δ_1 and δ_2 be c.r.n.g. with $\delta_1 \leq \delta_2$. Then *R10.2 asserts the existence of a certain fan. In the proof of the uniform continuity theorem *R10.3, we establish that the closed interval $[\delta_1, \delta_2]$ coincides (Heyting **1956** p. 42) with this fan. (Cf. Brouwer **1924** p. 192, **1928a** p. 5.)

*R10.2. $\vdash \delta_1,\delta_2 \in R' \;\&\; \delta_1 \leq \delta_2 \supset \exists\sigma\{\mathrm{Spr}(\sigma) \;\&\; \sigma(1)=0$
$\&\; \forall a[\sigma(a)=0 \supset \exists b \forall s(\sigma(a*2^{s+1})=0 \supset s \leq b)]$
$\&\; \forall\alpha[\alpha\in\sigma \sim \alpha\in R' \;\&\; \delta_1 \leq \alpha \leq \delta_2]\}$.

PROOF. Assume $\delta_1,\delta_2 \in R'$, $\delta_1 \leq \delta_2$. Introduce σ via Lemma 5.5 (c) (cf. the proof of *R0.8):

$$(A) \qquad \forall a\sigma(a) = \begin{cases} 0 \text{ if } a=1 \vee \\ \quad [\mathrm{Seq}(a) \;\&\; \mathrm{lh}(a)=1 \;\&\; \delta_1(0) \leq (a)_0 \dot{-} 1 \leq \delta_2(0)] \vee \\ \quad [\mathrm{Seq}(a) \;\&\; \mathrm{lh}(a)>1 \;\&\; |2((a)_{\mathrm{lh}(a)\dot{-}2} \dot{-} 1) - \\ \quad ((a)_{\mathrm{lh}(a)\dot{-}1} \dot{-} 1)| \leq 1 \;\&\; \sigma(\Pi_{i<\mathrm{lh}(a)\dot{-}1} p_i^{(a)_i})=0 \\ \quad \&\; \delta_1(\mathrm{lh}(a) \dot{-} 1) \leq (a)_{\mathrm{lh}(a)\dot{-}1} \dot{-} 1 \leq \delta_2(\mathrm{lh}(a) \dot{-} 1)], \\ 1 \text{ otherwise.} \end{cases}$$

IIb. Assume $\sigma(a)=0$. CASE 1: $a=1$. Then $\mathrm{Seq}(a*2^{\delta_1(0)+1})$ & $\mathrm{lh}(a*2^{\delta_1(0)+1})=1$ & $\delta_1(0)=(a*2^{\delta_1(0)+1})_0 \dot{-} 1 \leq \delta_2(0)$ [$\delta_1 \leq \delta_2$]. So by (A), $\sigma(a*2^{\delta_1(0)+1})=0$, and by $\exists$-introd., $\exists s\sigma(a*2^{s+1})=0$. CASE 2: $a \neq 1$. Using (A), $\mathrm{Seq}(a)$ & $\mathrm{lh}(a) \geq 1$ & $\delta_1(\mathrm{lh}(a) \dot{-} 1) \leq (a)_{\mathrm{lh}(a)\dot{-}1} \dot{-} 1 \leq \delta_2(\mathrm{lh}(a) \dot{-} 1)$. SUBCASE 2.1: $\delta_1(\mathrm{lh}(a) \dot{-} 1) = (a)_{\mathrm{lh}(a)\dot{-}1} \dot{-} 1$. Letting t be $a*2^{\delta_1(\mathrm{lh}(a))+1}$, we deduce $\mathrm{Seq}(t)$ & $\mathrm{lh}(t)>1$, $|2((t)_{\mathrm{lh}(t)\dot{-}2} \dot{-} 1) - ((t)_{\mathrm{lh}(t)\dot{-}1} \dot{-} 1)| = |2\delta_1(\mathrm{lh}(a) \dot{-} 1) - \delta_1(\mathrm{lh}(a))| \leq 1$ [$\delta_1 \in R'$], $\sigma(\Pi_{i<\mathrm{lh}(t)\dot{-}1} p_i^{(t)_i}) = \sigma(a) = 0$, $\delta_1(\mathrm{lh}(t) \dot{-} 1) = \delta_1(\mathrm{lh}(a)) = (t)_{\mathrm{lh}(t)\dot{-}1} \dot{-} 1 \leq \delta_2(\mathrm{lh}(a))$ [$\delta_1 \leq \delta_2$] $= \delta_2(\mathrm{lh}(t) \dot{-} 1)$. So by (A), $\sigma(t)=0$. By $\exists$-introd., $\exists s\sigma(a*2^{s+1})=0$. SUBCASE 2.2: $(a)_{\mathrm{lh}(a)\dot{-}1} \dot{-} 1 = \delta_2(\mathrm{lh}(a) \dot{-} 1)$. Similarly. SUBCASE 2.3: $\delta_1(\mathrm{lh}(a) \dot{-} 1) < (a)_{\mathrm{lh}(a)\dot{-}1} \dot{-} 1 < \delta_2(\mathrm{lh}(a) \dot{-} 1)$. Let t be $a*2^{2((a)_{\mathrm{lh}(a)\dot{-}1} \dot{-} 1)+1}$. Then $\mathrm{Seq}(t)$ & $\mathrm{lh}(t)>1$ & $\sigma(\Pi_{i<\mathrm{lh}(t)\dot{-}1} p_i^{(t)_i})=0$, and $|2((t)_{\mathrm{lh}(t)\dot{-}2} \dot{-} 1) - ((t)_{\mathrm{lh}(t)\dot{-}1} \dot{-} 1)| = |2((a)_{\mathrm{lh}(a)\dot{-}1} \dot{-} 1) - 2((a)_{\mathrm{lh}(a)\dot{-}1} \dot{-} 1)| = 0$. From subcase hyp., **(i)** $\delta_1(\mathrm{lh}(a) \dot{-} 1)+1 \leq (a)_{\mathrm{lh}(a)\dot{-}1} \dot{-} 1 \leq \delta_2(\mathrm{lh}(a) \dot{-} 1) \dot{-} 1$. So $\delta_1(\mathrm{lh}(t) \dot{-} 1) = \delta_1(\mathrm{lh}(a)) \leq 2\delta_1(\mathrm{lh}(a) \dot{-} 1)+1$ [$\delta_1 \in R'$, *R0.5a] $< 2(\delta_1(\mathrm{lh}(a) \dot{-} 1)+1) \leq 2((a)_{\mathrm{lh}(a)\dot{-}1} \dot{-} 1)$ [(i)] $= (t)_{\mathrm{lh}(t)\dot{-}1} \dot{-} 1 \leq 2(\delta_2(\mathrm{lh}(a) \dot{-} 1) \dot{-} 1)$ [(i)] $= 2\delta_2(\mathrm{lh}(a) \dot{-} 1) \dot{-} 2 \leq 2\delta_2(\mathrm{lh}(a) \dot{-} 1) \dot{-} 1$ [*6.18] $\leq \delta_2(\mathrm{lh}(a))$ [$\delta_2 \in R'$, *R0.5a–c] $= \delta_2(\mathrm{lh}(t) \dot{-} 1)$. So by (A), $\sigma(t)=0$. By $\exists$-introd., $\exists s\sigma(a*2^{s+1})=0$.

IIIa. Assume $\alpha\in\sigma$. As before, $\alpha \in R'$. Also by $\forall$-elim. with 1 for x, and (A), $\delta_1(0) \leq \alpha(0) \leq \delta_2(0)$; and by $\forall$-elim. with $x+2$ for x, and (A), $\delta_1(x+1) \leq \alpha(x+1) \leq \delta_2(x+1)$. Hence by induction cases (IM p. 186) and $\forall$-introd., $\delta_1 \leq \alpha \leq \delta_2$.

IIIb. Assume $\alpha \in R'$ & $\delta_1 \leq \alpha \leq \delta_2$. By ind. on x with a double basis, $\sigma(\bar{\alpha}(x))=0$. By $\forall$-introd., $\alpha\in\sigma$.

IV. Toward $\forall a[\sigma(a)=0 \supset \exists b\forall s(\sigma(a*2^{s+1})=0 \supset s \leq b)]$, assume **(i)** $\sigma(a)=0$ and **(ii)** $\sigma(a*2^{s+1})=0$. By (i) and IIa, $\mathrm{Seq}(a)$; and then from (ii) and (A) by cases ($\mathrm{lh}(a)=0$, $\mathrm{lh}(a)>0$), $s \leq \delta_2(\mathrm{lh}(a))$.

15.2. Let $F(\alpha, \beta)$ be a predicate expressed formally by the formula $\mathrm{F}(\alpha, \beta)$. Suppose δ_1 and δ_2 are r.n.g. A necessary condition that $F(\alpha, \beta)$ be the representing predicate of a function from real numbers, represented by r.n.g. α in $[\delta_1, \delta_2]$, to real numbers, represented by r.n.g. β, is given formally by the following formula (provided γ, ζ are free for α, β in $\mathrm{F}(\alpha, \beta)$ and do not occur free in $\mathrm{F}(\alpha, \beta)$).

$$\mathfrak{A}(\mathrm{F}, \delta_1, \delta_2): \quad \forall\alpha\forall\gamma\forall\beta\forall\zeta\{\alpha\in[\delta_1, \delta_2] \\ \& \ \alpha \doteq \gamma \ \& \ \mathrm{F}(\alpha, \beta) \ \& \ \mathrm{F}(\gamma, \zeta) \supset \beta, \zeta\in\mathrm{R} \ \& \ \beta\doteq\zeta\}.$$

REMARK 15.1. However, $\mathfrak{A}(\mathrm{F}, \delta_1, \delta_2)$ does not entail either of the replacement properties **(a)** $\alpha\doteq\gamma \ \& \ \mathrm{F}(\alpha, \beta) \supset \mathrm{F}(\gamma, \beta)$ and **(b)** $\beta\doteq\zeta \ \& \ \mathrm{F}(\alpha, \beta) \supset \mathrm{F}(\alpha, \zeta)$. For, letting $\mathrm{F}(\alpha, \beta)$ be $\alpha\in\mathrm{R}' \ \& \ \alpha\doteq\beta$, $\vdash \mathfrak{A}(\mathrm{F}, \delta_1, \delta_2)$ (by *R0.7, *R1.7, *R1.6 and *R1.5), but $\vdash \lambda x1 \doteq \lambda x2 \ \& \ \mathrm{F}(\lambda x1, \lambda x1)$ and $\vdash \neg\mathrm{F}(\lambda x2, \lambda x1)$ (so the closure of (a) is refutable). Similarly, letting $\mathrm{F}(\alpha, \beta)$ be $\beta\in\mathrm{R}' \ \& \ \alpha\doteq\beta$, $\vdash \mathfrak{A}(\mathrm{F}, \delta_1, \delta_2)$, but $\vdash \lambda x1 \doteq \lambda x2 \ \& \ \mathrm{F}(\lambda x1, \lambda x1)$ and $\vdash \neg\mathrm{F}(\lambda x1, \lambda x2)$.

The fourth member of the conjunction in the hypothesis of *R10.3 asserts that the function represented by $\mathrm{F}(\alpha, \beta)$ is completely defined in $[\delta_1, \delta_2]$. The conclusion corresponds under the interpretation to a familiar form of the definition of uniform continuity.

*R10.3. $\vdash \delta_1, \delta_2\in\mathrm{R} \ \& \ \delta_1 \circ\!\not> \delta_2 \ \& \ \mathfrak{A}(\mathrm{F}, \delta_1, \delta_2) \ \&$
$\forall\alpha\{\alpha\in[\delta_1, \delta_2] \supset \exists\beta\mathrm{F}(\alpha, \beta)\}$
$\supset \forall n\exists m\forall\alpha\forall\gamma\forall\beta\forall\zeta\{|\alpha-\gamma| <\!\circ\, 1\cdot 2^{-m}$
$\& \ \alpha, \gamma\in[\delta_1, \delta_2] \ \& \ \mathrm{F}(\alpha, \beta) \ \& \ \mathrm{F}(\gamma, \zeta) \supset |\beta-\zeta| <\!\circ\, 1\cdot 2^{-n}\}.$

PROOF. Assume $\delta_1, \delta_2\in\mathrm{R}$, **(i)** $\delta_1 \circ\!\not> \delta_2$, **(ii)** $\mathfrak{A}(\mathrm{F}, \delta_1, \delta_2)$, **(iii)** $\forall\alpha\{\alpha\in[\delta_1, \delta_2] \supset \exists\beta\mathrm{F}(\alpha, \beta)\}$. Using *R1.11, assume (prior to $\exists$-elims.): **(iv)** $\delta_1'\in\mathrm{R}' \ \& \ \delta_1'\doteq\delta_1$, **(v)** $\delta_2'\in\mathrm{R}' \ \& \ \delta_2'\doteq\delta_2$. By (i) and *R6.13–*R6.14, $\delta_1' \circ\!\not> \delta_2'$. Using *R6.22, assume: **(vi)** $\delta_1''\in\mathrm{R}' \ \& \ \delta_1''\doteq\delta_1' \ \& \ \delta_1''\leq\delta_2'$. By *R6.14, **(vii)** $\delta_1'' \circ\!\not> \delta_2'$. Using (vi), (v) and *R10.2, assume (the conjunction of): **(viii)** $\mathrm{Spr}(\sigma) \ \& \ \sigma(1)=0 \ \& \ \forall a[\sigma(a)=0 \supset \exists b\forall s(\sigma(a*2^{s+1})=0 \supset s\leq b)]$ and **(ix)** $\forall\alpha[\alpha\in\sigma \sim \alpha\in\mathrm{R}' \ \& \ \delta_1''\leq\alpha\leq\delta_2']$. Toward (x), assume $\alpha\in\sigma$. Then successively $\alpha\in\mathrm{R}' \ \& \ \delta_1''\leq\alpha\leq\delta_2'$ [(ix)], $\alpha\in\mathrm{R}' \ \& \ \alpha \not<\!\circ\, \delta_1'' \ \&$ $\alpha \circ\!\not> \delta_2'$ [*R6.15], $\alpha\in\mathrm{R}' \ \& \ \alpha\in[\delta_1'', \delta_2']$ [*R8.6, (v)–(vii)] and $\alpha\in\mathrm{R}' \ \&$ $\alpha\in[\delta_1, \delta_2]$ [*R8.3, (vi), (iv); *R8.4, (v)]. Using (iii), assume $\mathrm{F}(\alpha, \beta)$. From (ii) (and *R1.4), $\beta\in\mathrm{R}$. So, writing $b=2^x 3^{\beta(x)}$, assume $\forall p 2^{n+2}|2^p(b)_1-\beta((b)_0+p)|<2^{(b)_0+p}$. Writing "$\mathrm{G}(n, \beta, b)$" for this, and using &-, $\exists$-, $\supset$- and $\forall$-introds.: **(x)** $\forall n\forall\alpha_{\alpha\in\sigma}\exists b\exists\beta(\mathrm{G}(n, \beta, b) \ \& \ \mathrm{F}(\alpha, \beta))$. Consider *27.8 as of the form A & B & C $\supset$ D, and write "$\mathrm{A}(n, \alpha, b)$"

for its $A(\alpha, b)$ (so C, D become $C(n)$, $D(n)$); by *69 and *89, $A \,\&\, B \,\&\, \forall n C(n) \supset \forall n D(n)$. This with $A \,\&\, B$ given by (viii), and $\forall n C(n)$ by (x), gives $\forall n D(n)$. So assume: **(xi)** $\forall\alpha_{\alpha\in\sigma}\exists b\forall\gamma_{\gamma\in\sigma}\{\forall x_{x<z}\gamma(x)=\alpha(x) \supset \exists\beta(G(n, \beta, b) \,\&\, F(\gamma, \beta))\}$. Toward the conclusion of *R10.3, assume **(xii)** $|\alpha-\gamma|<\circ 1\cdot 2^{-(z+4)} \,\&\, \alpha,\gamma\in[\delta_1, \delta_2] \,\&\, F(\alpha, \beta) \,\&\, F(\gamma, \zeta)$. Using *R10.1 (and *R8.1), assume: **(xiii)** $\alpha',\gamma'\in R' \,\&\, \alpha'\doteq\alpha \,\&\, \gamma'\doteq\gamma \,\&\, \forall x_{x\leq z}\alpha'(x)=\gamma'(x)$. Then **(xiv)** $\forall x_{x<z}\alpha'(x)=\gamma'(x)$. Using *R8.2–*R8.4 and (iv)–(vi), $\alpha',\gamma'\in[\delta''_1, \delta'_2]$, whence by (vii) and *R8.6: **(xv)** $\alpha'\not<\circ\,\delta''_1$, **(xvi)** $\alpha'\circ\!\not>\delta'_2$, **(xvii)** $\gamma'\not<\circ\,\delta''_1$, **(xviii)** $\gamma'\circ\!\not>\delta'_2$. By (xiii) and (v), $\alpha',\gamma',\delta'_2\in R'$. So using *R6.21 with (xvi), (xviii) and (xiv): **(xix)** $\alpha'',\gamma''\in R'$, **(xx)** $\alpha''\doteq\alpha' \,\&\, \alpha''\leq\delta'_2 \,\&\, \forall x(\alpha'(x)\leq\delta'_2(x) \sim \alpha''(x)=\alpha'(x)) \,\&\, \forall x(\alpha'(x)\geq\delta'_2(x) \sim \alpha''(x)=\delta'_2(x))$, **(xxi)** $\gamma''\doteq\gamma' \,\&\, \gamma''\leq\delta'_2 \,\&\, \forall x(\gamma'(x)\leq\delta'_2(x) \sim\gamma''(x)=\gamma'(x)) \,\&\, \forall x(\gamma'(x)\geq\delta'_2(x)\sim\gamma''(x)=\delta'_2(x))$, **(xxii)** $\forall x_{x<z}\alpha''(x)=\gamma''(x)$. Using (vi), $\alpha'',\gamma'',\delta''_1\in R'$. By (xv) and (xvii) with (xx), (xxi) and *R6.13, $\alpha''\not<\circ\,\delta''_1 \,\&\, \gamma''\not<\circ\,\delta''_1$. So using *R6.23: **(xxiii)** $\alpha''',\gamma'''\in R'$, **(xxiv)** $\alpha'''\doteq\alpha'' \,\&\, \alpha'''\geq\delta''_1 \,\&\, \forall x(\alpha''(x)\geq\delta''_1(x) \sim \alpha'''(x)=\alpha''(x)) \,\&\, \forall x(\alpha''(x)\leq\delta''_1(x) \sim \alpha'''(x)= \delta''_1(x))$, **(xxv)** $\gamma'''\doteq\gamma'' \,\&\, \gamma'''\geq\delta''_1 \,\&\, \forall x(\gamma''(x)\geq\delta''_1(x) \sim \gamma'''(x)=\gamma''(x)) \,\&\, \forall x(\gamma''(x)\leq\delta''_1(x) \sim \gamma'''(x)=\delta''_1(x))$, **(xxvi)** $\forall x_{x<z}\alpha'''(x)=\gamma'''(x)$. By (xiii), (xx)–(xxi) and (xxiv)–(xxv): **(xxvii)** $\alpha\doteq\alpha''' \,\&\, \gamma\doteq\gamma'''$. We deduce $\alpha'''(x)\leq\delta'_2(x)$ thus. CASE 1: $\alpha''(x)\leq\delta''_1(x)$. Then $\alpha'''(x)=\delta''_1(x)$ [(xxiv)] $\leq \delta'_2(x)$ [(vi)]. CASE 2: $\alpha''(x)>\delta''_1(x)$. Then $\alpha'''(x)=\alpha''(x)$ [(xxiv)] $\leq \delta'_2(x)$ [(xx)]. So, using also (xxiv), $\delta''_1(x)\leq\alpha'''(x)\leq\delta'_2(x)$, whence by $\forall$-introd., $\delta''_1\leq\alpha'''\leq\delta'_2$. Similarly, $\delta''_1\leq\gamma'''\leq\delta'_2$. With (xxiii) and (ix), $\alpha'''\in\sigma$ and $\gamma'''\in\sigma$. So using (xi), assume (after $\forall$-elim. with γ''' for α and prior to $\exists$-elim.): $\forall\gamma_{\gamma\in\sigma}\{\forall x_{x<z}\gamma(x)=\gamma'''(x) \supset \exists\beta(G(n, \beta, b) \,\&\, F(\gamma, \beta))\}$. By $\forall$-elims. with α''' and γ''' respectively for γ, and using (xxvi), $\exists\beta(G(n, \beta, b) \,\&\, F(\alpha''', \beta))$ and $\exists\beta(G(n, \beta, b) \,\&\, F(\gamma''', \beta))$. Prior to $\exists$-elims., assume **(xxviii)** $G(n, \beta', b) \,\&\, F(\alpha''', \beta')$, **(xxix)** $G(n, \zeta', b) \,\&\, F(\gamma''', \zeta')$. Now $\forall p 2^{n+2}|2^p(b)_1-\beta'((b)_0+p)|<2^{(b)_0+p}$ and $\forall p 2^{n+2}|2^p(b)_1-\zeta'((b)_0+p)| <2^{(b)_0+p}$. Writing $x=(b)_0$ and using *11.5, **(xxx)** $\forall p 2^{n+1}|\beta'(x+p)-\zeta'(x+p)|<2^{x+p}$. Now $2^{n+1}((1\cdot 2^{-n})(x+n+p) \dot- |\beta'(x+n+p)-\zeta'(x+n+p)|) = 2^{n+1+x+p} \dot- 2^{n+1}|\beta'(x+n+p) \dot- \zeta'(x+n+p)|$ [*R9.1, etc.] $>$ $2^{n+1+x+p} \dot- 2^{x+n+p}$ [(xxx), *6.20 with its hyp. $a<c$ given by $c \dot- b>0$] $=$ $2^{x+n+p}(2 \dot- 1) \geq 2^{x+n+p}$. So by $\forall$- and $\exists$-introds., **(xxxi)** $|\beta'-\zeta'|<\circ 1\cdot 2^{-n}$. Using (xii), (xxvii) and (xxviii), $\alpha\in[\delta_1, \delta_2] \,\&\, \alpha\doteq\alpha''' \,\&\, F(\alpha, \beta) \,\&\, F(\alpha''', \beta')$. So by (ii), $\beta\doteq\beta'$. Similarly $\zeta\doteq\zeta'$. So with (xxxi) and *R5.7, *R5.4, *R6.13: $|\beta-\zeta|<\circ 1\cdot 2^{-n}$.

15.3. Using *R10.3, we shall establish that the continuum is "indivisible" ("unzerlegbar"; Brouwer **1927** p. 66, **1928a** p. 11 lines 4–9, Heyting **1956** p. 46), at least by any predicate $C(\alpha)$ expressible by a formula $\mathrm{C}(\alpha)$ of the system.

*R10.4. $\vdash \delta_1,\delta_2 \in \mathrm{R}\ \&\ \delta_1 \circ\!\not> \delta_2$
$\&\ \forall\alpha\forall\gamma\{\alpha\in[\delta_1, \delta_2]\ \&\ \alpha \doteq \gamma \supset (\mathrm{C}(\alpha) \sim \mathrm{C}(\gamma))\}$
$\&\ \forall\alpha\{\alpha\in[\delta_1, \delta_2] \supset \mathrm{C}(\alpha) \vee \neg\mathrm{C}(\alpha)\}$
$\supset \forall\alpha(\alpha\in[\delta_1, \delta_2] \supset \mathrm{C}(\alpha)) \vee \forall\alpha(\alpha\in[\delta_1, \delta_2] \supset \neg\mathrm{C}(\alpha))$.

Proof. Assume $\delta_1,\delta_2\in\mathrm{R}$ and the other hyps. of the implication. Using *R1.11, assume **(i)** $\delta_1'\in\mathrm{R}'\ \&\ \delta_1' \doteq \delta_1$. Now using *R6.14 and *R8.3: **(ii)** $\delta_1' \circ\!\not> \delta_2$, **(iii)** $\forall\alpha\forall\gamma\{\alpha\in[\delta_1', \delta_2]\ \&\ \alpha\doteq\gamma \supset (\mathrm{C}(\alpha) \sim \mathrm{C}(\gamma))\}$, **(iv)** $\forall\alpha\{\alpha\in[\delta_1', \delta_2] \supset \mathrm{C}(\alpha) \vee \neg\mathrm{C}(\alpha)\}$. Let $\mathrm{F}(\alpha, \beta)$ be $[\beta\doteq 0\ \&\ \mathrm{C}(\alpha)] \vee [\beta\doteq 1\ \&\ \neg\mathrm{C}(\alpha)]$. Toward (v), assume $\alpha\in[\delta_1', \delta_2]\ \&\ \alpha\doteq\gamma\ \&\ \mathrm{F}(\alpha, \beta)\ \&\ \mathrm{F}(\gamma, \zeta)$. By (iii), $\mathrm{C}(\alpha) \sim \mathrm{C}(\gamma)$. Now from $\mathrm{F}(\alpha, \beta)\ \&\ \mathrm{F}(\gamma, \zeta)$ by cases $(\mathrm{C}(\alpha), \neg\mathrm{C}(\alpha))$, $\beta\doteq\zeta$ and (using *R9.2, *R0.7, *R1.7) $\beta,\zeta\in\mathrm{R}$. By &-, ⊃- and ∀-introds., **(v)** $\mathfrak{A}(\mathrm{F}, \delta_1', \delta_2)$. Assuming $\alpha\in[\delta_1', \delta_2]$ and using (iv), $\mathrm{C}(\alpha) \vee \neg\mathrm{C}(\alpha)$, whence by cases (using *R1.4), $\exists\beta\mathrm{F}(\alpha, \beta)$. So **(vi)** $\forall\alpha\{\alpha\in[\delta_1', \delta_2] \supset \exists\beta\mathrm{F}(\alpha, \beta)\}$. Using $\delta_1'\in\mathrm{R}$ (from (i)), $\delta_2\in\mathrm{R}$, (ii), (v) and (vi) in *R10.3, assume: **(vii)** $\forall\alpha\forall\gamma\forall\beta\forall\zeta\{|\alpha-\gamma| <\circ 1\cdot 2^{-\mathrm{m}}\ \&\ \alpha,\gamma\in[\delta_1', \delta_2]\ \&\ \mathrm{F}(\alpha, \beta)\ \&\ \mathrm{F}(\gamma, \zeta) \supset |\beta-\zeta| <\circ 1\cdot 2^{-1}\}$. We shall deduce **(a)** $\forall\alpha\forall\gamma\{|\alpha-\gamma| <\circ 1\cdot 2^{-\mathrm{m}}\ \&\ \alpha,\gamma\in[\delta_1', \delta_2] \supset (\mathrm{C}(\alpha) \sim \mathrm{C}(\gamma))\}$. Assume $|\alpha-\gamma| <\circ 1\cdot 2^{-\mathrm{m}}\ \&\ \alpha,\gamma\in[\delta_1', \delta_2]$. By (vi), assume $\mathrm{F}(\alpha, \beta)$ and $\mathrm{F}(\gamma, \zeta)$. By (vii), $|\beta-\zeta| <\circ 1\cdot 2^{-1}$, whence by *R6.7, $|\beta-\zeta| \circ\!\not> 1\cdot 2^{-1}$. Toward $\mathrm{C}(\alpha) \supset \mathrm{C}(\gamma)$, assume $\mathrm{C}(\alpha)$. By $\mathrm{F}(\alpha, \beta)$, $\beta\doteq 0$. Now $\zeta\doteq 1$ would lead to $|\beta-\zeta| \doteq |0-1| \doteq 1$ [*R9.14] $\circ> 1\cdot 2^{-1}$ [*R9.7]. So $\neg\zeta\doteq 1$, whence from $\mathrm{F}(\gamma, \zeta)$, $\zeta\doteq 0\ \&\ \mathrm{C}(\gamma)$, whence $\mathrm{C}(\gamma)$. Thus by ⊃-introd., $\mathrm{C}(\alpha) \supset \mathrm{C}(\gamma)$. Similarly, $\mathrm{C}(\gamma) \supset \mathrm{C}(\alpha)$. By &-introd., $\mathrm{C}(\alpha) \sim \mathrm{C}(\gamma)$. By ⊃- and ∀-introds., (a). By *R8.5, **(viii)** $\delta_1'\in[\delta_1', \delta_2]$. Thence by (iv): **(ix)** $\mathrm{C}(\delta_1') \vee \neg\mathrm{C}(\delta_1')$.

Case 1: $\mathrm{C}(\delta_1')$. We deduce **(b)** $\alpha\in\mathrm{R}'\ \&\ \alpha(\mathrm{m}+2)=\delta_1'(\mathrm{m}+2)+\mathrm{a}\ \&\ \alpha\in[\delta_1', \delta_2] \supset \mathrm{C}(\alpha)$ by induction on a with a double basis (IM p. 193), thus: First basis. Assume $\alpha\in\mathrm{R}'$, **(i′)** $\alpha(\mathrm{m}+2)=\delta_1'(\mathrm{m}+2)+0$ and **(ii′)** $\alpha\in[\delta_1', \delta_2]$. Now $|\alpha-\delta_1'| \circ\!\not> |\alpha-\alpha(\mathrm{m}+2)2^{-(\mathrm{m}+2)}|+|\alpha(\mathrm{m}+2)2^{-(\mathrm{m}+2)}-\delta_1'|$ [*R6.17] $= |\alpha-\alpha(\mathrm{m}+2)2^{-(\mathrm{m}+2)}|+|\delta_1'(\mathrm{m}+2)2^{-(\mathrm{m}+2)}-\delta_1'|$ [(i′), *R9.5, etc.] $\circ\!\not> 1\cdot 2^{-(\mathrm{m}+2)}+1\cdot 2^{-(\mathrm{m}+2)}$ [*R9.15 twice, with $\alpha\in\mathrm{R}'$ or $\delta_1'\in\mathrm{R}'$, *R6.16 (twice)] $\doteq 1\cdot 2^{-(\mathrm{m}+1)}$ [*R9.10, *R9.3] $<\circ 1\cdot 2^{-\mathrm{m}}$ [*R9.7]. So, using also (ii′) and (viii), $|\alpha-\delta_1'| <\circ 1\cdot 2^{-\mathrm{m}}\ \&\ \alpha,\delta_1'\in[\delta_1', \delta_2]$. By (a), $\mathrm{C}(\alpha) \sim \mathrm{C}(\delta_1')$. Using case hyp., $\mathrm{C}(\alpha)$. Second basis. Assume $\alpha\in\mathrm{R}'$, **(i″)** $\alpha(\mathrm{m}+2)=\delta_1'(\mathrm{m}+2)+1$, **(ii″)** $\alpha\in[\delta_1', \delta_2]$. Now $|\alpha-\delta_1'| \circ\!\not>$

$|\alpha-\alpha(m+2)2^{-(m+2)}|+|\alpha(m+2)2^{-(m+2)}-\delta'_1(m+2)2^{-(m+2)}|+|\delta'_1(m+2)\cdot 2^{-(m+2)}-\delta'_1|$ [*R6.17] $\circ\!\not> 1\cdot 2^{-(m+2)}+|\alpha(m+2)2^{-(m+2)}-\delta'_1(m+2)2^{-(m+2)}| +1\cdot 2^{-(m+2)}$ [*R9.15] $= 1\cdot 2^{-(m+2)}+|(\delta'_1(m+2)+1)2^{-(m+2)}-\delta'_1(m+2)\cdot 2^{-(m+2)}|+1\cdot 2^{-(m+2)}$ [(i″)] $\doteq 1\cdot 2^{-(m+2)}+1\cdot 2^{-(m+2)}+1\cdot 2^{-(m+2)}$ [*R9.10, *R5.6] $\doteq 3\cdot 2^{-(m+2)}$ [*R9.10] $<\!\circ\, 1\cdot 2^{-m}$ [*R9.7]. So, using also (ii″) and (viii), $|\alpha-\delta'_1|<\!\circ 1\cdot 2^{-m}$ & $\alpha,\delta'_1\in[\delta'_1, \delta_2]$. By (a), $C(\alpha) \sim C(\delta'_1)$, and by Case 1 hyp., $C(\alpha)$. IND. STEP. Assume $\alpha\in R'$, **(i‴)** $\alpha(m+2)=\delta'_1(m+2)+a''$, **(ii‴)** $\alpha\in[\delta'_1, \delta_2]$, and the (second) hyp. ind. By *R9.2: **(iii‴)** $(\delta'_1(m+2)+a')2^{-(m+2)}\in R'$. By *R9.1, etc.: **(iv‴)** $((\delta'_1(m+2)+a')\cdot 2^{-(m+2)})(m+2)=\delta'_1(m+2)+a'$. Also: **(v‴)** $(\delta'_1(m+2)+a')2^{-(m+2)} \not<\!\circ\, (\delta'_1(m+2)+1)2^{-(m+2)}$ [*R9.8] $\not<\!\circ\, \delta'_1$ [*R9.17]. And: **(vi‴)** $(\delta'_1(m+2)+a')2^{-(m+2)} = ((\delta'_1(m+2)+a'')\dot{-}1)2^{-(m+2)}$ [*R9.5] $= (\alpha(m+2)\dot{-}1)\cdot 2^{-(m+2)}$ [(i‴)] $\circ\!\not> \alpha$ [*R9.16] $\circ\!\not> \delta_2$ [(ii‴), *R8.6 with (ii), and $\alpha,\delta'_1,\delta_2\in R$]. Combining (v‴) and (vi‴) by *R8.6 (with (ii), (iii‴), etc.): **(vii‴)** $(\delta'_1(m+2)+a')2^{-(m+2)}\in[\delta'_1, \delta_2]$. From (iii‴), (iv‴) and (vii‴) by the hyp. ind.: **(viii‴)** $C((\delta'_1(m+2)+a')2^{-(m+2)})$. Further: **(ix‴)** $|\alpha-(\delta'_1(m+2)+a')2^{-(m+2)}| \circ\!\not> |\alpha-(\delta'_1(m+2)+a'')2^{-(m+2)}|+|(\delta'_1(m+2)+a'')2^{-(m+2)}-(\delta'_1(m+2)+a')2^{-(m+2)}| \doteq |\alpha-(\delta'_1(m+2)+a'')2^{-(m+2)}|+1\cdot 2^{-(m+2)} = |\alpha-\alpha(m+2)2^{-(m+2)}|+1\cdot 2^{-(m+2)}$ [(i‴)] $\circ\!\not> 1\cdot 2^{-(m+2)}+1\cdot 2^{-(m+2)} \doteq 1\cdot 2^{-(m+1)} <\!\circ\, 1\cdot 2^{-m}$. By (a) with (ix‴), (ii‴) and (vii‴), $C(\alpha) \sim C((\delta'_1(m+2)+a')2^{-(m+2)})$, whence by (viii‴), $C(\alpha)$. Now from (b) we deduce **(c)** $\forall\alpha(\alpha\in[\delta'_1, \delta_2] \supset C(\alpha))$ thus. Assume $\alpha\in[\delta'_1, \delta_2]$. Using *R1.11, assume $\alpha'\in R'$ & $\alpha'\doteq\alpha$. By *R8.2, $\alpha'\in[\delta'_1, \delta_2]$, whence by (ii), etc. and *R8.6, $\alpha'\not<\!\circ\delta'_1$. Using *R6.24 assume $\alpha''\in R'$ & $\alpha''\doteq\alpha'$ & $\alpha''\geq\delta'_1$. By *R8.2, $\alpha''\in[\delta'_1, \delta_2]$. From $\alpha''\geq\delta'_1$, assume $\alpha''(m+2)= \delta'_1(m+2)+a$. By (b), $C(\alpha'')$. Also $\alpha''\doteq\alpha$. By (iii), $C(\alpha)$. By $\supset$- and $\forall$-introd., (c). From (c), using (i) and *R8.3, $\forall\alpha(\alpha\in[\delta_1, \delta_2] \supset C(\alpha))$. By $\vee$-introd., $\forall\alpha(\alpha\in[\delta_1, \delta_2] \supset C(\alpha)) \vee \forall\alpha(\alpha\in[\delta_1, \delta_2] \supset \neg C(\alpha))$.

CASE 2: Similarly, $\forall\alpha(\alpha\in[\delta_1, \delta_2] \supset \neg C(\alpha))$.

§ 16. The structure of the continuum. **16.1.** In "Die Struktur des Kontinuums" (Brouwer **1928a**), Brouwer discusses seven properties of the continuum (pp. 6–7), giving in most cases both intuitionistic counterexamples to classical theorems and (in general, without proof) intuitionistically true analogues of these classical results. For the property of discreteness, our *R9.23 corresponds to Brouwer's counter-example (bottom p. 7 item 1); no intuitionistic analogue is given. For the next property, that of ordering (item 2 pp. 7–9), see 14.6 above where Brouwer's axioms for virtual order are derived for $<$. For the remaining properties we concentrate on obtaining the intuitionistic theorems rather than the counterexamples. The results appear as *R12.2 (density in itself), *R12.4 (compactness), *R13.8 (everywhere density), *R14.12 with *R14.11 (separability in itself), *R14.13 (free connectedness). As remarked in the introduction, in the case of the latter two we simplify and (in the case of the last) amend Brouwer's formulation. (These are respectively Brouwer's items 3, 7, 6, 4, 5.)

16.2. The relations of inclusion and proper inclusion for closed intervals are expressed formally by the formulas on the right (abbreviated on the left) in *R11.1 and *R11.2, respectively. *R11.3 is proved in Brouwer **1924–7** II p. 454 Footnote 1.

*R11.1. $\vdash [\delta_1, \delta_2] \subseteq [\eta_1, \eta_2] \sim$
$\delta_1, \delta_2, \eta_1, \eta_2 \in R \ \& \ \forall\alpha(\alpha \in [\delta_1, \delta_2] \supset \alpha \in [\eta_1, \eta_2])$.

*R11.2. $\vdash [\delta_1, \delta_2] \subset [\eta_1, \eta_2] \sim$
$[\delta_1, \delta_2] \subseteq [\eta_1, \eta_2] \ \& \ \neg[\eta_1, \eta_2] \subseteq [\delta_1, \delta_2]$.

*R11.3. $\vdash [\delta_1, \delta_2] \subseteq [\eta_1, \eta_2] \sim \delta_1, \delta_2 \in [\eta_1, \eta_2]$.

*R11.4. $\vdash \delta_1, \delta_2 \in R \supset [\delta_1, \delta_2] \subseteq [\delta_1, \delta_2]$.

*R11.5. $\vdash \neg[\delta_1, \delta_2] \subset [\delta_1, \delta_2]$.

*R11.6. $\vdash [\delta_1, \delta_2] \subseteq [\eta_1, \eta_2] \ \& \ [\eta_1, \eta_2] \subseteq [\theta_1, \theta_2]$
$\supset [\delta_1, \delta_2] \subseteq [\theta_1, \theta_2]$.

*R11.7. $\vdash [\delta_1, \delta_2] \subset [\eta_1, \eta_2] \ \& \ [\eta_1, \eta_2] \subset [\theta_1, \theta_2]$
$\supset [\delta_1, \delta_2] \subset [\theta_1, \theta_2]$.

Proofs. *R11.3. I. Assume $[\delta_1, \delta_2] \subseteq [\eta_1, \eta_2]$, whence $\delta_1, \delta_2 \in R$ and $\delta_1, \delta_2 \in [\delta_1, \delta_2] \supset \delta_1, \delta_2 \in [\eta_1, \eta_2]$, whence, using *R8.5, $\delta_1, \delta_2 \in [\eta_1, \eta_2]$. II. Assume $\delta_1, \delta_2 \in [\eta_1, \eta_2]$, whence **(i)** $\delta_1, \delta_2, \eta_1, \eta_2 \in R$ and **(ii)** $\neg(\delta_1 \dot{<} \eta_1 \ \& \ \delta_1 \dot{<} \eta_2) \ \& \ \neg(\delta_1 \dot{>} \eta_1 \ \& \ \delta_1 \dot{>} \eta_2)$ and **(iii)** $\neg(\delta_2 \dot{<} \eta_1 \ \& \ \delta_2 \dot{<} \eta_2) \ \& \ \neg(\delta_2 \dot{>} \eta_1 \ \& \ \delta_2 \dot{>} \eta_2)$. Prior to $\supset$- and $\forall$-introds., assume **(iv)** $\alpha \in [\delta_1, \delta_2]$, whence **(v)** $\neg(\alpha \dot{<} \delta_1 \ \& \ \alpha \dot{<} \delta_2) \ \& \ \neg(\alpha \dot{>} \delta_1 \ \& \ \alpha \dot{>} \delta_2)$. Assume, for reductio ad absurdum, **(a)** $\alpha \dot{<} \eta_1 \ \& \ \alpha \dot{<} \eta_2$. Assuming $\alpha \dot{>} \delta_1 \lor \alpha \doteq \delta_1$, we deduce by cases using *R7.4 (with (i), (iv)) and *R7.2, $\delta_1 \dot{<} \eta_1 \ \& \ \delta_1 \dot{<} \eta_2$, contradicting (ii). Hence $\neg(\alpha \dot{>} \delta_1 \lor \alpha \doteq \delta_1)$, whence by *63, $\neg\alpha \dot{>} \delta_1$ & $\neg\alpha \doteq \delta_1$. By *R7.11 (and (i), (iv)), $\alpha \dot{<} \delta_1$. Similarly $\alpha \dot{<} \delta_2$, and by &-introd. $\alpha \dot{<} \delta_1 \ \& \ \alpha \dot{<} \delta_2$, contradicting (v). So, rejecting (a), $\neg(\alpha \dot{<} \eta_1 \ \& \ \alpha \dot{<} \eta_2)$. Similarly, $\neg(\alpha \dot{>} \eta_1 \ \& \ \alpha \dot{>} \eta_2)$. By &-introd., $\neg(\alpha \dot{<} \eta_1 \ \& \ \alpha \dot{<} \eta_2) \ \& \ \neg(\alpha \dot{>} \eta_1 \ \& \ \alpha \dot{>} \eta_2)$, whence with (i) and (iv), $\alpha \in [\eta_1, \eta_2]$.

*R11.6. Assume **(i)** $[\delta_1, \delta_2] \subseteq [\eta_1, \eta_2]$ and **(ii)** $[\eta_1, \eta_2] \subseteq [\theta_1, \theta_2]$.

By (i) and *R11.3, $\delta_1, \delta_2 \in [\eta_1, \eta_2]$. So by (ii), $\delta_1, \delta_2 \in [\theta_1, \theta_2]$, and by *R11.3, $[\delta_1, \delta_2] \subseteq [\theta_1, \theta_2]$.

*R11.7. Assume **(i)** $[\delta_1, \delta_2] \subset [\eta_1, \eta_2]$ and **(ii)** $[\eta_1, \eta_2] \subset [\theta_1, \theta_2]$. By *R11.6, $[\delta_1, \delta_2] \subseteq [\theta_1, \theta_2]$. Assuming $[\theta_1, \theta_2] \subseteq [\delta_1, \delta_2]$, $[\theta_1, \theta_2] \subseteq [\eta_1, \eta_2]$ follows by (i) and *R11.6, contradicting (ii). Hence $\neg[\theta_1, \theta_2] \subseteq [\delta_1, \delta_2]$.

16.3. We shall represent sequences of real numbers by sequences $\lambda x\varphi(2^0 3^x)$, $\lambda x\varphi(2^1 3^x)$, $\lambda x\varphi(2^2 3^x)$, ..., where $(n)\lambda x\varphi(2^n 3^x) \in R$.

^{x}R12.1. $\quad \varphi_{[n]}(x) = \varphi(2^n 3^x)$.

When "$\varphi_{[n]}$" is used without argument, it shall abbreviate $\lambda x\varphi_{[n]}(x)$ (i.e. $\lambda x f_i(n, x, \varphi)$ for the function symbol f_i introduced with the axiom ^{x}R12.1; cf. 5.1).

16.4. We now establish the property of the continuum of being "dense in itself" ("in sich dicht"; Brouwer **1928a** p. 9 line 6 from below to p. 10 line 2, with p. 7 lines 11–12). That is, we show that, for each r.n.g. α there is a sequence of properly nested closed intervals each containing α, such that each r.n.g. β contained in each of the intervals $\doteq \alpha$.

*R12.2. $\quad \vdash \alpha \in R \supset \exists\varphi\exists\psi\{\forall n[\varphi_{[n+1]}, \psi_{[n+1]}] \subset [\varphi_{[n]}, \psi_{[n]}]$
$\quad \& \ \forall n\alpha \in [\varphi_{[n]}, \psi_{[n]}] \ \& \ \forall\beta(\forall n\beta \in [\varphi_{[n]}, \psi_{[n]}] \supset \beta \doteq \alpha)\}$.

Proof. Assume $\alpha \in R$. Using *R1.11, assume: **(i)** $\alpha' \in R' \ \& \ \alpha' \doteq \alpha$. Using Lemma 5.3 (a), introduce φ and ψ: **(ii)** $\forall a\varphi(a) = ((\alpha'((a)_0) \dot{-} 1) \cdot 2^{-(a)_0})((a)_1)$, **(iii)** $\forall a\psi(a) = ((\alpha'((a)_0) + 2)2^{-(a)_0})((a)_1)$. Now $\varphi_{[n]}(x) = ((\alpha'(n) \dot{-} 1)2^{-n})(x)$, whence by (^{x}0.1 and) $\forall$-introds., **(iv)** $\forall n\varphi_{[n]} = (\alpha'(n) \dot{-} 1)2^{-n}$. Similarly, **(v)** $\forall n\psi_{[n]} = (\alpha'(n) + 2)2^{-n}$. By *R9.2 (and *R0.7), **(vi)** $\forall n\varphi_{[n]}, \psi_{[n]} \in R$. Using *R9.8, **(vii)** $\forall n\varphi_{[n]} <\circ \psi_{[n]}$, whence by *R6.7, **(viii)** $\forall n\varphi_{[n]} \circ\not> \psi_{[n]}$.

I. Using *R0.5a–c and (i): **(ix)** $2\alpha'(n) \dot{-} 2 \leq \alpha'(n+1) \dot{-} 1$. Now: **(x)** $\varphi_{[n]} = (\alpha'(n) \dot{-} 1)2^{-n}$ [(iv)] $= (2\alpha'(n) \dot{-} 2)2^{-(n+1)}$ [*R9.3] $\leq\circ (\alpha'(n+1) \dot{-} 1) \cdot 2^{-(n+1)}$ [(ix), *R9.8, *R9.5] $= \varphi_{[n+1]} <\circ \psi_{[n+1]}$ [(vii)] $= (\alpha'(n+1)+2) \cdot 2^{-(n+1)} <\circ (2\alpha'(n)+4)2^{-(n+1)}$ [*R0.5a, etc.] $= (\alpha'(n)+2)2^{-n} = \psi_{[n]}$. Using (x), *R7.7 (and (vi)), we readily obtain: **(xi)** $[\varphi_{[n+1]}, \psi_{[n+1]}] \subseteq [\varphi_{[n]}, \psi_{[n]}]$, **(xii)** $\psi_{[n]} \notin [\varphi_{[n+1]}, \psi_{[n+1]}]$. By *R8.5, $\psi_{[n]} \in [\varphi_{[n]}, \psi_{[n]}]$, which with (xii) gives **(xiii)** $\neg[\varphi_{[n]}, \psi_{[n]}] \subseteq [\varphi_{[n+1]}, \psi_{[n+1]}]$. From (xi) and (xiii) by &- and $\forall$-introd., **(xiv)** $\forall n[\varphi_{[n+1]}, \psi_{[n+1]}] \subset [\varphi_{[n]}, \psi_{[n]}]$.

II. Also $\alpha' \not<\circ (\alpha'(n) \dot{-} 1)2^{-n}$ [*R9.16, (i)] $= \varphi_{[n]}$. And $\alpha' \circ\not> (\alpha'(n)+1) \cdot$

2^{-n} [*R9.17] $\circ\!\not>$ $(\alpha'(n)+2)2^{-n}$ [*R9.8, *R6.7] $= \psi_{[n]}$. So $\alpha' \not<\!\circ\, \varphi_{[n]}$ & $\alpha' \circ\!\not> \psi_{[n]}$, whence by *R8.6 (with (i), (vi), (viii)), $\alpha' \in [\varphi_{[n]}, \psi_{[n]}]$, and by *R8.2, $\alpha \in [\varphi_{[n]}, \psi_{[n]}]$. By $\forall$-introd.: **(xv)** $\forall n \alpha \in [\varphi_{[n]}, \psi_{[n]}]$.

III. Assume $\forall n \beta \in [\varphi_{[n]}, \psi_{[n]}]$. Using *R1.11: **(xvi)** $\beta' \in R'$ & $\beta' \doteq \beta$. By *R8.2, **(xvii)** $\forall n \beta' \in [\varphi_{[n]}, \psi_{[n]}]$. Assume for reductio ad absurdum, **(a)** $\alpha' <\!\circ\, \beta'$. Assume $\forall p 2^k(\beta'(x+p) \dot{-} \alpha'(x+p)) \geq 2^{x+p}$. By $\forall$-elim., $2^k(\beta'(x+2+k) \dot{-} \alpha'(x+2+k)) \geq 2^{x+2+k}$, whence $(\beta'(x+2+k) \dot{-} \alpha'(x+2+k)) \geq 2^{x+2} > 3$. So $\beta'(x+2+k) > \alpha'(x+2+k)+3$, and $\beta'(x+2+k) \dot{-} 1 > \alpha'(x+2+k)+2$. Now $\beta' \not<\!\circ$ $(\beta'(x+2+k) \dot{-} 1)2^{-(x+2+k)}$ $\circ\!>$ $(\alpha'(x+2+k)+2)2^{-(x+2+k)} = \psi_{[x+2+k]}$. So by *R8.6 (with (viii), (vi)), $\beta' \notin [\varphi_{[x+2+k]}, \psi_{[x+2+k]}]$, contradicting (xvii). Rejecting (a): **(xviii)** $\alpha' \not<\!\circ\, \beta'$. Similarly (using *R9.17, (iv)): **(xix)** $\alpha' \circ\!\not> \beta'$. By (xviii), (xix) (with (i) and (xvi)) and *R6.5, $\beta' \doteq \alpha'$, whence $\beta \doteq \alpha$. By $\supset$- and $\forall$-introd.: **(xx)** $\forall \beta(\forall n \beta \in [\varphi_{[n]}, \psi_{[n]}] \supset \beta \doteq \alpha)$.

16.5. We establish a form of compactness defined in Brouwer **1928a** p. 12 lines 20–32. The assertion is that in the continuum there is no "hollow interval nest" ("hohle Intervallschachtelung"), that is, no sequence of nested closed intervals I_n such that, for each r.n.g. α, there is an integer n_α such that α is not in I_{n_α}. This is expressed formally by *R12.4. *R12.3 is a useful lemma.

*R12.3. $\vdash \forall n [\varphi_{[n+1]}, \psi_{[n+1]}] \subseteq [\varphi_{[n]}, \psi_{[n]}] \supset$
$\forall n \forall m_{m \leq n} [\varphi_{[n]}, \psi_{[n]}] \subseteq [\varphi_{[m]}, \psi_{[m]}]$.

*R12.4. $\vdash \neg \exists \varphi \exists \psi \{\forall n [\varphi_{[n+1]}, \psi_{[n+1]}] \subseteq [\varphi_{[n]}, \psi_{[n]}]$
& $\forall \alpha_{\alpha \in R} \exists b \alpha \notin [\varphi_{[b]}, \psi_{[b]}]\}$.

Proofs. *R12.3. Assume the hyp. **(i)**. Using *R11.1, $\varphi_{[m]}, \psi_{[m]} \in R$. We deduce $[\varphi_{[m+p]}, \psi_{[m+p]}] \subseteq [\varphi_{[m]}, \psi_{[m]}]$ by ind. on p, thus. Basis. Use *R11.4. Ind. step. Using *R11.6, $[\varphi_{[m+p']}, \psi_{[m+p']}] \subseteq [\varphi_{[m+p]}, \psi_{[m+p]}]$ [(i)] $\subseteq [\varphi_{[m]}, \psi_{[m]}]$ [hyp. ind.].

*R12.4. Assume (prior to $\exists$-elims.): **(i)** $\forall n [\varphi_{[n+1]}, \psi_{[n+1]}] \subseteq [\varphi_{[n]}, \psi_{[n]}]$, and **(ii)** $\forall \alpha_{\alpha \in R} \exists b \alpha \notin [\varphi_{[b]}, \psi_{[b]}]$. By (i): **(iii)** $\forall n \varphi_{[n]}, \psi_{[n]} \in R$. Using *R1.11, assume: **(iv)** $\varphi' \in R'$ & $\varphi' \doteq \varphi_{[0]}$, **(v)** $\psi' \in R'$ & $\psi' \doteq \psi_{[0]}$. Using Lemma 5.3 (a), introduce: **(vi)** $\delta_1 = 0$ and **(vii)** $\delta_2 = \varphi'(0) + \psi'(0) + 1$. By *R9.2: **(viii)** $\delta_1, \delta_2 \in R'$. By *R9.18: **(ix)** $\delta_1 \circ\!\not> \delta_2$. Using *R9.1, $\delta_2(x+1) = 2\delta_2(x)$; so by $\forall$-introd., **(x)** $\forall x \delta_2(x+1) = 2\delta_2(x)$. Further $\delta_1(x) = 0 \leq \delta_2(x)$; so by $\forall$-introd., **(xi)** $\delta_1 \leq \delta_2$. We deduce **(a)** $[\varphi_{[n]}, \psi_{[n]}] \subseteq [\delta_1, \delta_2]$ by induction on n, thus. Basis. By *R9.18, $\varphi' \not<\!\circ\, \delta_1$. Also, $\varphi' \circ\!\not>$ $(\varphi'(0)+1)2^{-0}$ [*R9.17, (iv)] $\circ\!\not>$ $(\varphi'(0)+\psi'(0)+1)2^{-0}$

[*R9.8, *R9.5; *R6.7, *R6.8] $= \delta_2$. So by *R8.6 (and (ix), etc.), $\varphi' \in [\delta_1, \delta_2]$, and by *R8.2 and (iv), $\varphi_{[0]} \in [\delta_1, \delta_2]$. Similarly, $\psi_{[0]} \in [\delta_1, \delta_2]$. Using *R11.3, $[\varphi_{[0]}, \psi_{[0]}] \subseteq [\delta_1, \delta_2]$. IND. STEP. Using *R11.6, $[\varphi_{[n+1]}, \psi_{[n+1]}] \subseteq [\varphi_{[n]}, \psi_{[n]}]$ [(i)] $\subseteq [\delta_1, \delta_2]$ [hyp. ind.]. Using (viii) and (xi) in *R10.2: **(xii)** $\mathrm{Spr}(\sigma)$ & $\sigma(1)=0$ & $\forall a[\sigma(a)=0 \supset \exists b \forall s(\sigma(a*2^{s+1})=0 \supset s \leq b)]$, **(xiii)** $\forall \alpha[\alpha \in \sigma \sim \alpha \in R'$ & $\delta_1 \leq \alpha \leq \delta_2]$. Assuming $\alpha \in \sigma$, (xiii) (and *R0.7) gives $\alpha \in R$, whence by (ii), $\exists b \alpha \notin [\varphi_{[b]}, \psi_{[b]}]$. By $\supset$- and $\forall$-introds., **(xiv)** $\forall \alpha_{\alpha \in \sigma} \exists b \alpha \notin [\varphi_{[b]}, \psi_{[b]}]$. Using (xii) and (xiv) in *27.8, assume: **(xv)** $\forall \alpha_{\alpha \in \sigma} \exists b \forall \gamma_{\gamma \in \sigma} \{\forall x_{x<z} \gamma(x)=\alpha(x) \supset \gamma \notin [\varphi_{[b]}, \psi_{[b]}]\}$. Toward (xx), assume **(xvi)** $\mathrm{Seq}(a)$ & $\mathrm{lh}(a)=z$ & $\forall x_{x<z}(a)_x \dot{-} 1 \leq \delta_2(x)$ & $\forall x_{x<z \dot{-} 1} |2((a)_x \dot{-} 1)-((a)_{x'} \dot{-} 1)| \leq 1$, which we abbreviate "$B(a, z, \delta_2)$". Introduce υ by:

$$\textbf{(xvii)} \qquad \forall x \upsilon(x) = \begin{cases} (a)_x \dot{-} 1 & \text{if } x<z, \\ 2^{x \dot{-} (z \dot{-} 1)}((a)_{z \dot{-} 1} \dot{-} 1) & \text{if } x \geq z. \end{cases}$$

Thus **(xviii)** $\forall x_{x<z} \upsilon(x)=(a)_x \dot{-} 1$. We shall deduce **(b)** $|2\upsilon(x)-\upsilon(x')| \leq 1$. CASE 1: $x<z \dot{-} 1$. Then $z>1$. So $x' < (z \dot{-} 1)+1 = z$. So $|2\upsilon(x)-\upsilon(x')| = |2((a)_x \dot{-} 1)-((a)_{x'} \dot{-} 1)| \leq 1$ [(xvi), case hyp.]. CASE 2: $x=z \dot{-} 1$. SUBCASE 2.1: $z=0$. Then by (xvi), $a=1$. So $|2\upsilon(x)-\upsilon(x')| = |0-0| \leq 1$. SUBCASE 2.2: $z \geq 1$. Then $x<z$ and $x'=z$; so $|2\upsilon(x)-\upsilon(x')| = |2((a)_x \dot{-} 1)-2((a)_x \dot{-} 1)| = 0 \leq 1$. CASE 3: $x>z \dot{-} 1$. Then $x \geq z$. So $|2\upsilon(x)-\upsilon(x')| = |2 \cdot 2^{x \dot{-} (z \dot{-} 1)}((a)_{z \dot{-} 1} \dot{-} 1)-2^{x' \dot{-} (z \dot{-} 1)}((a)_{z \dot{-} 1} \dot{-} 1)| = 0 \leq 1$. From (b) by $\forall$-introd., **(xix)** $\upsilon \in R'$. We next deduce **(c)** $\upsilon(x) \leq \delta_2(x)$ by ind. on x, thus. BASIS: By cases ($z=0$, $z>0$), $\upsilon(0) = (a)_0 \dot{-} 1$ [(xvii)] $\leq \delta_2(0)$ [(xvi)]. IND. STEP. CASE 1: $x'<z$. Use (xvii) and (xvi). CASE 2: $x' \geq z$. Then $\upsilon(x') = 2^{x' \dot{-} (z \dot{-} 1)}((a)_{z \dot{-} 1} \dot{-} 1) = 2 \cdot 2^{x \dot{-} (z \dot{-} 1)}((a)_{z \dot{-} 1} \dot{-} 1) = 2\upsilon(x)$ [by subcases: $x'>z$, $x'=z$] $\leq 2\delta_2(x)$ [hyp. ind.] $= \delta_2(x')$ [(x)]. From (c) by $\forall$-introd., $\upsilon \leq \delta_2$. Similarly to (xi), $\delta_1 \leq \upsilon$. With (xiii) and (xix), $\upsilon \in \sigma$. By (xv), $\exists b \forall \gamma_{\gamma \in \sigma} \{\forall x_{x<z} \gamma(x)= \upsilon(x) \supset \gamma \notin [\varphi_{[b]}, \psi_{[b]}]\}$. Using (xviii), $\exists b \forall \gamma_{\gamma \in \sigma} \{\forall x_{x<z} \gamma(x)=(a)_x \dot{-} 1 \supset \gamma \notin [\varphi_{[b]}, \psi_{[b]}]\}$. By $\supset$- and $\forall$-introd. (discharging (xvi)): **(xx)** $\forall a\{B(a, z, \delta_2) \supset \exists b \forall \gamma_{\gamma \in \sigma} \{\forall x_{x<z} \gamma(x)=(a)_x \dot{-} 1 \supset \gamma \notin [\varphi_{[b]}, \psi_{[b]}]\}\}$. By Remark 4.1 and 5.5 preceding *15.1, $B(a, z, \delta_2) \vee \neg B(a, z, \delta_2)$. So by cases, $\forall a \exists b\{[B(a, z, \delta_2)$ & $\forall \gamma_{\gamma \in \sigma} \{\forall x_{x<z} \gamma(x)=(a)_x \dot{-} 1 \supset \gamma \notin [\varphi_{[b]}, \psi_{[b]}]\}] \vee [\neg B(a, z, \delta_2)$ & $b=0]\}$. Using *2.2: **(xxi)** $\forall a\{[B(a, z, \delta_2)$ & $\forall \gamma_{\gamma \in \sigma} \{\forall x_{x<z} \gamma(x)=(a)_x \dot{-} 1 \supset \gamma \notin [\varphi_{[\beta(a)]}, \psi_{[\beta(a)]}]\}] \vee [\neg B(a, z, \delta_2)$ & $\beta(a)=0]\}$. Let $t=\max_{i \leq \bar{\delta}_2(z)} \beta(i)$. We deduce a contradiction thus. By (a) and *R11.3, $\varphi_{[t]} \in [\delta_1, \delta_2]$. Using *R1.11: **(xxii)** $\varphi'_t \in R'$ & $\varphi'_t \doteq \varphi_{[t]}$. Now $\varphi'_t \in [\delta_1, \delta_2]$, whence by *R8.6 (with (ix), etc.), $\varphi'_t{}^{\circ} \not> \delta_2$. So using *R6.22,

(xxiii) $\varphi''_t \in R'$ & $\varphi''_t \doteq \varphi_{[t]}$ & $\varphi''_t \leq \delta_2$. Similarly to (xi), $\delta_1 \leq \varphi''_t$. So using (xiii): **(xxiv)** $\varphi''_t \in \sigma$. Using (xxiii) with ˣ23.1, *B21, *3.12, etc., $\overline{\varphi''_t}(z) \leq \overline{\delta_2}(z)$. So by *H7, letting $s = \beta(\overline{\varphi''_t}(z))$: **(xxv)** $s \leq t$. Employing *23.5 and *23.2, we readily deduce $B(\overline{\varphi''_t}(z), z, \delta_2)$ from (xxiii), using cases ($z \leq 1$, $z > 1$) for the last part. So by (xxi) and *23.2: $\forall\gamma_{\gamma\in\sigma}\{\forall x_{x<z}\gamma(x) = \varphi''_t(x) \supset \gamma \notin [\varphi_{[s]}, \psi_{[s]}]\}$. Thence with (xxiv), $\varphi''_t \notin [\varphi_{[s]}, \psi_{[s]}]$, whence by (xxiii) and *R8.2, $\varphi_{[t]} \notin [\varphi_{[s]}, \psi_{[s]}]$. But by *R12.3, (i) and (xxv), $[\varphi_{[t]}, \psi_{[t]}] \subseteq [\varphi_{[s]}, \psi_{[s]}]$, whence by *R11.3, $\varphi_{[t]} \in [\varphi_{[s]}, \psi_{[s]}]$.

16.6. The formula on the right of *R13.1, abbreviated "$\alpha\in(\delta_1, \delta_2)$", expresses the assertion that α is an r.n.g. in the open interval (δ_1, δ_2), or that α is "between" δ_1 and δ_2 (Brouwer **1928a** p. 9 lines 13–10 from below; for *R13.5, cf. lines 8–6 from below).

*R13.1. $\vdash \alpha\in(\delta_1, \delta_2) \sim \alpha\in[\delta_1, \delta_2]\ \&\ \neg\alpha\doteq\delta_1\ \&\ \neg\alpha\doteq\delta_2$.
*R13.2. $\vdash \alpha\doteq\beta\ \&\ \alpha\in(\delta_1, \delta_2) \supset \beta\in(\delta_1, \delta_2)$.
*R13.3. $\vdash \delta_1\doteq\delta'_1\ \&\ \alpha\in(\delta_1, \delta_2) \supset \alpha\in(\delta'_1, \delta_2)$.
*R13.4. $\vdash \delta_2\doteq\delta'_2\ \&\ \alpha\in(\delta_1, \delta_2) \supset \alpha\in(\delta_1, \delta'_2)$.
*R13.5. $\vdash \alpha,\delta_1,\delta_2\in R\ \&\ \delta_1 \lessdot \delta_2 \supset \{\alpha\in(\delta_1, \delta_2) \sim \delta_1 \lessdot \alpha \lessdot \delta_2\}$.

PROOFS. *R13.5. Assume $\alpha,\delta_1,\delta_2\in R$ and **(i)** $\delta_1 \lessdot \delta_2$. I. Assume **(ii)** $\alpha\in[\delta_1, \delta_2]$, **(iii)** $\neg\alpha\doteq\delta_1\ \&\ \neg\alpha\doteq\delta_2$. Using (i) and (ii) (and *R7.4), $\neg\alpha\lessdot\delta_1\ \&\ \neg\alpha\gtrdot\delta_2$. Thence with (iii) and *R7.11, $\delta_1\lessdot\alpha\ \&\ \alpha\lessdot\delta_2$. II. Use *R7.11.

*R13.8 asserts that the continuum is "everywhere dense" ("überall dicht"; Brouwer **1928a** p. 12 lines 4–9). *R13.6 and *R13.7 are used in the proof.

*R13.6. $\vdash \forall x|2\alpha(x)-\alpha(x')| \leq 2 \supset \forall p \forall x|2^p\alpha(x)-\alpha(x+p)| < 2^{p+1}$.
*R13.7. $\vdash \forall x|2\alpha(x)-\alpha(x')| \leq 2 \supset \alpha\in R$.
*R13.8. $\vdash \alpha,\beta\in R\ \&\ \neg\alpha\doteq\beta \supset \exists\gamma_{\gamma\in R}\gamma\in(\alpha, \beta)$.

PROOFS. *R13.6. Assuming **(i)** $\forall x|2\alpha(x)-\alpha(x')| \leq 2$, we deduce $\forall x|2^p\alpha(x)-\alpha(x+p)| < 2^{p+1}$ by ind. on p, thus. IND. STEP. $|2^{p'}\alpha(x)-\alpha(x+p')| \leq 2^p|2\alpha(x)-\alpha(x')|+|2^p\alpha(x')-\alpha(x'+p)| < 2^{p+1}+2^{p+1}$ [(i), hyp. ind.] $= 2^{p'+1}$.

*R13.7. Assume $\forall x|2\alpha(x)-\alpha(x')| \leq 2$. Using *R13.6, $2^k|2^p\alpha(k+1)-\alpha(k+1+p)| < 2^{k+1+p}$, whence $\alpha\in R$.

*R13.8. Assume $\alpha,\beta\in R$, **(i)** $\neg\alpha\doteq\beta$. Using *R1.11: **(ii)** $\alpha'\in R'\ \&\ \alpha'\doteq\alpha$,

(iii) $\beta' \in R'$ & $\beta' \doteq \beta$. Introduce γ by Lemma 5.3 (a):

(A) $\forall x \gamma(x) = \min(\alpha'(x), \beta'(x)) +$
$$[\max(\alpha'(x), \beta'(x)) \dot{-} \min(\alpha'(x), \beta'(x))/2].$$

Thus we can write $\gamma(x) = b + [a \dot{-} b/2]$ [where $a \geq b$ by *8.8] $\leq b + [a \dot{-} b/1]$ [*13.11] $= b + (a \dot{-} b)$ [*13.6] $= a$ [*6.7]. So **(iv)** $\forall x \min(\alpha'(x), \beta'(x)) \leq \gamma(x) \leq \max(\alpha'(x), \beta'(x))$. We shall next deduce **(v)** $(\alpha'(x)+\beta'(x)) \dot{-} 1 \leq 2\gamma(x) \leq \alpha'(x)+\beta'(x)$ and **(vi)** $(\alpha'(x)+\beta'(x)) \dot{-} 1 \leq \gamma(x') \leq \alpha'(x)+\beta'(x)+1$. CASE 1: $\alpha'(x) \leq \beta'(x)$ & $\alpha'(x') \leq \beta'(x')$. Toward (v), we use cases from *12.3 with $r<2 \sim r=0 \vee r=1$ (adapting IM p. 198 line 7). CASE A: $\mathrm{rm}(\beta'(x) \dot{-} \alpha'(x), 2)=0$. Then by *13.4, $\beta'(x) \dot{-} \alpha'(x) = 2[\beta'(x) \dot{-} \alpha'(x)/2]$. So $2\gamma(x) = 2\alpha'(x) + 2[\beta'(x) \dot{-} \alpha'(x)/2]$ [(A), Case 1 hyp., *7.2, *8.2] $= 2\alpha'(x) + (\beta'(x) \dot{-} \alpha'(x)) = \alpha'(x)+\beta'(x)$ [*6.6, *6.3]. CASE B: $\mathrm{rm}(\beta'(x) \dot{-} \alpha'(x), 2)=1$. Then similarly $\beta'(x) \dot{-} \alpha'(x) = 2[\beta'(x) \dot{-} \alpha'(x)/2]+1$, so $\beta'(x) \geq \alpha'(x)+1$. Now $2\gamma(x) = 2\alpha'(x) + ((\beta'(x) \dot{-} \alpha'(x)) \dot{-} 1) = 2\alpha'(x) + (\beta'(x) \dot{-} (\alpha'(x)+1))$ [*6.5] $= (2\alpha'(x)+\beta'(x)) \dot{-} (\alpha'(x)+1)$ [*6.6] $= (\alpha'(x)+\beta'(x)) \dot{-} 1$ [*6.5, *6.3]. For (vi): CASE A′: $\alpha'(x') < 2\alpha'(x)$ & $\beta'(x') < 2\beta'(x)$. Then $2\alpha'(x) \geq 1$ & $2\beta'(x) \geq 1$ and $\alpha'(x') = 2\alpha'(x) \dot{-} 1$ & $\beta'(x') = 2\beta'(x) \dot{-} 1$ [*R0.5a–b, (ii), (iii)]. So $\gamma(x') = (2\alpha'(x) \dot{-} 1) + [(2\beta'(x) \dot{-} 1) \dot{-} (2\alpha'(x) \dot{-} 1)/2] = (2\alpha'(x) \dot{-} 1) + [2\beta'(x) \dot{-} 2\alpha'(x)/2]$ [*6.9, *6.7] $= (2\alpha'(x) \dot{-} 1) + (\beta'(x) \dot{-} \alpha'(x))$ [*6.14, *13.8, ×13.1] $= (2\alpha'(x)+\beta'(x)) \dot{-} (1+\alpha'(x))$ [*6.6 twice, *6.5] $= (\alpha'(x)+\beta'(x)) \dot{-} 1$ [*6.5, *6.3]. CASE B′: $\alpha'(x') < 2\alpha'(x)$ & $\beta'(x') = 2\beta'(x)$. Similarly $\gamma(x') = (2\alpha'(x) \dot{-} 1) + [2\beta'(x) \dot{-} (2\alpha'(x) \dot{-} 1)/2] = (2\alpha'(x) \dot{-} 1) + [(2\beta'(x)+2) \dot{-} (2\alpha'(x)+1)/2] = (2\alpha'(x) \dot{-} 1) + [((2\beta'(x)+2) \dot{-} 2\alpha'(x)) \dot{-} 1/2] = (2\alpha'(x) \dot{-} 1) + (((\beta'(x)+1) \dot{-} \alpha'(x)) \dot{-} 1)$ [*13.9, *13.7] $= (\alpha'(x)+\beta'(x)) \dot{-} 1$. CASE C′: $\alpha'(x') < 2\alpha'(x)$ & $\beta'(x') > 2\beta'(x)$. Similarly $\gamma(x') = (2\alpha'(x) \dot{-} 1) + [(2\beta'(x)+1) \dot{-} (2\alpha'(x) \dot{-} 1)/2] = (2\alpha'(x) \dot{-} 1) + [(2\beta'(x)+2) \dot{-} 2\alpha'(x)/2]$, etc. Cases D′–I′ are treated similarly. CASE 2: $\alpha'(x) \geq \beta'(x)$ & $\alpha'(x') \geq \beta'(x')$. Symmetric to Case 1. CASE 3: $\alpha'(x) \leq \beta'(x)$ & $\alpha'(x') \geq \beta'(x')$. SUBCASE 3.1: $\alpha'(x) < \beta'(x)$. But then $\alpha'(x') \leq 2\alpha'(x)+1$ [*R0.5a–c, (ii)] $\leq 2\beta'(x) \dot{-} 1 \leq \beta'(x')$; so this subcase comes under Case 1. SUBCASE 3.2: $\alpha'(x) = \beta'(x)$. Then $\alpha'(x) \geq \beta'(x)$, so this subcase comes under Case 2. CASE 4: $\alpha'(x) \geq \beta'(x)$ & $\alpha'(x') \leq \beta'(x')$. Symmetric to Case 3. From (v) and (vi), $|2\gamma(x) - \gamma(x')| \leq 2$, whence by $\forall$-introd. and *R13.7: **(vii)** $\gamma \in R$. Toward (viii), assume **(a)** $\gamma <_\circ \alpha'$, and for reductio ad absurdum, **(b)** $\gamma <_\circ \beta'$. Thence assume $\forall p 2^{k_1}(\alpha'(x_1+p) - \gamma(x_1+p)) \geq 2^{x_1+p}$ and $\forall p 2^{k_2}(\beta'(x_2+p) \dot{-} \gamma(x_2+p)) \geq 2^{x_2+p}$, and, using *R0.9: $\forall p \alpha'(x+p) > \gamma(x+p)$ and $\forall p \beta'(x+p) > \gamma(x+p)$. Thence $\alpha'(x) > \gamma(x)$ and $\beta'(x) > \gamma(x)$, so by *7.6, $\min(\alpha'(x), \beta'(x)) > \gamma(x)$, contradicting (iv). So, rejecting (b), $\gamma \not<_\circ \beta'$. By $\supset$-introd., discharging (a): **(viii)** $\gamma <_\circ \alpha' \supset \gamma \not<_\circ \beta'$, and by contraposition $\neg\gamma \not<_\circ \beta' \supset \neg\gamma <_\circ \alpha'$. By *R7.8, *R7.9, (ii), (iii) and (vii), $\gamma \prec \beta' \supset \neg\gamma \prec \alpha'$, whence by *58b, $\neg(\gamma \prec \alpha' \;\&\; \gamma \prec \beta')$. Symmetrically, $\neg(\gamma \succ \alpha' \;\&\; \gamma \succ \beta')$. By &-introd. with (ii), (iii) and (vii), $\gamma \in [\alpha', \beta']$, and thence by (ii), (iii) and *R8.3–*R8.4: **(ix)** $\gamma \in [\alpha, \beta]$. Toward $\neg\gamma \doteq \alpha$, assume **(c)** $\gamma \doteq \alpha'$. We shall first contradict **(d)** $\gamma <_\circ \beta'$. Assume from (c), (d) and *R0.9: **(e)** $\forall p 2^{k+1}|\gamma(x+p) - \alpha'(x+p)| < 2^{x+p}$, and $\forall p 2^{k_1}(\beta'(x+p) \dot{-} \gamma(x+p)) \geq 2^{x+p}$. Then $\forall p \beta'(x+p) > \gamma(x+p)$. Using also (iv) (since $\min(a, b) = a \vee \min(a, b) = b$): **(f)** $\forall p \alpha'(x+p) \leq \gamma(x+p) < \beta'(x+p)$. Letting t be $x+k+1+p$, we deduce **(g)** $\forall p 2^k(1+(\gamma(t) \dot{-} \alpha'(t))) < 2^t$ thus. Case 2: $\gamma(t) \dot{-} \alpha'(t) \neq 0$. Then $2^k(1+(\gamma(t) \dot{-} \alpha'(t))) \leq 2^{k+1}(\gamma(t) \dot{-} \alpha'(t)) < 2^t$ [(e)]. Now $\gamma(t) \geq ((\alpha'(t)+\beta'(t)) \dot{-} 1) \dot{-} \gamma(t)$ [(v), *6.17, *6.3] $= ((\alpha'(t)+\beta'(t)) \dot{-} \gamma(t)) \dot{-} 1$ [*6.5] $= (\alpha'(t)+(\beta'(t) \dot{-} \gamma(t))) \dot{-} 1$ [*6.6, (f)]. So **(h)** $(1+\gamma(t)) \dot{-} \alpha'(t) \geq \beta'(t) \dot{-} \gamma(t)$. Thence $2^k|\beta'(t) - \gamma(t)| = 2^k(\beta'(t) \dot{-} \gamma(t))$ [(f)] $\leq 2^k((1+\gamma(t)) \dot{-} \alpha'(t))$ [(h)] $= 2^k(1+(\gamma(t) \dot{-} \alpha'(t)))$ [*6.6, (f)] $< 2^t$ [(g)]. By $\forall$-, $\exists$- and $\forall$-introd., $\beta' \doteq \gamma$, contradicting (d) by *R6.4. Hence $\gamma \not<_\circ \beta'$. Similarly, $\gamma {}_\circ\!\not> \beta'$. By *R6.5, $\gamma \doteq \beta'$. By (c), $\alpha' \doteq \beta'$, and with (ii), (iii), $\alpha \doteq \beta$, contradicting (i). Hence, rejecting (c), $\neg\gamma \doteq \alpha'$. By (ii), $\neg\gamma \doteq \alpha$. Symmetrically, $\neg\gamma \doteq \beta$. By &-introd. with (ix): **(x)** $\gamma \in (\alpha, \beta)$.

16.7. To express the "sharp difference" of two r.n.g. (Brouwer **1928a** p. 10 lines 27–33), we use the formula on the right of *R14.1, abbreviated "$\alpha \neq_s \beta$". (For *R14.2, cf. **1928a** p. 10 Footnote 7.)

*R14.1. $\vdash \alpha \neq_s \beta \sim \neg\alpha \doteq \beta \;\&\; \forall\gamma_{\gamma \in R}\{\gamma \notin (\alpha, \beta) \supset (\neg\gamma \succ \alpha \;\&\; \neg\gamma \succ \beta) \vee (\neg\gamma \prec \alpha \;\&\; \neg\gamma \prec \beta)\}$.

*R14.2. $\vdash \alpha, \beta \in R \;\&\; \alpha \neq_s \beta \supset \alpha \prec \beta \vee \beta \prec \alpha$.

Proofs. *R14.2. Assume $\alpha, \beta \in R$ and $\alpha \neq_s \beta$. Then **(i)** $\neg\alpha \doteq \beta$ and **(ii)** $\alpha \notin (\alpha, \beta) \supset (\neg\alpha \succ \alpha \;\&\; \neg\alpha \succ \beta) \vee (\neg\alpha \prec \alpha \;\&\; \neg\alpha \prec \beta)$. Using *R1.4, $\alpha \notin (\alpha, \beta)$. Using also *R7.6, (ii) reduces by *41 and *45 to $\neg\alpha \succ \beta \vee \neg\alpha \prec \beta$. Thence by (i) and *R7.11, $\alpha \prec \beta \vee \beta \prec \alpha$.

For r.n.g. the sharp difference is equivalent to the apartness relation # of *R2.1, as we show in *R14.11. *R14.3–*R14.10 establish the existence of a spread with suitable properties, preparatory to using Brouwer's principle (*27.6) in the proof of *R14.11.

*R14.3. $\vdash \alpha, \beta \in R' \supset \forall x(\beta(x) \dot{-} \alpha(x) = 2 \supset \beta(x') \dot{-} \alpha(x') \geq 2)$.

*R14.4. $\vdash \alpha, \beta \in R' \supset \forall x(\beta(x) \dot{-} \alpha(x) > 2 \supset \beta(x') \dot{-} \alpha(x') > 2)$.

*R14.5. $\vdash \alpha,\beta\in R' \supset \forall x(\beta(x) \dot{-} \alpha(x)=2 \,\&\, \beta(x') \dot{-} \alpha(x')=2 \supset \alpha(x')=2\alpha(x)+1 \,\&\, \beta(x')=2\beta(x) \dot{-} 1)$.

PROOFS. *R14.3. Assume $\alpha,\beta\in R'$ and $\beta(x) \dot{-} \alpha(x)=2$. Now $2\beta(x) \dot{-} 1=2\alpha(x)+3$. So $\beta(x') \dot{-} \alpha(x') \geq (2\beta(x) \dot{-} 1) \dot{-} \alpha(x')$ [$\beta\in R'$, *R0.5a–c, *6.17] $\geq (2\beta(x) \dot{-} 1) \dot{-} (2\alpha(x)+1)$ [$\alpha\in R'$, *R0.5a–c, *6.18] $= (2\alpha(x)+3) \dot{-} (2\alpha(x)+1) = 2$.

*R14.4. Assume $\alpha,\beta\in R'$ and $\beta(x) \dot{-} \alpha(x)>2$. Now $2\beta(x) \dot{-} 1 > 2\alpha(x)+3 > 2\alpha(x)+1$. So $\beta(x') \dot{-} \alpha(x') \geq (2\beta(x) \dot{-} 1) \dot{-} (2\alpha(x)+1) > (2\alpha(x)+3) \dot{-} (2\alpha(x)+1)$ [*6.19] $= 2$.

*R14.5. Assume $\alpha,\beta\in R'$, $\beta(x) \dot{-} \alpha(x)=2$ and $\beta(x') \dot{-} \alpha(x')=2$. If $\alpha(x')\neq 2\alpha(x)+1$, then by *R0.5a, $\alpha(x')\leq 2\alpha(x)$, whence $2 = \beta(x') \dot{-} \alpha(x') \geq \beta(x') \dot{-} 2\alpha(x) \geq (2\beta(x) \dot{-} 1) \dot{-} 2\alpha(x) = 2(\beta(x) \dot{-} \alpha(x)) \dot{-} 1 = 3$. Hence $\alpha(x')=2\alpha(x)+1$. Similarly, $\beta(x')=2\beta(x) \dot{-} 1$.

*R14.6. $\vdash \alpha,\beta\in R' \,\&\, \alpha\leq\beta \supset \exists\alpha'_{\alpha'\in R'}\exists\beta'_{\beta'\in R'}\{\alpha' \stackrel{\circ}{=} \alpha \,\&\, \beta' \stackrel{\circ}{=} \beta \,\&\, \alpha'\leq\beta' \,\&\, \forall x(\beta'(x) \dot{-} \alpha'(x)=2 \supset \beta'(x') \dot{-} \alpha'(x')>2)\}$.

PROOF. Assume **(i)** $\alpha\in R'$, **(ii)** $\beta\in R'$, **(iii)** $\alpha\leq\beta$. Introduce α' and β', using Lemma 5.5 (b) and (a), and letting $A(\alpha, \beta, x)$ be $\beta(x) \dot{-} \alpha(x)=2 \,\&\, \beta(x') \dot{-} \alpha(x')=2$.

(A1) $$\alpha'(0)=\begin{cases}\alpha(0)+1 \text{ if } A(\alpha, \beta, 0),\\ \alpha(0) \text{ otherwise.}\end{cases}$$

(A2) $$\forall x\alpha'(x')=\begin{cases}\alpha(x')+1 \text{ if } A(\alpha, \beta, x') \,\&\, \alpha(x')\neq 2\alpha'(x)+1,\\ \alpha(x') \text{ otherwise.}\end{cases}$$

(B) $$\beta'(x)=\begin{cases}\alpha(x)+1 \text{ if } A(\alpha, \beta, x) \,\&\, \alpha(x)=2\alpha'(x \dot{-} 1)+1,\\ \beta(x) \text{ otherwise.}\end{cases}$$

Then easily: **(iv)** $\forall x(\alpha(x)\leq\alpha'(x)\leq\alpha(x)+1)$, **(v)** $\forall x(A(\alpha, \beta, x) \,\&\, \alpha(x)\neq 2\alpha'(x \dot{-} 1)+1 \supset \alpha'(x)=\alpha(x)+1)$ and **(vi)** $\forall x(\neg A(\alpha, \beta, x) \supset \alpha'(x)=\alpha(x) \,\&\, \beta'(x)=\beta(x))$.

I. We shall deduce $|2\alpha'(x)-\alpha'(x')|\leq 1$, whence (by $\forall$-introd.) $\alpha'\in R'$. CASE 1: $A(\alpha, \beta, x) \,\&\, \alpha(x)\neq 2\alpha'(x \dot{-} 1)+1$. SUBCASE 1.1: $A(\alpha, \beta, x') \,\&\, \alpha(x')\neq 2\alpha'(x)+1$. Then by *R14.5, $\alpha(x')=2\alpha(x)+1$. By (v), $\alpha'(x)=\alpha(x)+1$ and $\alpha'(x')=\alpha(x')+1$. So $|2\alpha'(x)-\alpha'(x')| = |2(\alpha(x)+1)-(\alpha(x')+1)| = |(2\alpha(x)+2)-(2\alpha(x)+2)| = 0 \leq 1$. SUBCASE 1.2: $\neg A(\alpha, \beta, x') \vee \alpha(x')=2\alpha'(x)+1$. By case hyp. and (v), $\alpha'(x)=\alpha(x)+1$. So $|2\alpha'(x)-\alpha'(x')| = |2(\alpha(x)+1)-\alpha(x')|$ [(A2)] $= |(2\alpha(x)+4)-(\alpha(x')+2)|$ [*11.7] $= |2\beta(x)-\beta(x')|$ [$A(\alpha, \beta, x)$] ≤ 1 [(ii)]. CASE 2: $\neg A(\alpha, \beta, x) \vee$

$\alpha(x)=2\alpha'(x \dot{-} 1)+1$. We deduce **(a)** $\alpha'(x)=\alpha(x)$ thus. CASE A: $x=0$. By (iv), $\alpha'(0)\geq\alpha(0)$. Hence $\alpha(0)\neq 2\alpha'(0)+1$, whence by Case 2 hyp., $\neg A(\alpha, \beta, 0)$. By (A1), $\alpha'(0)=\alpha(0)$. CASE B: $x>0$. Use Case 2 hyp. SUBCASE 2.1: $A(\alpha, \beta, x')$ & $\alpha(x')\neq 2\alpha'(x)+1$. Then by (v): **(i')** $\alpha'(x')=\alpha(x')+1$. By (a), $\alpha(x')\neq 2\alpha(x)+1$. By (i) and *R0.5a–b, $\alpha(x')+1=2\alpha(x) \vee \alpha(x')=2\alpha(x)$. We deduce **(b)** $\alpha'(x')=2\alpha(x) \vee \alpha'(x')=2\alpha(x)+1$. CASE A: $\alpha(x')+1=2\alpha(x)$. Then $\alpha'(x') = \alpha(x')+1$ [(i')] $= 2\alpha(x)$. CASE B: $\alpha(x')=2\alpha(x)$. Then $\alpha'(x') = \alpha(x')+1 = 2\alpha(x)+1$. Now by cases from (b), using (a): $|2\alpha'(x)-\alpha'(x')|\leq 1$. SUBCASE 2.2: $\neg A(\alpha, \beta, x') \vee \alpha(x')=2\alpha'(x)+1$. Then $|2\alpha'(x)-\alpha'(x')| = |2\alpha(x)-\alpha(x')|$ [(a), (A2)] ≤ 1 [(i)].

II. We shall next deduce $|2\beta'(x)-\beta'(x')|\leq 1$, whence (by $\forall$-introd.) $\beta'\epsilon R'$. CASE 1: $A(\alpha, \beta, x)$ & $\alpha(x)=2\alpha'(x \dot{-} 1)+1$. SUBCASE 1.1: $A(\alpha, \beta, x')$ & $\alpha(x')=2\alpha'(x)+1$. By *R14.5, $\alpha(x')=2\alpha(x)+1$. Now $|2\beta'(x)-\beta'(x')| = |2(\alpha(x)+1)-(\alpha(x')+1)|$ [(B)] $= |(2\alpha(x)+2)-(2\alpha(x)+2)| = 0 \leq 1$. SUBCASE 1.2: $\neg A(\alpha, \beta, x') \vee \alpha(x')\neq 2\alpha'(x)+1$. Then $\alpha(x')=2\alpha(x)+1$ (using *R14.5) and $\beta(x')=\alpha(x')+2$ (using case hyp.). So $|2\beta'(x)-\beta'(x')| = |2(\alpha(x)+1)-\beta(x')|$ [(B)] $= |(\alpha(x')+1)-(\alpha(x')+2)| = 1$. CASE 2: $\neg A(\alpha, \beta, x) \vee \alpha(x)\neq 2\alpha'(x \dot{-} 1)+1$. SUBCASE 2.1: $A(\alpha, \beta, x')$ & $\alpha(x')=2\alpha'(x)+1$. If $\alpha'(x)\neq\alpha(x)$, then by (iv), $\alpha'(x)=\alpha(x)+1$, $\alpha(x') = 2\alpha'(x)+1$ [subcase hyp.] $= 2(\alpha(x)+1)+1 = 2\alpha(x)+3$, contradicting (i) by *R0.5a. So **(i'')** $\alpha'(x)=\alpha(x)$. So by subcase hyp.: **(ii'')** $\alpha(x')=2\alpha(x)+1$, and also using *R14.4 (with $A(\alpha, \beta, x')$): **(iii'')** $\beta(x) \dot{-} \alpha(x)\leq 2$. Assume for reductio ad absurdum **(a')** $\beta(x) \dot{-} \alpha(x)=2$. Then, using $A(\alpha, \beta, x')$ (from subcase hyp.), $A(\alpha, \beta, x)$. So by case hyp., $\alpha(x)\neq 2\alpha'(x \dot{-} 1)+1$. Now by (v), $\alpha'(x)=\alpha(x)+1$, contradicting (i''). Thus, rejecting (a'), **(iv'')** $\beta(x) \dot{-} \alpha(x)\neq 2$. If $\beta(x) \dot{-} \alpha(x)=0$, then $\beta(x') = \alpha(x')+2$ [subcase hyp.] $= 2\alpha(x)+3$ [(ii'')] $\geq 2\beta(x)+3$ [*6.11], contradicting (ii) by *R0.5a. Thus $\beta(x) \dot{-} \alpha(x)\neq 0$. Then by (iii'') and (iv''), $\beta(x) \dot{-} \alpha(x)=1$. So $|2\beta'(x)-\beta'(x')| = |2\beta(x)-(\alpha(x')+1)|$ [(B)] $= |2(\alpha(x)+1)-(\alpha(x')+1)|$ $= |(2\alpha(x)+2)-(2\alpha(x)+2)|$ [(ii'')] $= 0 \leq 1$. SUBCASE 2.2: $\neg A(\alpha, \beta, x') \vee \alpha(x')\neq 2\alpha'(x)+1$. Then $|2\beta'(x)-\beta'(x')| = |2\beta(x)-\beta(x')| \leq 1$ [(ii)].

III. From (iv) by *11.15a, $|\alpha'(k'+p)-\alpha(k'+p)|\leq 1$. Thence $2^{k}|\alpha'(k'+p)-\alpha(k'+p)|<2^{k'+p}$, whence by $\forall$-, $\exists$- and $\forall$-introd., $\alpha'\doteq\alpha$.

IV. We shall next deduce $|\beta'(k'+p)-\beta(k'+p)|\leq 1$, whence $\beta'\doteq\beta$. CASE 1: $A(\alpha, \beta, k'+p)$ & $\alpha(k'+p)=2\alpha'((k'+p) \dot{-} 1)+1$. Then $\beta(k'+p)$ $=\alpha(k'+p)+2$. So $|\beta'(k'+p)-\beta(k'+p)| = |(\alpha(k'+p)+1)-(\alpha(k'+p)+2)|$ [(B)] $= 1$. CASE 2: $\neg A(\alpha, \beta, k'+p) \vee \alpha(k'+p)\neq 2\alpha'((k'+p) \dot{-} 1)+1$. Then $|\beta'(k'+p)-\beta(k'+p)| = |\beta(k'+p)-\beta(k'+p)| = 0 \leq 1$.

V. We shall deduce $\alpha'(x) \leq \beta'(x)$, whence $\alpha' \leq \beta'$. CASE 1: $A(\alpha, \beta, x)$. SUBCASE 1.1: $\alpha(x) = 2\alpha'(x \dot{-} 1)+1$. Then $\alpha'(x) \leq \alpha(x)+1$ [(iv)] $= \beta'(x)$ [(B)]. SUBCASE 1.2: $\alpha(x) \neq 2\alpha'(x \dot{-} 1)+1$. Then $\alpha'(x) = \alpha(x)+1$ [(v)] $< \alpha(x)+2 = \beta(x)$ [case hyp.] $= \beta'(x)$ [(B)]. CASE 2: $\neg A(\alpha, \beta, x)$. Then $\alpha'(x) = \alpha(x)$ [(vi)] $\leq \beta(x)$ [(iii)] $= \beta'(x)$ [(vi)].

VI. Toward (x), assume **(vii)** $\beta'(x) \dot{-} \alpha'(x) = 2$. Assume for reductio ad absurdum, **(a″)** $A(\alpha, \beta, x)$. CASE 1: $\alpha(x) = 2\alpha'(x \dot{-} 1)+1$. Then $\beta'(x) \dot{-} \alpha'(x) = (\alpha(x)+1) \dot{-} \alpha'(x)$ [(B)] $\leq (\alpha(x)+1) \dot{-} \alpha(x)$ [(iv)] $= 1$, contradicting (vii). CASE 2: $\alpha(x) \neq 2\alpha'(x \dot{-} 1)+1$. Then $\beta'(x) \dot{-} \alpha'(x) = \beta(x) \dot{-} \alpha'(x)$ [(B)] $= \beta(x) \dot{-} (\alpha(x)+1)$ [(v)] $= (\beta(x) \dot{-} \alpha(x)) \dot{-} 1 = 2 \dot{-} 1$ [(a″)] $= 1$, contradicting (vii). Thus $\neg A(\alpha, \beta, x)$, whence **(viii)** $\beta(x) \dot{-} \alpha(x) \neq 2 \vee \beta(x') \dot{-} \alpha(x') \neq 2$. By (vi), $\alpha'(x) = \alpha(x) \;\&\; \beta'(x) = \beta(x)$. So, using (vii) and (viii): **(ix)** $\beta(x') \dot{-} \alpha(x') \neq 2$. Thus $\neg A(\alpha, \beta, x')$. By (vi), $\alpha'(x') = \alpha(x') \;\&\; \beta'(x') = \beta(x')$. So by (ix), $\beta'(x') \dot{-} \alpha'(x') \neq 2$. So by *R14.3, I, II and (vii): $\beta'(x') \dot{-} \alpha'(x') > 2$. By $\supset$- and $\forall$-introd., **(x)** $\forall x(\beta'(x) \dot{-} \alpha'(x) = 2 \supset \beta'(x') \dot{-} \alpha'(x') > 2)$.

In *R14.8 $A(\gamma, \alpha, \beta, y)$ is $\forall x[(x \leq y \supset \gamma(x) = \alpha(x)) \;\&\; (x > y \supset \gamma(x) = \beta(x))]$.

*R14.7. $\vdash \alpha, \beta \in R' \;\&\; \beta(x) \dot{-} \alpha(x) > 2 \supset |\beta - \alpha| \not<_\circ 1 \cdot 2^{-x}$.

*R14.8. $\vdash \alpha, \beta \in R' \;\&\; \alpha \leq \beta \;\&$
$\forall x(\beta(x) \dot{-} \alpha(x) = 2 \supset \beta(x') \dot{-} \alpha(x') > 2) \;\&\; |\beta - \alpha| <_\circ 1 \cdot 2^{-(y+2)}$
$\supset \exists \gamma_{\gamma \in R'} A(\gamma, \alpha, \beta, y) \vee \exists \gamma_{\gamma \in R'} A(\gamma, \beta, \alpha, y)$.

PROOFS. *R14.7. Assume **(i)** $\alpha, \beta \in R'$, **(ii)** $\beta(x) \dot{-} \alpha(x) > 2$. Then $3 \cdot 2^{-x} \leq_\circ |\beta(x) - \alpha(x)| 2^{-x}$ [(ii), *R9.8, *R9.5] $\doteq |\beta(x)2^{-x} - \alpha(x)2^{-x}|$ [*R9.14] $\circ\!\not> |\beta(x)2^{-x} - \beta| + |\beta - \alpha| + |\alpha - \alpha(x)2^{-x}|$ [*R6.17] $\circ\!\not> 1 \cdot 2^{-x} + |\beta - \alpha| + 1 \cdot 2^{-x}$ [*R9.15, (i); *R6.16] $\doteq 2 \cdot 2^{-x} + |\beta - \alpha|$. Thence, using *R6.16, $|\beta - \alpha| \not<_\circ 1 \cdot 2^{-x}$.

*R14.8. Assume **(i)** $\alpha \in R'$, **(ii)** $\beta \in R'$, **(iii)** $\alpha \leq \beta$, **(iv)** $\forall x(\beta(x) \dot{-} \alpha(x) = 2 \supset \beta(x') \dot{-} \alpha(x') > 2)$, **(v)** $|\beta - \alpha| <_\circ 1 \cdot 2^{-(y+2)}$. Toward (vi), assume **(a)** $x \leq y+1$, and for reductio ad absurdum, **(b)** $\beta(x) \dot{-} \alpha(x) \geq 2$. By cases from (b), using *R14.4 and (iv): $\beta(x') \dot{-} \alpha(x') > 2$. Now $|\beta - \alpha| \not<_\circ 1 \cdot 2^{-x'}$ [*R14.7] $\circ\!\geq 1 \cdot 2^{-(y+2)}$ [(a), with *R9.5, *R9.7, *3.12], contradicting (v). By $\supset$- and $\forall$-introd.: **(vi)** $\forall x(x \leq y+1 \supset \beta(x) \dot{-} \alpha(x) \leq 1)$. By (vi), $\beta(y) \dot{-} \alpha(y) \leq 1$. We shall deduce **(c)** $\exists \gamma_{\gamma \in R'} A(\gamma, \alpha, \beta, y) \vee \exists \gamma_{\gamma \in R'} A(\gamma, \beta, \alpha, y)$. CASE 1: $\beta(y) \dot{-} \alpha(y) = 0$. Introduce γ by Lemma 5.5 (a):

$$\text{(A)} \qquad \forall x \gamma(x) = \begin{cases} \alpha(x) & \text{if } x \leq y, \\ \beta(x) & \text{if } x > y. \end{cases}$$

Then $A(\gamma, \alpha, \beta, y)$. We shall deduce $|2\gamma(x)-\gamma(x')| \leq 1$. CASE A: $x<y$. Use (i). CASE B: $x=y$. Then $|2\gamma(x)-\gamma(x')| = |2\alpha(y)-\beta(y')| = |2\beta(y)-\beta(y')|$ [Case 1 hyp., (iii)] ≤ 1 [(ii)]. CASE C: $x>y$. Use (ii). By $\forall$-introd., $\gamma \in R'$. By &-, $\exists$- and $\vee$-introd., (c). CASE 2: $\beta(y) \dot{-} \alpha(y)=1$. Then **(i')** $2\beta(y) \dot{-} 1=2\alpha(y)+1$. Using (vi), $\beta(y') \dot{-} \alpha(y') \leq 1$. SUBCASE 2.1: $\beta(y') \dot{-} \alpha(y')=0$. Introduce γ by (A) above, whence $A(\gamma, \alpha, \beta, y)$. We shall deduce $|2\gamma(x)-\gamma(x')| \leq 1$. CASE A: $x<y$. Use (i). CASE B: $x=y$. Then $|2\gamma(x)-\gamma(x')| = |2\alpha(y)-\beta(y')| = |2\alpha(y)-\alpha(y')|$ [subcase hyp., (iii)] ≤ 1 [(i)]. CASE C: $x>y$. Use (ii). By $\forall$-introd., $\gamma \in R'$. By &-, $\exists$- and $\vee$-introd., (c). SUBCASE 2.2: $\beta(y') \dot{-} \alpha(y')=1$. Then **(ii')** $\beta(y')>\alpha(y')$. If $2\alpha(y)>\alpha(y')$, then by (i'), $2\beta(y) \dot{-} 1>\alpha(y')$, and thence $1 = \beta(y') \dot{-} \alpha(y')$ [subcase hyp.] $\geq (2\beta(y) \dot{-} 1) \dot{-} \alpha(y')$ [(ii), *R0.5a–c, *6.17] $> (2\beta(y) \dot{-} 1) \dot{-} 2\alpha(y)$ [*6.20] $= (2\alpha(y)+1) \dot{-} 2\alpha(y)$ [(i')] $= 1$. Hence $2\alpha(y) \leq \alpha(y')$. SUB²CASE 2.2.1: $2\alpha(y)=\alpha(y')$. If $\beta(y')>2\beta(y) \dot{-} 1$, then $1 = \beta(y') \dot{-} \alpha(y')$ [subcase hyp.] $> (2\beta(y) \dot{-} 1) \dot{-} \alpha(y')$ [(ii'), *6.19] $= (2\alpha(y)+1) \dot{-} 2\alpha(y)$ [(i'), sub²case hyp.] $= 1$. Hence $\beta(y') \leq 2\beta(y) \dot{-} 1$, whence by *R0.5a–c and (i'): **(iii')** $\beta(y')=2\alpha(y)+1$. Introduce γ by (A) above, whence $A(\gamma, \alpha, \beta, y)$. We shall deduce $|2\gamma(x)-\gamma(x')| \leq 1$. CASE B: $x=y$. Then $|2\gamma(x)-\gamma(x')| = |2\alpha(y)-\beta(y')| = |2\alpha(y)-(2\alpha(y)+1)|$ [(iii')] $= 1$. By $\forall$-introd., $\gamma \in R'$. By &-, $\exists$- and $\vee$-introd., (c). SUB²CASE 2.2.2: $2\alpha(y)<\alpha(y')$. By (i) and *R0.5a: **(i'')** $2\alpha(y)+1=\alpha(y')$. Introduce γ thus:

$$\text{(B)} \qquad \forall x \gamma(x) = \begin{cases} \beta(x) & \text{if } x \leq y, \\ \alpha(x) & \text{if } x>y. \end{cases}$$

Then $A(\gamma, \beta, \alpha, y)$. We shall deduce $|2\gamma(x)-\gamma(x')| \leq 1$. CASE A: $x<y$. Use (ii). CASE B: $x=y$. Then $|2\gamma(x)-\gamma(x')| = |2\beta(y)-\alpha(y')| = |2\beta(y)-(2\alpha(y)+1)|$ [(i'')] $= |(2\alpha(y)+2)-(2\alpha(y)+1)|$ [Case 2 hyp.] $= 1$. CASE C: $x>y$. Use (i). By $\forall$-introd., $\gamma \in R'$. By &-, $\exists$- and $\vee$-introd., (c).

*R14.9. $\vdash \alpha, \beta \in R' \supset \exists \sigma \{\mathrm{Spr}(\sigma) \mathbin{\&} \sigma(1)=0 \mathbin{\&} \forall \gamma [\gamma \in \sigma \sim \gamma \in R' \mathbin{\&} \forall x(\gamma(x)=\alpha(x) \vee \gamma(x)=\beta(x))]\}$.

*R14.10. $\vdash \forall x(\gamma(x)=\alpha(x) \vee \gamma(x)=\beta(x)) \supset \gamma \notin (\alpha, \beta)$.

PROOFS. *R14.9. Assume **(i)** $\alpha \in R'$, **(ii)** $\beta \in R'$. Introduce σ via Lemma 5.5 (c) (cf. the proofs of *R0.8, *R10.2):

$$\text{(A)} \qquad \forall a \sigma(a) = \begin{cases} 0 \text{ if } a=1 \vee \{\mathrm{Seq}(a) \mathbin{\&} \mathrm{lh}(a)=1 \mathbin{\&} [(a)_0 \dot{-} 1=\alpha(0) \\ \qquad \vee (a)_0 \dot{-} 1=\beta(0)]\} \vee \{\mathrm{Seq}(a) \mathbin{\&} \mathrm{lh}(a)>1 \\ \qquad \mathbin{\&} |2((a)_{\mathrm{lh}(a) \dot{-} 2} \dot{-} 1)-((a)_{\mathrm{lh}(a) \dot{-} 1} \dot{-} 1)| \leq 1 \\ \qquad \mathbin{\&} \sigma(\Pi_{i<\mathrm{lh}(a) \dot{-} 1} p_i^{(a)_i})=0 \mathbin{\&} [(a)_{\mathrm{lh}(a) \dot{-} 1} \dot{-} 1= \\ \qquad \alpha(\mathrm{lh}(a) \dot{-} 1) \vee (a)_{\mathrm{lh}(a) \dot{-} 1} \dot{-} 1=\beta(\mathrm{lh}(a) \dot{-} 1)]\}, \\ 1 \text{ otherwise.} \end{cases}$$

IIb. Assume $\sigma(a)=0$, toward deducing $\exists s\sigma(a*2^{s+1})=0$. CASE 1: $a=1$. Then by (A), $\sigma(a*2^{\alpha(0)+1})=0$. CASE 2: $a\neq 1$. Then by (A), $\mathrm{Seq}(a)\ \&\ \mathrm{lh}(a)\geq 1\ \&\ [(a)_{\mathrm{lh}(a)\dot{-}1}\dot{-}1=\alpha(\mathrm{lh}(a)\dot{-}1)\vee(a)_{\mathrm{lh}(a)\dot{-}1}\dot{-}1=\beta(\mathrm{lh}(a)\dot{-}1)]$. SUBCASE 2.1: $(a)_{\mathrm{lh}(a)\dot{-}1}\dot{-}1=\alpha(\mathrm{lh}(a)\dot{-}1)$. Using (A) and (i), $\sigma(a*2^{\alpha(\mathrm{lh}(a))+1})=0$. SUBCASE 2.2: $(a)_{\mathrm{lh}(a)\dot{-}1}\dot{-}1=\beta(\mathrm{lh}(a)\dot{-}1)$. Using (A) and (ii), $\sigma(a*2^{\beta(\mathrm{lh}(a))+1})=0$.

IIIa. Assume **(a)** $\gamma\in\sigma$. By $\forall$-elim., $\sigma(\bar{\gamma}(x''))=0$, whence using (A), $|2\gamma(x)-\gamma(x')|\leq 1$. By $\forall$-introd., $\gamma\in R'$. By $\forall$-elim. from (a), $\sigma(\bar{\gamma}(x'))=0$, whence from (A) by cases ($x=0$, $x>0$): $\gamma(x)=\alpha(x)\vee\gamma(x)=\beta(x)$. By $\forall$-introd., $\forall x(\gamma(x)=\alpha(x)\vee\gamma(x)=\beta(x))$.

IIIb. Assume $\gamma\in R'\ \&\ \forall x(\gamma(x)=\alpha(x)\vee\gamma(x)=\beta(x))$. By ind. on x with a double basis, $\sigma(\bar{\gamma}(x))=0$. By $\forall$-introd., $\gamma\in\sigma$.

*R14.10. Assume **(i)** $\forall x(\gamma(x)=\alpha(x)\vee\gamma(x)=\beta(x))$, and for reductio ad absurdum **(a)** $\gamma\in(\alpha,\beta)$, whence $\alpha\in R$ and $\gamma\in R$. If $\gamma\circ\!>\alpha$, then assuming $\forall p2^k(\gamma(x+p)\dot{-}\alpha(x+p))\geq 2^{x+p}$, we deduce successively $\gamma(x+p)\neq\alpha(x+p)$, $\gamma(x+p)=\beta(x+p)$ [(i)], and $\gamma\doteq\beta$ [$\forall$-introd., *R1.2], contradicting (ii). Thus $\gamma\circ\!\not>\alpha$. Similarly, $\gamma\not<\!\circ\alpha$. By *R6.5 (with $\alpha,\gamma\in R$), $\gamma\doteq\alpha$, contradicting (ii). So, rejecting (a), $\gamma\notin(\alpha,\beta)$.

*R14.11. $\vdash\alpha,\beta\in R\supset(\alpha\neq_s\beta\sim\alpha\#\beta)$.

PROOF. Assume $\alpha,\beta\in R$.

I. Assume **(i)** $\alpha\neq_s\beta$. Using *R1.11, assume **(ii)** $\alpha'\in R'\ \&\ \alpha'\doteq\alpha$, **(iii)** $\beta'\in R'\ \&\ \beta'\doteq\beta$. By (i), *R14.2 and *R7.2–*R7.3, $\alpha'<\beta'\vee\beta'<\alpha'$.

CASE 1: $\alpha'<\beta'$. Then using (*R7.1, (ii), (iii) and) *R6.22, assume: **(iv)** $\alpha''\in R'\ \&\ \alpha''\doteq\alpha'\ \&\ \alpha''\leq\beta'$. Using *R14.6 with (iv) and (iii), assume: **(v)** $\alpha'''\in R'\ \&\ \beta''\in R'\ \&\ \alpha'''\doteq\alpha''\ \&\ \beta''\doteq\beta'\ \&\ \alpha'''\leq\beta''\ \&\ \forall x(\beta''(x)\dot{-}\alpha'''(x)=2\supset\beta''(x')\dot{-}\alpha'''(x')>2)$. By (ii)–(v): **(vi)** $\alpha'''\doteq\alpha''\doteq\alpha'\doteq\alpha\ \&\ \beta''\doteq\beta'\doteq\beta$. Applying *R14.9 (with (v)), assume: **(vii)** $\mathrm{Spr}(\sigma)\ \&\ \sigma(1)=0\ \&\ \forall\gamma[\gamma\in\sigma\sim\gamma\in R'\ \&\ \forall x(\gamma(x)=\alpha'''(x)\vee\gamma(x)=\beta''(x))]$. Toward (viii), assume $\gamma\in\sigma$. Then $\gamma\in R'\ \&\ \forall x(\gamma(x)=\alpha'''(x)\vee\gamma(x)=\beta''(x))$. By *R14.10, $\gamma\notin(\alpha''',\beta'')$, whence by (vi) with *R13.3–*R13.4, $\gamma\notin(\alpha,\beta)$. So by (i) with *R14.1 and $\gamma\in R'$, $(\neg\gamma>\alpha\ \&\ \neg\gamma>\beta)\vee(\neg\gamma<\alpha\ \&\ \neg\gamma<\beta)$. By $\supset$- and $\forall$-introd., **(viii)** $\forall\gamma_{\gamma\in\sigma}\{(\neg\gamma>\alpha\ \&\ \neg\gamma>\beta)\vee(\neg\gamma<\alpha\ \&\ \neg\gamma<\beta)\}$. Applying *27.6 with (vii) and (viii), assume: **(ix)** $\forall\gamma_{\gamma\in\sigma}\exists y\{\forall x[\tau(\bar{\gamma}(x))>0\supset y=x]\ \&\ \{(\neg\gamma>\alpha\ \&\ \neg\gamma>\beta\ \&\ \tau(\bar{\gamma}(y))=1)\vee(\neg\gamma<\alpha\ \&\ \neg\gamma<\beta\ \&\ \tau(\bar{\gamma}(y))=2)\}\}$. By (vii) and (v), $\alpha'''\in\sigma$. So using (ix), assume: **(x)** $\forall x[\tau(\overline{\alpha'''}(x))>0\supset y_1=x]\ \&\ \{(\neg\alpha'''>\alpha\ \&\ \neg\alpha'''>\beta\ \&\ \tau(\overline{\alpha'''}(y_1))=1)\vee(\neg\alpha'''<\alpha\ \&\ \neg\alpha'''<\beta\ \&\ \tau(\overline{\alpha'''}(y_1))=2)\}$. By case hyp. and (vi), $\alpha'''<\beta$. Hence by (x): **(xi)**

$\tau(\overline{\alpha'''}(y_1))=1$. Similarly to (x) and (xi): **(xii)** $\forall x[\tau(\overline{\beta''}(x))>0 \supset y_2=x]$ & $\{(\neg\beta''\gtrdot\alpha \,\&\, \neg\beta''\gtrdot\beta \,\&\, \tau(\overline{\beta''}(y_2))=1) \vee (\neg\beta''\lessdot\alpha \,\&\, \neg\beta''\lessdot\beta \,\&\, \tau(\overline{\beta''}(y_2))=2)\}$, **(xiii)** $\tau(\overline{\beta''}(y_2))=2$. Let $y=\max(y_1, y_2)$. Assume for reductio ad absurdum: **(xiv)** $|\beta''-\alpha'''|<\circ 1\cdot 2^{-(y+2)}$. Thence by *R14.8 with (v), $\exists\gamma_{\gamma\in R'}A(\gamma, \alpha''', \beta'', y) \vee \exists\gamma_{\gamma\in R'}A(\gamma, \beta'', \alpha''', y)$. Case A: $\exists\gamma_{\gamma\in R'}A(\gamma, \alpha''', \beta'', y)$. Assume **(xv)** $\gamma\in R'$ & $A(\gamma, \alpha''', \beta'', y)$. Then $\forall x(x<y_1 \supset \gamma(x)=\alpha'''(x))$, whence **(xvi)** $\tau(\bar{\gamma}(y_1)) = \tau(\overline{\alpha'''}(y_1)) = 1$ [(xi)]. By (vii) with (xv), $\gamma\in\sigma$, so using (ix): **(xvii)** $\forall x[\tau(\bar{\gamma}(x))>0 \supset z=x]$ & $\{(\neg\gamma\gtrdot\alpha \,\&\, \neg\gamma\gtrdot\beta \,\&\, \tau(\bar{\gamma}(z))=1) \vee (\neg\gamma\lessdot\alpha \,\&\, \neg\gamma\lessdot\beta \,\&\, \tau(\bar{\gamma}(z))=2)\}$. Using (xvi) in (xvii), $z=y_1$. So by (xvi), $\tau(\bar{\gamma}(z))=1$, whence by (xvii): **(xviii)** $\neg\gamma\gtrdot\alpha$ & $\neg\gamma\gtrdot\beta$. But using (xv), $2^k|\gamma(y'+p)-\beta''(y'+p)| = 0 < 2^{y'+p}$, whence $\gamma \doteq \beta'' \doteq \beta'$ [(vi)] $\gtrdot \alpha'$ [Case 1 hyp.] $\doteq \alpha$ [(vi)], contradicting (xviii). Case B: $\exists\gamma_{\gamma\in R'}A(\gamma, \beta'', \alpha''', y)$. Similarly. Thus rejecting (xiv), $|\beta''-\alpha'''| \not<\circ 1\cdot 2^{-(y+2)} \circ> 1\cdot 2^{-(y+3)}$ [*R9.7]. So assume $\forall p 2^k((|\beta''-\alpha'''|)(x+p) \dot{-} (1\cdot 2^{-(y+3)})(x+p)) \geq 2^{x+p}$. Thence by *6.15, etc., $2^k(|\beta''-\alpha'''|)(x+p)\geq 2^{x+p}$, whence by *R5.1 and $\forall$- and $\exists$-introds., $\beta''\#\alpha'''$. Thence by (vi) with *R2.4 and *R2.3, $\alpha\#\beta$.

Case 2: $\beta'\lessdot\alpha'$. Similarly. (The assumptions are symmetric in α, β.)

II. Assume **(i)** $\alpha\#\beta$. By *R2.5, **(ii)** $\neg\alpha\doteq\beta$. Assume **(iii)** $\gamma\in R$, **(iv)** $\gamma\notin(\alpha, \beta)$. We shall deduce **(a)** $(\neg\gamma\gtrdot\alpha \,\&\, \neg\gamma\gtrdot\beta) \vee (\neg\gamma\lessdot\alpha \,\&\, \neg\gamma\lessdot\beta)$. By *R6.2, (i) (with $\alpha,\beta\in R$): $\alpha<\circ\beta \vee \beta<\circ\alpha$. Case 1: $\alpha<\circ\beta$. By *R6.7: **(v)** $\alpha\circ\not>\beta$. By *R6.9 (and $\alpha, \beta\in R$, (iii)), $\alpha<\circ\gamma \vee \gamma<\circ\beta$. Subcase 1.1: $\alpha<\circ\gamma$. By *R6.7, **(vi)** $\gamma\not<\circ\alpha$, and by *R6.4, **(vii)** $\neg\gamma\doteq\alpha$. For reductio ad absurdum, assume **(b)** $\gamma<\circ\beta$. By *R6.7, $\gamma\circ\not>\beta$. Thence using *R8.6, (v) and (vi): **(c)** $\gamma\in[\alpha, \beta]$. By (b) and *R6.4, $\neg\gamma\doteq\beta$. Thence using (vii) and (c), $\gamma\in(\alpha, \beta)$, contradicting (iv). So rejecting (b), $\gamma\not<\circ\beta$. With (vi) and *R7.8, $\neg\gamma\lessdot\alpha \,\&\, \neg\gamma\lessdot\beta$, whence by $\vee$-introd., (a). Subcase 1.2: $\gamma<\circ\beta$. Similarly. Case 2: $\beta<\circ\alpha$. Similarly. By $\supset$-, $\forall$- and $\supset$-introd., $\forall\gamma_{\gamma\in R}(\gamma\notin(\alpha, \beta) \supset (\neg\gamma\gtrdot\alpha \,\&\, \neg\gamma\gtrdot\beta) \vee (\neg\gamma\lessdot\alpha \,\&\, \neg\gamma\lessdot\beta))$. Then with (ii), $\alpha\neq_s\beta$.

In our formulation of "separability in itself" ("Separabilität in sich"; Brouwer **1928a** p. 10 lines 33–37), we use $\alpha\#\beta$ rather than $\alpha\neq_s\beta$ (which puts the nontrivial part of Brouwer's proposition with *R14.11).

*R14.12. $\vdash \alpha,\beta\in R \,\&\, \alpha\#\beta \supset \exists a\exists m(a2^{-m}\in(\alpha, \beta))$.

Proof. Assume $\alpha,\beta\in R$ and $\alpha\#\beta$. By *R6.2, $\alpha<\circ\beta \vee \beta<\circ\alpha$. Case 1: $\alpha<\circ\beta$. Using *R9.19, assume (preceding $\exists$-elims.) $\alpha<\circ a2^{-m}<\circ\beta$.

By *R7.7, $\alpha \lessdot a2^{-m} \lessdot \beta$. Then using *R13.5 (with *R9.2, etc.), $a2^{-m} \in (\alpha, \beta)$. CASE 2: $\beta <\circ \alpha$. Similarly.

16.8. We define the unit continuum [0, 1] to be "freely connected" ("freizusammenhängend") if, for each pair $A(\gamma)$ and $B(\gamma)$ of species of r.n.g. such that (1) $A(\gamma) \,\&\, B(\zeta) \rightarrow \gamma \lessdot \zeta$ and (2) for each two r.n.g. α and β in [0, 1] with $\alpha \neq_s \beta \,\&\, \alpha \lessdot \beta$, either, for each γ in [0, 1], $\overline{\gamma \gtrdot \alpha} \rightarrow A(\gamma)$, or, for each γ in [0, 1], $\overline{\gamma \lessdot \beta} \rightarrow B(\gamma)$, the following is the case. There is a r.n.g. η in [0, 1] such that, for each γ in [0, 1], $\gamma \lessdot \eta \rightarrow A(\gamma)$, and for each γ in [0, 1], $\gamma \gtrdot \eta \rightarrow B(\gamma)$. (Brouwer **1928a** p. 11 lines 14–3 from below.) Brouwer required the above only of $A(\gamma)$ and $B(\gamma)$ out of which [0, 1] is "composed" ("zusammengesetzt", p. 10 bottom), i.e. such that there does not exist a γ in [0, 1] for which neither $A(\gamma)$ nor $B(\gamma)$. We do not need to make this restriction in establishing the free connectedness of [0, 1] in *R14.14.

In *R14.13 we consider a generalization of Brouwer's result to the whole continuum $[0, \infty)$. We add the hypothesis that $B(\gamma)$ has an element. (For $[0, \infty)$ the theorem would not hold otherwise.) Instead of $\alpha \neq_s \beta \,\&\, \alpha \lessdot \beta$ we use the simpler equivalent $\alpha <\circ \beta$ ($\supset$ by *R14.11 with *R6.2, *R7.1; $\subset$ by* R6.3, *R14.11, *R7.7), and instead of $\neg \gamma > \alpha$ we use $\gamma \circ \not> \alpha$ (equivalent by *R7.8). Further, in the conclusion instead of $\lessdot$ we put $<\circ$, for a reason to be given in Remark 16.1.

*R14.13. $\vdash \forall\gamma_{\gamma\in R}\forall\zeta_{\zeta\in R}\{A(\gamma) \,\&\, B(\zeta) \supset \gamma \lessdot \zeta\}$
$\quad \& \; \forall\alpha_{\alpha\in R}\forall\beta_{\beta\in R}\{\alpha <\circ \beta \supset \forall\gamma_{\gamma\in R}[\gamma\circ\not>\alpha \supset A(\gamma)]$
$\quad \vee \; \forall\gamma_{\gamma\in R}[\gamma \not<\circ \beta \supset B(\gamma)]\} \,\&\, \exists b B(b2^{-0})$
$\quad \supset \exists\eta_{\eta\in R}\{\forall\gamma_{\gamma\in R}[\gamma <\circ \eta \supset A(\gamma)] \,\&\, \forall\gamma_{\gamma\in R}[\gamma\circ > \eta \supset B(\gamma)]\}.$

PROOF. Assume the hyps. **(i)–(iii)**. Using (ii), *R9.2 and *R9.8, $\forall\gamma_{\gamma\in R}[\gamma\circ\not> a2^{-m} \supset A(\gamma)] \vee \forall\gamma_{\gamma\in R}[\gamma \not<\circ (a+1)2^{-m} \supset B(\gamma)]$. By cases and $\forall$-introd., $\forall a \forall m \exists y\{[\{\forall\gamma_{\gamma\in R}[\gamma\circ\not> a2^{-m} \supset A(\gamma)] \,\&\, y=0\} \vee \{\forall\gamma_{\gamma\in R}[\gamma \not<\circ (a+1)2^{-m} \supset B(\gamma)] \,\&\, y=1\}] \,\&\, y \leq 1\}$. Applying *25.7, assume: **(iv)** $\forall a \forall m\{[\{\forall\gamma_{\gamma\in R}[\gamma\circ\not> a2^{-m} \supset A(\gamma)] \,\&\, \varkappa(\langle a,m\rangle)=0\} \vee \{\forall\gamma_{\gamma\in R}[\gamma \not<\circ (a+1)2^{-m} \supset B(\gamma)] \,\&\, \varkappa(\langle a, m\rangle)=1\}] \,\&\, \varkappa(\langle a, m\rangle) \leq 1\}$. Assume from (iii): **(v)** $B(b2^{-0})$. For reductio ad absurdum, assume $\varkappa(\langle b, 0\rangle)=0$. By (iv), $\forall\gamma_{\gamma\in R}[\gamma\circ\not> b2^{-0} \supset A(\gamma)]$. Using *R9.2 and *R9.8, $A(b2^{-0})$. So by (i) and (v), $b2^{-0} \lessdot b2^{-0}$, contradicting *R7.6. Hence $\varkappa(\langle b, 0\rangle) \neq 0$. So by (iv): **(vi)** $\varkappa(\langle b, 0\rangle)=1$. Using #E, let M be $\mu y_{y\leq b}\varkappa(\langle y, 0\rangle)=1$. Introduce η by Lemma 5.5 (b) (with cases exhaustive by (iv)):

$$\eta(0)=\mathrm{M},$$

$$\text{(A)}\qquad \forall x\eta(x')=\begin{cases} 2\eta(x)\dot{-}1 \text{ if } \varkappa(\langle 2\eta(x)\dot{-}1, x'\rangle)=1, \\ 2\eta(x) \text{ if } \varkappa(\langle 2\eta(x)\dot{-}1, x'\rangle)=0 \\ \qquad\qquad \& \ \varkappa(\langle 2\eta(x), x'\rangle)=1, \\ 2\eta(x)+1 \text{ if } \varkappa(\langle 2\eta(x)\dot{-}1, x'\rangle)=0 \\ \qquad\qquad \& \ \varkappa(\langle 2\eta(x), x'\rangle)=0. \end{cases}$$

Using *R0.5b–a, **(vii)** $\eta\in\mathrm{R}'$, whence **(viii)** $\eta\in\mathrm{R}$. We shall deduce **(ix)** $\forall\gamma_{\gamma\in\mathrm{R}}[\gamma<_\circ(\eta(x)\dot{-}1)2^{-x}\supset A(\gamma)]\ \&\ \forall\gamma_{\gamma\in\mathrm{R}}[\gamma\,{}_\circ\!>(\eta(x)+1)2^{-x}\supset B(\gamma)]$ by ind. on x. BASIS. By (A), $\eta(0)=\mathrm{M}$. Toward (a) assume $\gamma\in\mathrm{R}$ and $\gamma<_\circ(\mathrm{M}\dot{-}1)2^{-0}$. Then by *R9.18 (with *6.4), $\mathrm{M}>0$. So $\mathrm{M}\dot{-}1<\mathrm{M}$. Hence by (vi) and *E5, $\varkappa(\langle\mathrm{M}\dot{-}1, 0\rangle)\neq 1$. So by (iv) (and using *R6.7), $A(\gamma)$. By $\supset$- and $\forall$-introd., **(a)** $\forall\gamma_{\gamma\in\mathrm{R}}[\gamma<_\circ(\eta(0)\dot{-}1)2^{-0}\supset A(\gamma)]$. Toward (b), assume $\gamma\in\mathrm{R}$ and $\gamma\,{}_\circ\!>(\mathrm{M}+1)2^{-0}$. By (vi) and *E5, $\varkappa(\langle\mathrm{M}, 0\rangle)=1$. So by (iv), $B(\gamma)$. By $\supset$- and $\forall$-introd.: **(b)** $\forall\gamma_{\gamma\in\mathrm{R}}[\gamma\,{}_\circ\!>(\eta(0)+1)2^{-0}\supset B(\gamma)]$. IND. STEP. We use cases (the same as in (A)) to deduce **(c)** $\forall\gamma_{\gamma\in\mathrm{R}}[\gamma<_\circ(\eta(x')\dot{-}1)2^{-x'}\supset A(\gamma)]$ and **(d)** $\forall\gamma_{\gamma\in\mathrm{R}}[\gamma\,{}_\circ\!>(\eta(x')+1)2^{-x'}\supset B(\gamma)]$. CASE 1: $\varkappa(\langle 2\eta(x)\dot{-}1, x'\rangle)=1$. By (A), $\eta(x')=2\eta(x)\dot{-}1$. Toward (c), assume $\gamma\in\mathrm{R}$ and $\gamma<_\circ(\eta(x')\dot{-}1)2^{-x'}=(2\eta(x)\dot{-}2)2^{-x'}=(\eta(x)\dot{-}1)2^{-x}$ [*R9.3]. By hyp. ind., $A(\gamma)$. By $\supset$- and $\forall$-introd., (c). Toward (d), assume $\gamma\in\mathrm{R}$ and $\gamma\,{}_\circ\!>(\eta(x')+1)2^{-x'}=((2\eta(x)\dot{-}1)+1)2^{-x'}$. By (iv) and case hyp., $B(\gamma)$. By $\supset$- and $\forall$-introd., (d). CASE 2: $\varkappa(\langle 2\eta(x)\dot{-}1, x'\rangle)=0\ \&\ \varkappa(\langle 2\eta(x), x'\rangle)=1$. Similarly. CASE 3: $\varkappa(\langle 2\eta(x)\dot{-}1, x'\rangle)=0\ \&\ \varkappa(\langle 2\eta(x), x'\rangle)=0$. Similarly. By $\forall$-introd. (and *87) from (ix): **(x)** $\forall x\forall\gamma_{\gamma\in\mathrm{R}}[\gamma<_\circ(\eta(x)\dot{-}1)2^{-x}\supset A(\gamma)]\ \&\ \forall x\forall\gamma_{\gamma\in\mathrm{R}}[\gamma\,{}_\circ\!>(\eta(x)+1)2^{-x}\supset B(\gamma)]$. Toward (xi), assume $\gamma\in\mathrm{R}$ and $\gamma<_\circ\eta$. Using *R1.11, assume $\gamma'\in\mathrm{R}'\ \&\ \gamma'\doteq\gamma$. Then $\gamma'<_\circ\eta$. Assume $\forall p2^{k}(\eta(x+p)\dot{-}\gamma'(x+p))\geq 2^{x+p}$, whence $2^{k}(\eta(x+k+2)\dot{-}\gamma'(x+k+2))\geq 2^{x+k+2}$, whence $\eta(x+k+2)\dot{-}\gamma'(x+k+2)\geq 4$. Now $\gamma\doteq\gamma'\ {}_\circ\!\not>\ (\gamma'(x+k+2)+1)2^{-(x+k+2)}$ [*R9.17] $<_\circ(\eta(x+k+2)\dot{-}1)2^{-(x+k+2)}$ [*R9.8]. Using (x), $A(\gamma)$. By $\supset$- and $\forall$-introd.: **(xi)** $\forall\gamma_{\gamma\in\mathrm{R}}[\gamma<_\circ\eta\supset A(\gamma)]$. Similarly: **(xii)** $\forall\gamma_{\gamma\in\mathrm{R}}[\gamma\,{}_\circ\!>\eta\supset B(\gamma)]$. Combining (viii), (xi) and (xii): $\exists\eta_{\eta\in\mathrm{R}}\{\forall\gamma_{\gamma\in\mathrm{R}}[\gamma<_\circ\eta\supset A(\gamma)]\ \&\ \forall\gamma_{\gamma\in\mathrm{R}}[\gamma\,{}_\circ\!>\eta\supset B(\gamma)]\}$.

Using *R14.13, we prove the free connectedness of [0, 1]. We let $\alpha\in\mathrm{I}$ abbreviate $\alpha\in\mathrm{R}\ \&\ \alpha\,{}_\circ\!\not>1$; the latter is equivalent to $\alpha\in[0, 1]$ by *R8.6, *R9.18, etc.

*R14.14. $\vdash\forall\gamma\{A(\gamma)\vee B(\gamma)\supset\gamma\in\mathrm{I}\}\ \&\ \forall\gamma\forall\zeta\{A(\gamma)\ \&\ B(\zeta)\supset\gamma<\zeta\}$
$\&\ \forall\alpha_{\alpha\in\mathrm{I}}\forall\beta_{\beta\in\mathrm{I}}\{\alpha<_\circ\beta\supset\forall\gamma_{\gamma\in\mathrm{I}}[\gamma\,{}_\circ\!\not>\alpha\supset A(\gamma)]\vee$
$\forall\gamma_{\gamma\in\mathrm{I}}[\gamma\not<_\circ\beta\supset B(\gamma)]\}\supset$
$\exists\eta_{\eta\in\mathrm{I}}\{\forall\gamma_{\gamma\in\mathrm{I}}[\gamma<_\circ\eta\supset A(\gamma)]\ \&\ \forall\gamma_{\gamma\in\mathrm{I}}[\gamma\,{}_\circ\!>\eta\supset B(\gamma)]\}.$

PROOF. Assume the hyp. **(i)–(iii)**. Let $B'(\gamma)$ be $\neg\neg(B(\gamma) \vee \gamma \dot{>} 1)$. Toward (iv), assume $\gamma,\zeta \in R$ and $A(\gamma)$. By (i), $\gamma \in I$. Assuming $B(\zeta) \vee \zeta \dot{>} 1$, we deduce $\gamma \dot{<} \zeta$ by cases (using (ii) in the first case). So by *12 and *R7.9: $B'(\zeta) \supset \gamma \dot{<} \zeta$. Hence **(iv)** $\forall\gamma_{\gamma \in R}\forall\zeta_{\zeta \in R}\{A(\gamma) \,\&\, B'(\zeta) \supset \gamma \dot{<} \zeta\}$. Toward (v), assume $\alpha,\beta \in R$ and $\alpha <\circ \beta$. By *R6.9, $\alpha <\circ 1 \vee 1 <\circ \beta$. We shall deduce **(a)** $\forall\gamma_{\gamma \in R}[\gamma \circ\!\not> \alpha \supset A(\gamma)] \vee \forall\gamma_{\gamma \in R}[\gamma \not<\!\circ \beta \supset B'(\gamma)]$. CASE 1: $\alpha <\circ 1$. Using *R9.19, assume $\alpha <\circ a_1 2^{-m_1} <\circ 1$ and $\alpha <\circ a_2 2^{-m_2} <\circ \beta$. By *R9.4, *R9.7, etc.: $a_1 2^{-m_1} \leq\circ a_2 2^{-m_2} \vee a_1 2^{-m_1} \circ\!> a_2 2^{-m_2}$. Thence by cases we deduce a formula from which we can assume: **(i′)** $\alpha <\circ a2^{-m} <\circ 1$, **(ii′)** $\alpha <\circ a2^{-m} <\circ \beta$. By (iii) (with α, $a2^{-m}$ for α, β) and (i′): $\forall\gamma_{\gamma \in I}[\gamma \circ\!\not> \alpha \supset A(\gamma)] \vee \forall\gamma_{\gamma \in I}[\gamma \not<\!\circ a2^{-m} \supset B(\gamma)]$. SUBCASE 1.1: $\forall\gamma_{\gamma \in I}[\gamma \circ\!\not> \alpha \supset A(\gamma)]$. Then $\forall\gamma_{\gamma \in R}[\gamma \circ\!\not> \alpha \supset A(\gamma)]$, whence (a). SUBCASE 1.2: $\forall\gamma_{\gamma \in I}[\gamma \not<\!\circ a2^{-m} \supset B(\gamma)]$. Assume $\gamma \in R$ and $\gamma \not<\!\circ a2^{-m}$. If $\gamma \in I$, then $B(\gamma)$, whence $B(\gamma) \vee \gamma \dot{>} 1$. If $\gamma \notin I$, then $\neg\gamma \circ\!\not> 1$, whence by *R7.9, $\gamma \dot{>} 1$, whence $B(\gamma) \vee \gamma \dot{>} 1$. So $\gamma \in I \vee \gamma \notin I \supset B(\gamma) \vee \gamma \dot{>} 1$. Thence by *12 and *63, $\neg(B(\gamma) \vee \gamma \dot{>} 1) \supset \gamma \notin I \,\&\, \neg\gamma \notin I$. Thus $\neg\neg(B(\gamma) \vee \gamma \dot{>} 1)$, i.e. $B'(\gamma)$. So $\forall\gamma_{\gamma \in R}[\gamma \not<\!\circ a2^{-m} \supset B'(\gamma)]$. Using (ii′), $\forall\gamma_{\gamma \in R}[\gamma \not<\!\circ \beta \supset B'(\gamma)]$, whence (a). CASE 2: $1 <\circ \beta$. We deduce $\forall\gamma_{\gamma \in R}[\gamma \not<\!\circ \beta \supset B'(\gamma)]$. Thus **(v)** $\forall\alpha_{\alpha \in R}\forall\beta_{\beta \in R}\{\alpha <\circ \beta \supset \forall\gamma_{\gamma \in R}[\gamma \circ\!\not> \alpha \supset A(\gamma)] \vee \forall\gamma_{\gamma \in R}[\gamma \not<\!\circ \beta \supset B'(\gamma)]\}$. Also we easily deduce $B'(2)$, whence **(vi)** $\exists b B'(b2^{-0})$. Using (iv), (v), (vi) in *R14.13, assume: **(vii)** $\eta \in R$, **(viii)** $\forall\gamma_{\gamma \in R}[\gamma <\circ \eta \supset A(\gamma)] \,\&\, \forall\gamma_{\gamma \in R}[\gamma \circ\!> \eta \supset B'(\gamma)]$. For reductio ad absurdum, assume **(b)** $\eta \circ\!> 1$. Using *R9.19, assume $1 <\circ b2^{-n} <\circ \eta$. From $1 <\circ b2^{-n}$, $B'(b2^{-n})$. From $b2^{-n} <\circ \eta$ and (viii), $A(b2^{-n})$. Then by (iv), $b2^{-n} \dot{<} b2^{-n}$. So, rejecting (b), $\eta \circ\!\not> 1$, whence with (vii), $\eta \in I$.

REMARK 16.1. It is impossible here to establish the modification of *R14.13 in which $<\circ$ ($\circ>$) in the conclusion is replaced by $\dot{<}$ ($\dot{>}$), and Brouwer's hypothesis $\neg\exists\gamma_{\gamma \in R}[\neg A(\gamma) \,\&\, \neg B(\gamma)]$ is added. For, we show that if the instance **(A)** of this for $\gamma \leq\circ \rho$, $\gamma \circ\!> \rho$ as the $A(\gamma)$, $B(\gamma)$ were provable, so would be the formula $\rho,\gamma \in R \supset (\rho <\circ \gamma \sim \rho \dot{<} \gamma)$ expressing the equivalence of $<\circ$ to $\dot{<}$ for r.n.g. This equivalence is denied by Brouwer in **1949**, **1951**, and shown to be unprovable by Kleene in 18.2 of Chapter IV below (cf. Theorem 18.2). So assume **(i)** $\rho \in R$. I. We first deduce the hypotheses of (A). Ia. Assume $\gamma,\zeta \in R$, $\gamma \leq\circ \rho$, $\zeta \circ\!> \rho$. Then $\gamma <\circ \zeta$, whence $\gamma \dot{<} \zeta$. Ib. Assume $\alpha,\beta \in R$ and $\alpha <\circ \beta$. By *R6.9 and (i), $\alpha <\circ \rho \vee \rho <\circ \beta$. CASE 1: $\alpha <\circ \rho$. Assume $\gamma \in R$ and $\gamma \circ\!\not> \alpha$. Then $\gamma <\circ \rho$, whence $\gamma \leq\circ \rho$. By $\supset$-, $\forall$- and $\vee$-introds., $\forall\gamma_{\gamma \in R}[\gamma \circ\!\not> \alpha \supset \gamma \leq\circ \rho] \vee \forall\gamma_{\gamma \in R}[\gamma \not<\!\circ \beta \supset \gamma \circ\!> \rho]$. CASE 2: $\rho <\circ \beta$. Similarly, assuming $\gamma \in R$ and $\gamma \not<\!\circ \beta$, and deducing $\gamma \circ\!> \rho$. Ic. Using (i) and *R1.11, assume $\rho'\in R'$ & $\rho'\doteq\rho$. Then $\rho \doteq \rho'\circ\!\not>(\rho'(0)+1)2^{-0}$ [*R9.17] $<\circ$ $(\rho'(0)+2)2^{-0}$, whence $\exists b(b2^{-0}\circ\!>\rho)$. Id. Prior to $\exists$-elim. and $\neg$-introd., assume $\gamma\in R$ & $\neg\gamma\leq\circ\rho$ & $\gamma\circ\!\not>\rho$. Using *63, $\gamma\not<\circ\rho$ & $\neg\gamma\doteq\rho$. Using *R6.5, $\gamma\doteq\rho$. II. Now using (A), after $\supset$-elim. and prior to $\exists$-elim.: **(ii)** $\eta\in R$ & $\forall\gamma_{\gamma\in R}[\gamma\lessdot\eta \supset \gamma\leq\circ\rho]$ & $\forall\gamma_{\gamma\in R}[\gamma\gtrdot\eta\supset\gamma\circ\!>\rho]$. Assume for reductio ad absurdum, **(a)** $\rho<\circ\eta$. Using *R13.8, assume $\gamma\in R$ & $\gamma\in(\rho,\eta)$. Then by *R8.6 and (a), **(b)** $\gamma\not<\circ\rho$ & $\gamma\circ\!\not>\eta$ & $\neg\gamma\doteq\rho$ & $\neg\gamma\doteq\eta$. So $\gamma\lessdot\eta$. By (ii), $\gamma\leq\circ\rho$, contradicting (b). Rejecting (a): **(iii)** $\rho\not<\circ\eta$. Assume for reductio ad absurdum, $\rho\circ\!>\eta$. By *R7.7, $\rho\gtrdot\eta$. By (ii), $\rho\circ\!>\rho$, contradicting *R6.8. Hence $\rho\circ\!\not>\eta$. Now by *R6.5, $\rho\doteq\eta$. So by (ii), $\forall\gamma_{\gamma\in R}(\gamma\gtrdot\rho\supset\gamma\circ\!>\rho)$. This with *R7.7 gives $\forall\gamma_{\gamma\in R}(\rho<\circ\gamma \sim \rho\lessdot\gamma)$.

Similarly, the corresponding modification of *R14.14 leads to the equivalence of $<\circ$ and $\lessdot$. For, using instead of (A) the like instance **(B)** of this, we obtain **(i)** $\rho\in I\supset\forall\gamma_{\gamma\in I}(\gamma\gtrdot\rho\supset\gamma\circ\!>\rho)$. Assume **(ii)** $\alpha,\beta\in R$ and **(iii)** $\alpha\gtrdot\beta$. Putting $\gamma=1\dot{-}(1\dot{-}(\alpha\dot{-}\beta))$ (cf. ˣ7.1), we can establish **(iv)** $\gamma\in I$, **(v)** $\gamma\gtrdot 0$ [using (iii)] and **(vi)** $\gamma\circ\!>0\supset\alpha\circ\!>\beta$. Using $0\in I$, (iv) and (v) in (i), and the result in (vi), we have after $\supset$-introds.: $\alpha,\beta\in R\supset(\alpha\gtrdot\beta\supset\alpha\circ\!>\beta)$.

CHAPTER IV

ON ORDER IN THE CONTINUUM

by S. C. KLEENE

§ 17. Introduction and preliminaries. 17.1. Brouwer in his paper "On order in the continuum, and the relation of truth to non-contradictoricity" **1951** considers five pairs of properties of real numbers, and makes various claims about the relations between the properties in the pairs. (There he cites his Dutch articles **1948b** and **1949a** for the proofs. But three of the pairs are also discussed in his English paper **1948** p. 1248.)

We shall list these claims, and determine (independently of **1948b** and **1949a**, or of **1948**) exactly which ones can be established on the basis of the formal system of Chapters I and III above. (Our results here, except those (in 18.2) based on §§ 10, 11 of Chapter II, were obtained in preliminary form in May 1955, before Vesley's Chapter III was written.)

We chose Brouwer's paper **1951** for this investigation (but could have chosen **1948**), because it contains accessible examples of results which Brouwer reached only by a new method, which he introduced in **1948**, and used in **1948a**, **1948b**, **1949**, **1949a** and later papers. This method rests on "defining" a choice sequence $\alpha(0), \alpha(1), \alpha(2), \ldots$ in such a way that $\alpha(x+1)$ depends on whether or not the "creating subject" between the choices of $\alpha(x)$ and $\alpha(x+1)$ will have solved a certain problem, the solution to which we do not now know how to obtain. The method is analyzed and discussed by van Dantzig **1949** and Heyting **1956** 8.1.1, in connection with an application of it in Brouwer's Dutch paper **1948a**, which is claimed to show the unprovability of $\overline{\alpha \doteq 0} \to \alpha \# 0$ for real numbers α.

To give this example, essentially as in Heyting **1956**, say P is a proposition such that no method is known which will lead to a proof of $\overline{P}$ or of $\overline{\overline{P}}$. Then the "creating subject" can choose successively the integral part and digits in the dual expansion $\alpha_0.\alpha_1\alpha_2\alpha_3\ldots$ of a real number α according to the following instruction: always choose 0, unless between the choices of α_x and α_{x+1} the truth of $\overline{P}$ or of $\overline{\overline{P}}$ first becomes evident to you, in which case choose that $\alpha_{x+1} = 1$. Now $\alpha \doteq 0$ would mean that $\overline{P}$ will never become known to the (any!) creating subject, so $\overline{\overline{P}}$, and also that $\overline{\overline{P}}$ will never become known, so $\overline{\overline{\overline{P}}}$, contradicting $\overline{\overline{P}}$; thus $\overline{\alpha \doteq 0}$. But we are in no position to infer $\alpha \# 0$ (as would follow if $\overline{\alpha \doteq 0} \to \alpha \# 0$ were proved), since $\alpha \# 0$ would mean that we can find an interval separating α from 0, which we would know to be the case only if we already knew how to find the solution to the problem whether $\overline{P}$ or $\overline{\overline{P}}$. Thus $\alpha \# 0$ is unproved, although $\overline{\alpha \doteq 0}$ has been proved. (But $\alpha \# 0$ is not absurd, i.e. not $\overline{\alpha \# 0}$, or then we would have $\alpha \doteq 0$; cf. *R2.7 in Chapter III. Likewise, not $\overline{\overline{\alpha \doteq 0} \to \alpha \# 0}$, or we would have $\overline{\alpha \# 0}$; cf. IM *60d.)

An immediate objection to Brouwer's argument is that a particular real number α has not been defined mathematically; the alleged "definition" makes "α" depend on the unpredeterminate activity of a "creating subject".

van Dantzig **1949** attempts to deal with this objection by giving an "objectivistic" or "formal" version, as contrasted to Brouwer's "subjectivistic" version. In this, van Dantzig introduces as parameter a sequence $\omega = (\omega_0, \omega_1, \omega_2, \ldots)$ of finite sets of "deductions", where the deductions in ω_x are performed between the choices of α_x and α_{x+1}, with $\alpha_{x+1} = 1$ if a deduction of $\overline{P}$ or of $\overline{\overline{P}}$ occurs in ω_x but no such deduction occurs in ω_y for any $y < x$ ($= 0$ otherwise; and $\alpha_0 = 0$). Thereby α becomes α^ω (notation ours). van Dantzig, after alluding p. 955 to "the existence of propositions which are undecidable within a definite formal system" (Gödel), says that the "deductions" should be "according to the rules of a given semantical metasystem". To the present writer, there is a fundamental vagueness of concept here, despite some detailed assumptions which van Dantzig lists. But if we attempt to proceed notwithstanding this vagueness, van Dantzig's reformulation of Brouwer's argument seems to go as follows. One can quantify over all such sequences ω of "deductions". If in every such sequence there is no "deduction" of $\overline{P}$ (of $\overline{\overline{P}}$), then $\overline{\overline{P}}$ (then $\overline{\overline{\overline{P}}}$); therefore, **(i)** $\overline{(\omega)}\alpha^\omega \doteq 0$. Also, if $\alpha^\omega \# 0$, then there is a "deduction" of $\overline{P}$ or of $\overline{\overline{P}}$ in ω; thus, so long as we do not know how to solve the problem whether $\overline{P}$ or $\overline{\overline{P}}$, **(ii)** $(E\omega)\alpha^\omega \# 0$ is unprovable.

But, as we see it, the unprovability of $\overline{\alpha \doteq 0} \rightarrow \alpha \# 0$ does not follow now. We can quantify this to obtain **(iii)** $(E\omega)\overline{\alpha^{\omega} \doteq 0} \rightarrow (E\omega)\alpha^{\omega} \# 0$. But to utilize (iii) with (i) to contradict (ii), we would first need to transform (i) into $(E\omega)\overline{\alpha^{\omega} \doteq 0}$, which we are only able to do classically.

Heyting affirms that it is not very important whether we express the result in Brouwer's words or in those of van Dantzig, or whether we call it a mathematical result or not, provided we understand what it means. Then he says that it shows it would be foolish to seek a proof of the equivalence of the relations $\overline{\alpha \doteq \beta}$ and $\alpha \# \beta$ between real numbers α and β. (The implication $\alpha \# \beta \rightarrow \overline{\alpha \doteq \beta}$ is known; cf. *R2.5.)

In informal intuitionism, care is necessary to identify the grounds for statements of the form "A does not hold (or is unprovable)", as contrasted to "A is absurd", which is simply $\bar{A}$. In a case like "the law of the excluded middle $A \vee \bar{A}$ does not hold", a ground is simply that $A \vee \bar{A}$, although not absurd itself (cf. *51a), has absurd consequences, e.g. ones of the form $(\alpha)(A(\alpha) \vee \bar{A}(\alpha))$ (cf. *27.17). If we admit quantification of proposition variables, its closure $(A)(A \vee \bar{A})$ is absurd. In the case of $\overline{\alpha \doteq \beta} \rightarrow \alpha \# \beta$, no absurd consequence by intuitionistic deductions of the older kind has been exhibited.

In formal intuitionism, "unprovable" takes on its metamathematical meaning, so there is no problem here unless one is proposing to vary the formal system under consideration. The semantical methods we have available will enable us to show that $\forall\beta_{\beta \in R}\forall\alpha_{\alpha \in R}(\neg\alpha \doteq \beta \supset \alpha \# \beta)$ is unprovable, but (classically) has no absurd consequences, in the formal system of intuitionistic analysis of Chapters I and III. (A related observation is made by Kreisel in **1962c**, but his $=$, $\#$ are not the $\doteq$, $\#$ for real number generators; cf. Remark 18.6.)

17.2. We list in Table 1 opposite the five pairs of properties (A_i, B_i) considered by Brouwer **1951**, in the symbolism of Vesley's Chapter III (extending IM and Chapter I) used for the moment informally. It is to be understood that α, β range over R (cf. *R0.1). The first four properties are from Brouwer's table p. 358, and the fifth (A_5, B_5) is from his Footnote 3.

Brouwer's continuum includes all the real numbers, while Vesley in Chapter III includes only the non-negative real numbers. Thereby Vesley simplifies the formalization, without evading any of the

TABLE 1

		$\underline{\beta\in R \,\&\, \beta\dot{>}0 \supset \neg\forall\alpha_{\alpha\in R} B_1}$	
			d
A_1: $\alpha\doteq\alpha$	B_1: $\alpha\circ\!\not>\beta \vee \alpha\not<\!\circ\beta$	$\beta\in R \,\&\, \beta\dot{>}0 \supset \neg\forall\alpha_{\alpha\in R}(A_1 \supset B_1)$	
		$\underline{\forall\beta_{\beta\in R}\forall\alpha_{\alpha\in R}(\neg\alpha\doteq\beta \supset \alpha\#\beta)}$	
			e
		$\forall\beta_{\beta\in R}\forall\alpha_{\alpha\in R}(\neg\alpha\doteq\beta \supset \alpha<\!\circ\beta \vee \alpha\circ\!>\beta)$	
			f
A_2: $\neg\alpha\doteq\beta$	B_2: $\alpha\dot{<}\beta \vee \alpha\dot{>}\beta$	$\underline{\forall\beta_{\beta\in R}\forall\alpha_{\alpha\in R}(A_2 \supset B_2)}$	
			g
A_3: $\alpha\not<\!\circ\beta$	B_3: $\alpha\doteq\beta \vee \alpha\circ\!>\beta$	$\beta\in R \supset \neg\forall\alpha_{\alpha\in R}(A_3 \supset B_3)$	
A_4: $\alpha\dot{>}\beta$	B_4: $\alpha\circ\!>\beta$	$\underline{\forall\beta_{\beta\in R}\forall\alpha_{\alpha\in R}(A_4 \supset B_4)}$	
			h
A_5: $\alpha\doteq\beta \vee \alpha\dot{>}\beta$	B_5: $\alpha\doteq\beta \vee \alpha\circ\!>\beta$	$\forall\beta_{\beta\in R}\forall\alpha_{\alpha\in R}(A_5 \supset B_5)$	j
		$\beta\in R \supset \neg\forall\alpha_{\alpha\in R}(\neg\neg B_5 \supset A_5)$	
			i
		$\underline{\beta\in R \supset \neg\forall\alpha_{\alpha\in R}(\neg\alpha\dot{<}\beta \supset A_5)}$	

fundamental problems; the inclusion of the negative reals would have added only a little uninteresting detail. Consequently in Vesley's formalization 0 (i.e. $0 \cdot 2^{-0}$) is a boundary point with a somewhat special status. For this reason, in rendering one of Brouwer's claims involving B_1, it would not do to use 0 as Brouwer did. We could simply substitute 1 for 0 throughout the properties; but we shall generalize to any $\beta\in R$ (with $\beta\dot{>}0$ when required). In place of $\neg\alpha\dot{<}\beta$ (as direct translation of Brouwer's into our symbols), we usually write $\alpha\not<\!\circ\beta$, which is equivalent to it by *R7.8.

For the first four pairs, Brouwer claims that $\neg\neg B_i$ is equivalent to A_i, but that $\neg\neg B_i$ is not equivalent to B_i. The claimed equivalence we write

a_i: $$\neg\neg B_i \sim A_i,$$

which (with $\beta,\alpha\in R \supset$ prefixed) we shall show in 17.3 to be provable. By *51b, *49a and *25, $\neg\neg(\neg\neg B_i \sim B_i)$. So the claimed non-equivalence is not to be rendered by $\neg(\neg\neg B_i \sim B_i)$ or $\forall\alpha_{\alpha\in R}\neg(\neg\neg B_i \sim B_i)$, which leaves us $\neg\forall\alpha_{\alpha\in R}(\neg\neg B_i \sim B_i)$ as the plausible rendering. Using *49a with *45, and a_i, the latter simplifies to $\neg\forall\alpha_{\alpha\in R}(A_i \supset B_i)$. Brouwer's **1948** p. 1248 Footnote 1 says that by "non-equivalence" he means the absurdity of equivalence (not just the unprovability of equivalence). For the second and fourth pairs, we shall find that his claims of non-equivalence (as rendered above) are not tenable in this sense.

For the fifth pair, if we understand correctly Brouwer's language (which speaks of the "converse" of a proposition not explicitly written as an implication), he claims non-equivalence of B_5 to A_5 (untenable), equivalence of $\neg B_5$ to $\neg A_5$, and non-equivalence of $\neg\neg B_5$ to A_5. Using

b: $$B_5 \supset A_5$$

(proved below), the equivalence simplifies to

c: $$\neg B_5 \supset \neg A_5.$$

Using b, c, *13 and *45, the claimed non-equivalences rendered in the plausible way simplify to $\neg\forall\alpha_{\alpha\in R}(A_5 \supset B_5)$ and $\neg\forall\alpha_{\alpha\in R}(\neg\neg B_5 \supset A_5)$. (The renderings $\neg(B_5 \sim A_5)$ and $\neg(\neg\neg B_5 \sim A_5)$ are refuted thus: By *51b, etc., $\neg\neg(\neg\neg A_5 \sim A_5)$. Thence by the said equivalence, $\neg\neg(\neg\neg B_5 \sim A_5)$. From this and $\neg\neg(\neg\neg B_5 \sim B_5)$, by *25 and *24, $\neg\neg(B_5 \sim A_5)$.)

Thus, to substantiate Brouwer's claims concerning the five pairs of properties in **1951**, we should prove six positive formulas a_1, a_2, a_3, a_4, b, c and six "non-implications" (more precisely, negated universalized implications).

The six positive formulas we shall prove in short order in 17.3.

Of the six non-implications, we shall prove only the three shown in the right column of the table with their negation signs retained ("$\neg\forall$"). The other three we have written omitting the $\neg$ before $\forall$ and in closed form; the resulting formulas we shall show to be formally undecidable. The work will be arranged as follows. Four other formulas have been inserted into the right column of Table 1, including one expressing the implication $\overline{\alpha \doteq \beta} \rightarrow \alpha \# \beta$ (cf. 17.1). The deducibilities d–j shown by arrows in the table we shall quickly establish in 17.3. The first and last (underlined) formulas we shall prove in 18.1. The second underlined formula we shall show (classically) to be realizable, and the third and fourth underlined formulas to be un$_{\mathrm{s}}$realizable, in 18.2.

17.3. a_1: $\beta,\alpha\epsilon R \supset (\neg\neg B_1 \sim A_1)$. Assume $\beta,\alpha\epsilon R$. I. By *R1.4, $\alpha \doteq \alpha$, i.e. A_1, whence by *11 $\neg\neg B_1 \supset A_1$. II. Assume $\alpha \circ > \beta$ & $\alpha < \circ \beta$. Thence by *R6.6 $\alpha < \circ \alpha$, contradicting *R6.8. So $\neg(\alpha \circ > \beta \,\&\, \alpha < \circ \beta)$, whence by *58c–d $\neg\neg(\alpha \circ \not> \beta \vee \alpha \not< \circ \beta)$, i.e. $\neg\neg B_1$.

a_2. Assume $\beta,\alpha\epsilon R$. Now $\neg\neg(\alpha < \beta \vee \alpha > \beta) \sim \neg(\neg\alpha < \beta \,\&\, \neg\alpha > \beta)$ [*63] $\sim \neg\alpha \doteq \beta$ [*R7.10].

a_3. Assume $\beta,\alpha\epsilon R$. Now $\neg\neg(\alpha \doteq \beta \vee \alpha \circ > \beta) \sim \neg(\neg\alpha \doteq \beta \,\&\, \alpha \circ \not> \beta)$ [*63] $\sim \neg\alpha < \beta$ [*R7.1] $\sim \alpha \not< \circ \beta$ [*R7.8].

a_4. *R7.9.

b. *R7.7 (without assuming $\beta,\alpha\epsilon R$).

c. Assume $\beta,\alpha\epsilon R$. Using *R7.8, $\neg\alpha \doteq \beta \,\&\, \alpha \circ \not> \beta \supset \neg\alpha \doteq \beta \,\&\, \neg\alpha > \beta$. Now use *63.

e. I. *R6.2. II. *R6.3, *R2.3.

i. We need that $\beta,\alpha\epsilon R$, $\neg\neg B_5 \supset A_5 \vdash \neg\alpha < \beta \supset A_5$. But by *R7.8 and a_3, $\beta,\alpha\epsilon R \vdash \neg\alpha < \beta \supset \neg\neg B_5$.

§ 18. Refutation or proof of independence, of certain classical order properties. 18.1. After we have *R15.2 and *R15.3, three similar results will follow by the arrows d, i, j in Table 1 (and three others *R9.23–*R9.25 were established in Chapter III).

The following proofs of *R15.1–*R15.3 are by Vesley. They employ some of the later results in Chapter III, and are shorter than the author's original proofs (which used nothing after *R9.2).

*R15.1. $\vdash \beta\epsilon R' \,\&\, \beta > 0 \supset \neg\forall\alpha_{\alpha\epsilon R'}(\alpha \circ \not> \beta \vee \alpha \not< \circ \beta)$.

*R15.2. $\vdash \beta\epsilon R \,\&\, \beta > 0 \supset \neg\forall\alpha_{\alpha\epsilon R}(\alpha \circ \not> \beta \vee \alpha \not< \circ \beta)$.

PROOFS. *R15.1. Assume **(a)** $\beta\epsilon R' \,\&\, \beta > 0$ and $\forall\alpha_{\alpha\epsilon R'}(\alpha \circ \not> \beta \vee \alpha \not< \circ \beta)$. Applying *27.6 with *R0.8, and omitting $\exists\tau$ prior to $\exists$-elim., assume **(b)** $\forall\alpha_{\alpha\epsilon R'}\exists y\{\forall x[\tau(\bar{\alpha}(x)) > 0 \supset y = x] \,\&\, \{(\alpha \circ \not> \beta \,\&\, \tau(\bar{\alpha}(y)) = 1) \vee (\alpha \not< \circ \beta \,\&\, \tau(\bar{\alpha}(y)) = 2)\}\}$. Using this with (a) and omitting $\exists y$ prior to $\exists$-elim.: **(c)** $(\beta \circ \not> \beta \,\&\, \tau(\bar{\beta}(y)) = 1) \vee (\beta \not< \circ \beta \,\&\, \tau(\bar{\beta}(y)) = 2)$. We begin with the slightly more complicated CASE 2: $\tau(\bar{\beta}(y)) = 2$. Using *R9.20 with (a), assume **(d)** $\alpha\epsilon R'$, **(e)** $\bar{\alpha}(y) = \bar{\beta}(y)$, **(f)** $\alpha \doteq (\beta(y) \dot{-} 1)2^{-y}$. By *R9.16, (a) and (f): **(g)** $\beta \not< \circ \alpha$. Using (d), (e) and case hyp. in (b): **(h)** $\alpha \not< \circ \beta$. Now **(i)** $\beta \doteq \alpha$ [(g), (h), *R6.5] $\doteq (\beta(y) \dot{-} 1)2^{-y}$ [(f)]. Thence by (a) and *R7.1: **(j)** $\beta(y) > 1$. Let **(k)** $\gamma = (\beta(y) \dot{-} 1)2^{-y}$ [Lemma 5.3 (a)]. By *R9.2: **(ł)** $\gamma\epsilon R'$. So assume from (b): $\forall x[\tau(\bar{\gamma}(x)) > 0 \supset z = x] \,\&\, \{(\gamma \circ \not> \beta \,\&\, \tau(\bar{\gamma}(z)) = 1) \vee (\gamma \not< \circ \beta \,\&\, \tau(\bar{\gamma}(z)) = 2)\}$. SUBCASE 2.2: $\tau(\bar{\gamma}(z)) = 2$. Let **(m)** $w = \max(z, y)$. Using (ł) and $\exists$-elim. from *R9.20 with γ, w for β, y, assume **(n)** $\delta \in R'$, **(o)** $\bar{\delta}(w)=\bar{\gamma}(w)$, **(p)** $\delta \doteq (\gamma(w) \dot{-} 1)2^{-w}$. Using (m), (o) and *23.4 $\bar{\delta}(z)=\bar{\gamma}(z)$, whence by subcase hyp. $\tau(\bar{\delta}(z))=2$. Using this and (n) in (b), $\delta \not\lessdot \circ\beta$. But $\delta \doteq ((\beta(y) \dot{-} 1)2^{-y}(w) \dot{-} 1)2^{-w}$ [(p), (k)] $= ((\beta(y) \dot{-} 1)2^{w \dot{-} y} \dot{-} 1)2^{-w}$ [*R9.1, (m), etc.] $<\circ$ $(\beta(y) \dot{-} 1)2^{w \dot{-} y} \cdot 2^{-w}$ [*R9.8, (j), *6.16] $= (\beta(y) \dot{-} 1)2^{w \dot{-} y}2^{-(y+(w \dot{-} y))}$ [(m), *6.7] $= (\beta(y) \dot{-} 1)2^{-y}$ [*R9.3] $\doteq \beta$ [(i)]. SUBCASE 2.1: $\tau(\bar{\gamma}(z))=1$. Similarly, using *R9.21 instead. CASE 1: $\tau(\bar{\beta}(y))=1$. Similarly to Case 2, using *R9.21 first (and not deducing (j)).

*R15.2. From *R15.1 (using *R1.11, *R7.3, *R1.5, *R0.7, *R6.13, *R6.14) as *R9.23 from *R9.22.

*R15.3. $\vdash \beta \in R \supset \neg\forall\alpha_{\alpha\in R}(\neg\alpha \lessdot \beta \supset \alpha \doteq \beta \vee \alpha \gtrdot \beta)$.

PROOF. Assume $\beta \in R$ and **(a)** $\forall\alpha_{\alpha\in R}(\neg\alpha \lessdot \beta \supset \alpha \doteq \beta \vee \alpha \gtrdot \beta)$. Now **(b)** $\beta, \beta+1 \in R$ [*R3.2, *R9.2, *R0.7]. Also $\beta = \beta+0$ [*R9.1, *13.1] $<\circ$ $\beta + 1$ [*R9.8, *R6.16, *R3.4], so **(c)** $\beta \circ\not\gtrdot \beta+1$ [*R6.7] and **(d)** $\neg\beta \doteq \beta+1$ [*R6.4]. Using *R1.5 and *R1.6, **(e)** $\forall\alpha\forall\gamma[\alpha \in [\beta, \beta+1]$ & $\alpha \doteq \gamma \supset (\alpha \doteq \beta \sim \gamma \doteq \beta)]$. Toward (f), assume $\alpha \in [\beta, \beta+1]$. By *R8.1, $\alpha \in R$. So by *R8.6, (b) and (c) $\alpha \not\lessdot \circ\beta$, whence by *R7.8 $\neg\alpha \lessdot \beta$, whence by (a) $\alpha \doteq \beta \vee \alpha \gtrdot \beta$, whence by *R7.1 $\alpha \doteq \beta \vee \neg\alpha \doteq \beta$. By $\supset$- and $\forall$-introd., **(f)** $\forall\alpha(\alpha \in [\beta, \beta+1] \supset \alpha \doteq \beta \vee \neg\alpha \doteq \beta)$. By *R10.4 (for $\alpha \doteq \beta$ as the C(α)) with (b), (c), (e) and (f), $\forall\alpha(\alpha \in [\beta, \beta+1] \supset \alpha \doteq \beta) \vee \forall\alpha(\alpha \in [\beta, \beta+1] \supset \neg\alpha \doteq \beta)$. CASE 1: $\forall\alpha(\alpha \in [\beta, \beta+1] \supset \alpha \doteq \beta)$. Then using *R8.5 with (b), $\beta+1 \doteq \beta$, contradicting (d). CASE 2: $\forall\alpha(\alpha \in [\beta, \beta+1] \supset \neg\alpha \doteq \beta)$. Similarly, $\neg\beta \doteq \beta$, contradicting *R1.4.

18.2. When we have established Theorems 18.1, 18.2 and 18.4, the formal undecidability of the five formulas without $\neg\forall$ in the table of 17.2 will follow (the unprovability intuitionistically, the irrefutability classically) by Theorem 11.3 (a), Theorem 9.3 (a) and Corollary 9.4, and the arrows e, f, g, h in Table 1 (with proofs in 17.3).

*R15.4. $\vdash \alpha, \beta \in R' \supset (\alpha \circ\!> \beta \sim \exists x\, \alpha(x) \dot{-} \beta(x) > 2)$.
*R15.5. $\vdash \alpha, \beta \in R' \supset (\alpha \# \beta \sim \exists x |\alpha(x) - \beta(x)| > 2)$.
*R15.6. $\vdash \alpha, \beta \in R' \supset (\alpha \doteq \beta \sim \forall x |\alpha(x) - \beta(x)| \leq 2)$.

PROOFS. *R15.4. Assume $\alpha, \beta \in R'$. I. Prior to $\exists$-elims. from $\alpha \circ\!> \beta$, assume $\forall p 2^k(\alpha(x+p) \dot{-} \beta(x+p)) \geq 2^{x+p}$. Thence $2^k(\alpha(x+k+2) \dot{-} \beta(x+k+2)) \geq 2^{x+k+2} \geq 2^{k+2}$, whence $\alpha(x+k+2) \dot{-} \beta(x+k+2) \geq 2^2 > 2$, whence $\exists x\, \alpha(x) \dot{-} \beta(x) > 2$. II. Prior to $\exists$-elim., assume $\alpha(x) \dot{-} \beta(x) > 2$. Thence $\alpha(x) \dot{-} \beta(x) \geq 3$, so **(a)** $3 \cdot 2^p \leq 2^p(\alpha(x) \dot{-} \beta(x)) =$

$2^p\alpha(x) \dot{-} 2^p\beta(x)$. Now $\alpha(x+p)+2^p > 2^p\alpha(x)$ [*R0.6, *11.15] $\geq 3\cdot 2^p+2^p\beta(x)$ [(a)] $= 2\cdot 2^p+2^p\beta(x)+2^p > 2\cdot 2^p+\beta(x+p)$ [*R0.6, *11.15], whence $\alpha(x+p) \dot{-} \beta(x+p) \geq 2^p$. So $2^x(\alpha(x+p) \dot{-} \beta(x+p)) \geq 2^{x+p}$, whence by $\forall$-, $\exists$- and $\exists$-introd., $\alpha \circ\!> \beta$.

*R15.5. Assume $\alpha,\beta \in R'$. I. Assume $\alpha \# \beta$. By *R6.2 (with *R0.7), $\beta \circ\!> \alpha \vee \alpha \circ\!> \beta$. CASE 1: $\beta \circ\!> \alpha$. By *R15.4, $\exists x \beta(x) \dot{-} \alpha(x) > 2$, whence $\exists x|\alpha(x)-\beta(x)|>2$. II. Prior to $\exists$-elim., assume $|\alpha(x)-\beta(x)|>2$. By *11.14a, $\alpha(x)>\beta(x)+2 \vee \beta(x)>\alpha(x)+2$. CASE 1: $\alpha(x)>\beta(x)+2$. Then $\alpha(x) \dot{-} \beta(x)>2$, whence by $\exists$-introd. and *R15.4 $\alpha \circ\!> \beta$, whence by *R6.3 (with *R2.3) $\alpha \# \beta$.

*R15.6. Assume $\alpha,\beta \in R'$. Then $\alpha \doteq \beta \sim \neg \alpha \# \beta$ [*R2.8] $\sim \neg \exists x|\alpha(x)-\beta(x)|>2$ [*R15.5] $\sim \forall x \neg |\alpha(x)-\beta(x)|>2$ [*86] $\sim \forall x|\alpha(x)-\beta(x)| \leq 2$ [*139–*141].

THEOREM 18.1^C. *Classically,*

(1) $\forall \beta_{\beta \in R} \forall \alpha_{\alpha \in R}(\neg \alpha \doteq \beta \supset \alpha \# \beta)$

is realizable.

PROOF. Using *R1.11 with *R1.6, *R1.5, *R2.4 and *R2.3, (1) is deducible from

(2) $\forall \beta_{\beta \in R'} \forall \alpha_{\alpha \in R'}(\neg \alpha \doteq \beta \supset \alpha \# \beta)$.

Using *R15.6 and *R15.5, (2) is deducible from

(3) $\forall \beta \forall \alpha[\neg \forall x|\alpha(x)-\beta(x)| \leq 2 \supset \exists x|\alpha(x)-\beta(x)|>2]$.

Using #15 and #D in 5.5, we can take the scopes of $\forall x$ and $\exists x$ in (3) to be prime. Classically, (3) is true. Hence by Lemma 8.4b (ii), (3) is realizable. By Theorem 9.3 (a), so is (1).

THEOREM 18.2. *The formula*

(4) $\forall \beta_{\beta \in R} \forall \alpha_{\alpha \in R}(\alpha > \beta \supset \alpha \circ\!> \beta)$

is un$_S$realizable.

PROOF. From (4) by *R0.7 and *R7.1, we can deduce

(5) $\beta,\alpha \in R'$ & $\alpha \not\lessdot\!\circ \beta$ & $\neg \alpha \doteq \beta \supset \alpha \circ\!> \beta$,

and thence by *R6.15, *R15.6 and *R15.4,

(6) $\beta,\alpha \in R'$ & $\alpha \geq \beta$ & $\neg \forall y|\alpha(y)-\beta(y)| \leq 2 \supset \exists y \alpha(y) \dot{-} \beta(y)>2$.

Thence we shall further deduce

(7) $\forall x[\neg \forall y \neg T(x, y) \supset \exists y T(x, y)]$

for the T(x, y) of Theorem 11.7 (b).

Accordingly, we assume

(a) $\neg\forall y\neg T(x, y)$,

and set out to deduce $\exists y T(x, y)$ from (a) and (6) with x held constant. By Lemma 8.8 $T_1(z, x, y) \vee \neg T_1(z, x, y)$, whence by substitution

(b) $T(x, y) \vee \neg T(x, y)$.

So, for any natural number b, we can assume, preparatory to $\exists$-elims. from Lemmas 5.3 (b) and 5.5 (b),

(c) $\beta(0)=\boldsymbol{b}$, $\forall y \beta(y')=2\beta(y)$,

(d) $\alpha(0)=\boldsymbol{b}+1$, $\forall y \alpha(y')=\begin{cases} 2\alpha(y) \dot{-} 1 & \text{if } \neg T(x, y), \\ 2\alpha(y) & \text{if } T(x, y). \end{cases}$

Using respectively *R0.5a–b, ind. on y, and (f):

(e) $\beta, \alpha \in R'$. (f) $\alpha(y) \geq \beta(y)+1$. (g) $\alpha \geq \beta$.

Next we establish

(h) $\alpha(y)>\beta(y)+1 \supset \alpha(y')>\beta(y')+2$.

Assume $\alpha(y)>\beta(y)+1$. Then $\alpha(y') \geq 2\alpha(y) \dot{-} 1 \geq 2(\beta(y)+2) \dot{-} 1 = 2\beta(y)+3 = \beta(y')+3 > \beta(y')+2$. — Next,

(i) $\forall y|\alpha(y)-\beta(y)| \leq 2 \supset \forall y \alpha(y)=\beta(y)+1$.

Assume **(A)** $\forall y|\alpha(y)-\beta(y)| \leq 2$. Now $\alpha(y)=\beta(y)+1$, or else by (f) $\alpha(y)>\beta(y)+1$, whereupon by (h) $\alpha(y')>\beta(y')+2$, so $|\alpha(y')-\beta(y')|>2$, contradicting (A). By $\forall$-introd., $\forall y \alpha(y)=\beta(y)+1$. — Next we establish, by ind. on y,

(j) $\{\alpha(y)=\beta(y)+1 \sim \forall z_{z<y}\neg T(x, z)\}$ & $\{\alpha(y)>\beta(y)+1 \sim \exists z_{z<y} T(x, z)\}$.

Using *140 [*58b, *86], the left [right] members of the two equivalences are mutually exclusive. So if we deduce both members of one equivalence, that one will follow by *11, *16, the other by *10a, *16. IND. STEP. By (f), we have two cases. CASE 1: $\alpha(y)=\beta(y)+1$. By hyp. ind., $\forall z_{z<y}\neg T(x, z)$. By (b), we have two subcases. SUBCASE 1.1: $T(x, y)$. Then $\exists z_{z<y'} T(x, z)$. Also $\alpha(y') = 2\alpha(y)$ [(d)] $= 2(\beta(y)+1)$ [Case 1 hyp.] $= 2\beta(y)+2 = \beta(y')+2 > \beta(y')+1$. CASE 2: $\alpha(y)>$

$\beta(y)+1$. By hyp. ind., $\exists z_{z<y}T(x, z)$, whence $\exists z_{z<y'}T(x, z)$. Using (h), $\alpha(y')>\beta(y')+1$. — Using (j), $\forall y \alpha(y)=\beta(y)+1 \supset \forall y \neg T(x, y)$. Combining this by *2 with (i), and using contraposition *12 and (a),

(k) $\neg \forall y |\alpha(y)-\beta(y)| \leq 2$.

If $\exists y \alpha(y) \dot{-} \beta(y)>2$, then $\exists y \alpha(y)>\beta(y)+1$, so by (j) $\exists y T(x, y)$. So

(ł) $\exists y \alpha(y) \dot{-} \beta(y)>2 \supset \exists y T(x, y)$.

Using (e), (g) and (k) in (6), and the result in (ł),

(m) $\exists y T(x, y)$.

Using $\exists$-elim. from Lemmas 5.3 (b) and 5.5 (b) to discharge the assumptions (c) and (d), $\supset$-introd. to discharge (a), and $\forall$-introd., we have the deducibility of (7) from (6), and thence from (4).

By Theorem 11.7 (b) (7) is un$_{\mathrm{s}}$realizable. By Theorem 11.3 (a), so is (4).

Remark 18.3. By (c), $\beta = \boldsymbol{b}$ [$= \boldsymbol{b}2^{-0}$, ${}^{\times}$R9.1], so the proof shows that $\forall \alpha_{\alpha \in R}(\alpha > \boldsymbol{b} \supset \alpha^{\circ} > \boldsymbol{b})$ is un$_{\mathrm{s}}$realizable. Modifying the proof inessentially, $\exists x \forall p 2\beta(x+p) \leq \beta(x+p') \leq 2\beta(x+p)+1 \supset \forall \alpha_{\alpha \in R}(\alpha > \beta \supset \alpha^{\circ} > \beta)$ is un$_{\mathrm{s}}$realizable.

Theorem 18.4. *The formula*

(8) $\forall \beta_{\beta \in R} \forall \alpha_{\alpha \in R}(\neg \alpha \doteq \beta \supset \alpha < \beta \vee \alpha > \beta)$

is un$_{\mathrm{s}}$realizable.

Proof. From (8) by *R0.7 and *R7.1, we can deduce

(9) $\beta, \alpha \in R' \ \& \ \neg \alpha \doteq \beta \supset \alpha^{\circ} \not> \beta \vee \alpha \not<^{\circ} \beta$,

and thence by *R15.6 and *R15.4

(10) $\beta, \alpha \in R' \ \& \ \neg \forall y |\alpha(y)-\beta(y)| \leq 2 \supset$
$\neg \exists y \alpha(y) > \beta(y)+2 \vee \neg \exists y \beta(y) > \alpha(y)+2$.

Thence we shall further deduce

(11) $\forall x[\neg(\forall y \neg W_0(x, y) \ \& \ \forall y \neg W_1(x, y)) \supset$
$\neg \forall y \neg W_0(x, y) \vee \neg \forall y \neg W_1(x, y)]$

for the $W_0(x, y)$ and $W_1(x, y)$ of Theorem 11.7 (d).

Accordingly, we assume

(a) $\neg(\forall y \neg W_0(x, y) \ \& \ \forall y \neg W_1(x, y))$.

By Lemma 8.8 and substitution, and middle Remark 4.1,

(b_0) $W_0(x, y) \vee \neg W_0(x, y)$, ($b_1$) $W_1(x, y) \vee \neg W_1(x, y)$.

The proof of IM p. 308 (51) is readily formalized to give

(c) $\neg(\exists y W_0(x, y) \,\&\, \exists y W_1(x, y))$,

whence

(d) $\neg(W_0(x, y) \,\&\, W_1(x, y))$.

So, for any natural number $b > 0$, we can assume, preparatory to $\exists$-elims. from Lemmas 5.3 (b) and 5.5 (b),

(e) $\beta(0) = \boldsymbol{b}$, $\beta(y') = 2\beta(y)$,

(f) $\alpha(0) = \boldsymbol{b}$,
$$\forall y\alpha(y') = \begin{cases} 2\alpha(y) \dot{-} 1 \text{ if } W_0(x, y) \,\&\, \neg W_1(x, y), \\ 2\alpha(y) \text{ if } \neg W_0(x, y) \,\&\, \neg W_1(x, y), \\ 2\alpha(y)+1 \text{ if } \neg W_0(x, y) \,\&\, W_1(x, y). \end{cases}$$

Using respectively *R0.5a–b and ind. on y:

(g) $\beta, \alpha \in R'$. (h) $\beta(y) = 2^y \boldsymbol{b} > 0$.

Next we establish, by ind. on y,

(i) $\{\alpha(y) < \beta(y) \sim \exists z_{z<y} W_0(x, z)\} \,\&$
$\{\alpha(y) = \beta(y) \sim \forall z_{z<y}(\neg W_0(x, z) \,\&\, \neg W_1(x, z))\} \,\&$
$\{\alpha(y) > \beta(y) \sim \exists z_{z<y} W_1(x, z)\}$.

Using *140, *141 [(c) etc.] the left [right] members are mutually exclusive; etc. (cf. (j) for Theorem 18.2). IND. STEP. CASE 1: $\alpha(y) < \beta(y)$. By hyp. ind., $\exists z_{z<y} W_0(x, z)$, whence $\exists z_{z<y'} W_0(x, z)$. Also $\alpha(y') \leq 2\alpha(y)$ [because (c) excludes the third case in (f)] $< 2\beta(y)$ [Case 1 hyp.] $= \beta(y')$. CASE 2: $\alpha(y) = \beta(y)$. SUBCASE 2.1: $W_0(x, y) \,\&\, \neg W_1(x, y)$. Then $\exists z_{z<y'} W_0(x, z)$. Also $\alpha(y') = 2\alpha(y) \dot{-} 1$ [(f)] $= 2\beta(y) \dot{-} 1$ [Case 2 hyp.] $< 2\beta(y)$ [*6.16, with $2\beta(y) > 0$ from (h)] $= \beta(y')$. SUBCASE 2.2: $\neg W_0(x, y) \,\&\, \neg W_1(x, y)$. By hyp. ind., $\forall z_{z<y}(\neg W_0(x, z) \,\&\, \neg W_1(x, z))$. So $\forall z_{z<y'}(\neg W_0(x, z) \,\&\, \neg W_1(x, z))$; etc. — Next,

(j_0) $\beta(y) > \alpha(y) \supset \beta(y'') \geq \alpha(y'') + 4$,
(j_1) $\alpha(y) > \beta(y) \supset \alpha(y'') \geq \beta(y'') + 4$.

For, assume $\beta(y) > \alpha(y)$. Then by (i) and (c), $\neg\exists y W_1(x, y)$. Hence $\alpha(y') + 2 \leq 2\alpha(y) + 2 = 2(\alpha(y)+1) \leq 2\beta(y) = \beta(y')$, and $\alpha(y'') + 4 \leq$

$2\alpha(y')+4 = 2(\alpha(y')+2) \leq 2\beta(y') = \beta(y'')$. —

(k) $\forall y|\alpha(y)-\beta(y)|\leq 2 \supset \forall y\alpha(y)=\beta(y)$.

For, assuming $\alpha(y)<\beta(y)$, by (j_0) we would have $|\alpha(y'')-\beta(y'')|\geq 4$. — By (k) with (i), $\forall y|\alpha(y)-\beta(y)|\leq 2 \supset \forall y(\neg W_0(x, y) \& \neg W_1(x, y))$, whence by *87, contraposition and (a),

(l) $\neg\forall y|\alpha(y)-\beta(y)|\leq 2$.

Next we establish

(m_0) $\neg\exists y\alpha(y)>\beta(y)+2 \supset \neg\forall y\neg W_0(x, y)$,
(m_1) $\neg\exists y\beta(y)>\alpha(y)+2 \supset \neg\forall y\neg W_1(x, y)$.

Assume $\neg\exists y\alpha(y)>\beta(y)+2$, whence $\forall y\alpha(y)\leq\beta(y)+2$, and by ($j_1$): **(A)** $\forall y\alpha(y)\leq\beta(y)$. Also assume $\forall y\neg W_0(x, y)$, whence by (i) $\forall y\neg\alpha(y)<\beta(y)$. So by (A) $\forall y\alpha(y)=\beta(y)$, whence by (i) $\forall y(\neg W_0(x, y) \& \neg W_1(x, y))$. This with *87 contradicts (a). — Using (g) and (l) in (10), and (m_0) and (m_1) with the result,

(n) $\neg\forall y\neg W_0(x, y) \vee \neg\forall y\neg W_1(x, y)$.

REMARK 18.5. By the proof, or for $m > 0$ an easy modification, $\forall\alpha_{\alpha\in R}(\neg\alpha\doteq \boldsymbol{b}2^{-m} \supset \alpha<\boldsymbol{b}2^{-m} \vee \alpha>\boldsymbol{b}2^{-m})$ is un$_s$realizable for any $b > 0$ and any m.

REMARK 18.6. Combining several deductions (namely, e+g in 17.3, the deduction of (7) from (4) in the proof of Theorem 18.2, beginning proof of Corollary 11.10 (a)), from

(1) $\forall\beta_{\beta\in R}\forall\alpha_{\alpha\in R}(\neg\alpha\doteq\beta \supset \alpha\#\beta)$

(in Theorem 18.1) we can deduce Markov's principle

M_1: $\forall z\forall x[\neg\forall y\neg T_1(z, x, y) \supset \exists yT_1(z, x, y)]$

(beginning 10.1). — Modifying inessentially the proof of Theorem 18.2, we can instead deduce from (4), and so by e+g from (1),

(3′) $\forall\beta\forall\alpha[\neg\forall y\neg|\alpha(y)-\beta(y)|>2 \supset \exists y|\alpha(y)-\beta(y)|>2]$.

Conversely, from (3′) we can deduce (3), and thence (by the proof of Theorem 18.1) (1). Thus: *Each two of* (1), (4), (3′) *are interdeducible.* — Formula (3′) is similar in form to M_1, but involves function variables. A direct extension of Markov's principle from algorithms for one-place number-theoretic functions to algorithms for type-2 functionals is

M_1': $\forall z\forall\alpha[\neg\forall y\neg T_0^1(\tilde{\alpha}(y), z, y) \supset \exists yT_0^1(\tilde{\alpha}(y), z, y)]$

where $T_0^1(w, z, y)$ is chosen by (the method of proof of) Lemma 8.5 to numeralwise express $T_0^1(w, z, y)$ IM p. 292. This formula M_1' is deducible from (1) essentially as M_1 and (3′). Conversely, by informal reasoning (which is presumably formalizable) in the theory of recursive functionals, (3′) follows from M_1'. — Thus, since the converses of (1) and (4) are provable (*R2.5, *R7.7): *Markov's principle extended to functionals as* (3′), *and presumably as* M_1', *is interdeducible with* (i) *the equivalence of inequality* $\neg\alpha \doteq \beta$ *to apartness* $\alpha \# \beta$, *and with* (ii) *the equivalence of the virtual ordering* $\alpha < \beta$ *to the pseudoordering* $\alpha <\circ \beta$, *in Brouwer's continuum.* — From (3′) or

$$(3'') \qquad \forall\beta\forall\alpha[\neg\forall y\alpha(y)=\beta(y) \supset \exists y\alpha(y)\neq\beta(y)]$$

(cf. van Rootselaar **1960**, Kreisel **1962c**), we can deduce

$$(3''') \qquad \forall\alpha[\neg\forall y\neg\alpha(y)=0 \supset \exists y\alpha(y)=0],$$

using $\forall$-elims. with $\beta_1=\lambda y0$, and $\alpha_1=\lambda y\overline{sg}(\alpha(y))+2$ or $\alpha_1=\lambda y\overline{sg}(\alpha(y))$ resp.; and conversely, using $\alpha_1=\lambda yt(\alpha, \beta, y)$ for a standard formula $t(\alpha, \beta, y)=0$ equivalent to $|\alpha(y)-\beta(y)|>2$ or to $\alpha(y)\neq\beta(y)$, resp. (cf. 5.5 preceding *14.1). Thus: *Each two of* (3′), (3″), (3‴) *are interdeducible.*

BIBLIOGRAPHY

by S. C. KLEENE

A date in gothic numerals used in conjunction with a name (e.g. Brouwer **1907**) constitutes a reference to this bibliography; in medieval tailciphers (e.g. Brouwer 1908), to the bibliography of Kleene's **Introduction to metamathematics** (**1952b**).

BELINFANTE, M. J.

1929. *Zur intuitionistischen Theorie der unendlichen Reihen.* **Sitzungsberichte der Preussische Akademie der Wissenschaften, Physikalisch-mathematische Klasse,** 1929, pp. 639-660.

BETH, EVERT W.

1947. *Semantical considerations on intuitionistic mathematics.* **Koninklijke Nederlandsche Akademie van Wetenschappen, Proceedings of the Section of Sciences,** vol. 50, pp. 1246-1251 (= **Indagationes mathematicae,** vol. 9, pp. 572-577).

1955. Les fondements logiques des mathématiques (2e éd. revue et augmentée). Paris (Gauthier-Villars) and Louvain (E. Nauwelaerts), xv+241 pp.

1955a. *Semantic entailment and formal derivability.* **Mededelingen der Koninklijke Nederlandse Akademie van Wetenschappen, Afd. letterkunde,** n.s., vol. 18, no. 13, pp. 309-342.

1956. *Semantic construction of intuitionistic logic.* Ibid., vol. 19, no. 11. Cf. **1959a**, Kleene **1957a**, Kreisel **1958b** pp. 380-383, **1960**, **1962b**, Dyson-Kreisel **1961**.

1959. The foundations of mathematics. Amsterdam (North-Holland Pub. Co.), XXVI+741 pp.

1959a. *Remarks on intuitionistic logic.* **Constructivity in mathematics,** Amsterdam (North-Holland Pub. Co.), pp. 15-25. Cf. Kreisel **1962b**.

BIRKHOFF, GARRETT

1940. Lattice theory. American Mathematical Society Colloquium publications, vol. 25, New York, v+155 pp. 2nd ed., 1948, xiv+283 pp.

BOREL, ÉMILE

1898. Leçons sur la théorie des fonctions. Paris (Gauthier-Villars), 8+136 pp. 4th ed. (Leçons sur la théorie des fonctions; principes de la théorie des ensembles en vue des applications à la théorie des fonctions). Paris (Gauthier-Villars), 1950, xii+295 pp.

1922. Méthodes et problèmes de la théorie des fonctions. Paris (Gauthier-Villars), ix+148 pp.

BROUWER, L. E. J.

1907. Over de grondslagen der wiskunde (On the foundations of mathematics). Dissertation. Amsterdam and Leipzig. 183 pp.

1912. Intuitionisme en formalisme. Groningen, 32 pp. Eng. tr. by Arnold Dresden, *Intuitionism and formalism,* **Bulletin of the American Mathematical Society,** vol. 20 (1913-4), pp. 81-96.

1918-9. *Begründung der Mengenlehre unabhängig vom logischen Satz vom ausgeschlossenen Dritten.* [I] *Erster Teil: Algemeine Mengenlehre.* **Verhandelingen der Koninklijke Nederlandsche Akademie van Wetenschappen te Amsterdam (Eerste sectie),** vol. 12, no. 5 (Amsterdam 1918), 43 pp. (errata at end of II). [II] *Zweiter Teil: Theorie der Punktmengen,* ibid., vol. 12, no. 7 (Amsterdam 1919), 33 pp.

1919. *Intuitionistische Mengenlehre.* **Jahresbericht der Deutschen Mathematiker-Vereinigung,** vol. 28, pp. 203-208. Also **Kon. Ned. Akad. Wet. Amsterdam, Proc. Sect. Sci.,** vol. 23, pp. 949-954.

1921. *Besitzt jede reele Zahl eine Dezimalbruchentwicklung?* **Mathematische Annalen,** vol. 83, pp. 201-210. Also **Kon. Ned. Akad. Wet. Amsterdam, Proc. Sect. Sci.,** vol. 23 (1921-2), pp. 955-964.

1923a. *Begründung der Funktionenlehre unabhängig vom logischen Satz vom ausgeschlossenen Dritten.* **Verhand. Kon. Ned. Akad. Wet. Amsterdam (Eerste sectie),** vol. 13, no. 2, 24 pp. (Errata in **1924** Footnote 1.)

1924. *Beweis, dass jede volle Funktion gleichmässig stetig ist.* **Kon. Ned. Akad. Wet. Amsterdam, Proc. Sect. Sci.,** vol. 27, pp. 189-193. Cf. **1924a.**

1924a. *Bemerkungen zum Beweis der gleichmässigen Stetigkeit voller Funktionen.* Ibid., pp. 644-646.

1924-7. *Zur Begründung der intuitionistischen Mathematik.* I. **Math. Ann.,** vol. 93 (1924-5), pp. 244-257 (erratum, vol. 95, p. 472). II. Ibid., vol. 95 (1925-6), pp. 453-472 (erratum, vol. 96, p. 488). III. Ibid., vol. 96 (1926-7), pp. 451-488.

1925. *Intuitionistische Zerlegung mathematischer Grundbegriffe.* **Jahresb. Deutsch. Math. Verein.,** vol. 33, pp. 251-256.

1927. *Über Definitionsbereiche von Funktionen.* **Math. Ann.,** vol. 97, pp. 60-75.

1928a. Die Struktur des Kontinuums. Vienna (Gottlieb Gistel), 12 pp.

1929. *Mathematik, Wissenschaft und Sprache.* **Monatshefte für Mathematik und Physik,** vol. 36, pp. 153-164.

1948. *Consciousness, philosophy, and mathematics.* **Proceedings of the Xth International Congress of Philosophy (Amsterdam, Aug. 11-18, 1948),** Amsterdam (North-Holland Pub. Co.) 1949, pp. 1235-1249.

1948a. *Essentieel negatieve eigenschappen* (Essentially negative properties). **Kon. Ned. Akad. Wet. (Amsterdam), Proc. Sect. Sci.,** vol. 51, pp. 963-964 (= **Indagationes mathematicae,** vol. 10, pp. 322-323).

1948b. *Opmerkingen over het beginsel van het uitgesloten derde en over negatieve asserties* (Remarks on the principle of the excluded third and on negative assertions). Ibid., vol. 51, pp. 1239-1243 (= **Indag. math.,** vol. 10, pp. 383-387).

1949. *De non-aequivalentie van de constructieve en de negatieve orderelatie in het continuum* (The non-equivalence of the constructive and the negative order relation in the continuum). Ibid., vol. 52, pp. 122-124 (= **Indag. math.,** vol. 11, pp. 37-39).

1949a. *Contradictoriteit der elementaire meetkunde* (Contradictoriness of elementary geometry). Ibid., vol. 52, pp. 315-316 (= **Indag. math.,** vol. 11, pp. 89-90).

1951. *On order in the continuum, and the relation of truth to non-contradictority.* Ibid., vol. 54 (or **Indag. math.,** vol. 13), pp. 357-358.

1952. *Historical background, principles and methods of intuitionism.* **South African journal of science,** vol. 49, pp. 139-146.

1954. *Points and spaces.* **Canadian journal of mathematics,** vol. 6, pp. 1-17.

CHURCH, ALONZO and ROSSER, J. B.

1936. *Some properties of conversion.* **Transactions of the American Mathematical Society,** vol. 39, pp. 472-482.

CURRY, HASKELL B. and FEYS, ROBERT

1958. Combinatory logic, vol. 1. Amsterdam (North-Holland Pub. Co.), XVI+417 pp.

DALEN, DIRK VAN

1963. Extension problems in intuitionistic plane projective geometry. Thesis Amsterdam (Drukkerij Holland), 36 pp. (with Dutch summary).

DANTZIG, DAVID VAN

1942. *On the affirmative content of Peano's theorem on differential equations.* **Kon. Ned. Akad. Wet. (Amsterdam), Proc. Sect. Sci.,** vol. 45, pp. 367-373 (= **Indag. math.,** vol. 4, pp. 140-146).

1949. *Comments on Brouwer's theorem on essentially-negative predicates.* Ibid., vol. 52, pp. 949-957 (= **Indag. math.,** vol. 11, pp. 347-355).

DAVIS, MARTIN

1958. Computability and unsolvability. New York, Toronto, London (McGraw-Hill), xxv+210 pp.

DEKKER, J. C. E.

1962. (editor) **Recursive function theory** [Symposium at New York, April 6-7, 1961]. Proceedings of the Symposia in Pure Mathematics, vol. 5, American Mathematical Society, Providence, R.I., 1962, vii+247 pp.
See Dekker and Myhill.

DEKKER, J. C. E. and MYHILL, JOHN

1960. Recursive equivalence types. Univ. of California publications in mathematics, n.s., vol. 3, no. 3, pp. 67-214, Berkeley and Los Angeles (U. of Calif. Press).

DE LOOR, B. See Loor, B. de

DIJKMAN, JACOBUS GERHARDUS

1948. *Recherche de la convergence négative dans les mathématiques intuitionistes.*

Kon. Ned. Akad. Wet. (Amsterdam), Proc. Sect. Sci., vol. 51, pp. 681-682 (= **Indag. math.,** vol. 10, pp. 232-243).

1952. **Convergentie en divergentie in de intuitionistische wiskunde (Convergence and divergence in intuitionistic mathematics).** Thesis Amsterdam, 's-Gravenhage, x+98 pp. (with English summary).

DYSON, V. H. and KREISEL, G.

1961. *Analysis of Beth's semantic construction of intuitionistic logic.* Part II of **Technical Report No. 3, Applied Mathematics and Statistics Laboratories, Stanford Univ., Stanford, Calif.,** 65 pp.

FREUDENTHAL, HANS

1936. *Zum intuitionistischen Raumbegriff.* **Compositio mathematica,** vol. 4, pp. 82-111.

GÁL, I. L., ROSSER, J. B. and SCOTT, D.

1958. *Generalization of a lemma of G. F. Rose.* **Journal of symbolic logic,** vol. 23, pp. 137-138.

GLIVENKO, V.

1928. *Sur la logique de M. Brouwer.* **Académie Royale de Belgique, Bulletins de la classe des sciences,** ser. 5, vol. 14, pp. 225-228.

GÖDEL, KURT

1933. *Eine Interpretation des intuitionistischen Aussagenkalküls.* **Ergebnisse eines mathematischen Kolloquiums,** Heft 4 (for 1931-2, pub. 1933), pp. 39-40.

1958. *Über eine bisher noch nicht benützte Erweiterung des finiten Standpunktes.* **Dialectica,** vol. 12 (nos. 47/48), pp. 280-287.

GRISS, G. F. C.

1944. *Negatieloze intuitionistische wiskunde* (Negationless intuitionistic mathematics). **Koninklijke Nederlandsche Akademie van Wetenschappen (Amsterdam), Verslagen van de gewone vergaderingen der Afdeeling Natuurkunde,** vol. 53, no. 5, pp. 261-268.

1946-51. *Negationless intuitionistic mathematics,* and ibid. II, III, IVa, IVb. **Kon. Ned. Akad. Wet. (Amsterdam), Proc., ser. A,** vol. 49 (1946) pp. 1127-1133, vol. 53 (1950) pp. 456-463, vol. 54 (1951) pp. 193-199, 452-471 (= **Indag. math.,** vol. 8 pp. 675-681, vol. 12 pp. 108-115, vol. 13 pp. 193-199, 452-471).

1955. *La mathématique intuitioniste sans négation.* **Nieuw archief voor wiskunde,** ser. 3, vol. 3, no. 3, pp. 134-142.

HARDY, G. H. and WRIGHT, E. M.

1954. **An introduction to the theory of numbers.** Oxford (Clarendon), xvi+419 pp.

HARROP, RONALD

1956. *On disjunctions and existential statements in intuitionistic systems of logic.* **Math. Ann.,** vol. 132, pp. 347-361.

1960. *Concerning formulas of the types $A \rightarrow B \vee C$, $A \rightarrow (Ex)B(x)$ in intuitionistic formal systems.* **Jour. symbolic logic,** vol. 25, pp. 27-32.

HERMES, HANS

1961. **Aufzählbarkeit, Entscheidbarkeit, Berechenbarkeit: Einführung in die Theorie der rekursiven Funktionen.** Die Grundlehren der mathematischen Wissenschaften, vol. 109, Springer-Verlag, Berlin-Göttingen-Heidelberg, x+246 pp.

HEYTING, AREND

1925. *Intuitionistische axiomatiek der projectieve meetkunde* (Intuitionistic axiomatics for projective geometry). Thesis Amsterdam. Groningen, 95 pp. Cf. **1927a.**

1927. *Die Theorie der linearen Gleichungen in einer Zahlenspezies mit nichtkommutativer Multiplikation.* **Math. Ann.**, vol. 98, pp. 465-490.

1927a. *Zur intuitionistischen Axiomatik der projektiven Geometrie.* Ibid., pp. 491-538.

1929. *De telbaarheidspraedicaten van Prof. Brouwer* (The enumerability predicates of Prof. Brouwer). **Nieuw archief voor wiskunde,** 2 s., vol. 16, no. 2, pp. 47-58.

1941. *Untersuchungen über intuitionistische Algebra.* **Verhand. Kon. Ned. Akad. Wet. Amsterdam (Eerste sectie),** vol. 18, no. 2, 36 pp.

1952-3. **Inleiding tot de intuitionistische wiskunde** (Introduction to intuitionistic mathematics). Mimeographed notes by J. J. de Iongh on lectures in 1952-53, Amsterdam, ii+98 pp.

1953. *Espace de Hilbert et intuitionnisme.* **Les méthodes formelles en axiomatique (à Paris** [7-9] **Décembre 1950),** pp. 59-64. Colloques Internationaux du Centre National de la Recherche Scientifique, XXXVI. Editions du Centre National de la Recherche Scientifique, Paris.

1953a. *Sur la théorie intuitionniste de la mesure.* **Bulletin de la Société Mathématique de Belgique,** vol. 6, pp. 70-78.

1955. **Les fondements des mathématiques. Intuitionnisme. Théorie de la démonstration.** Paris (Gauthier-Villars) and Louvain (E. Nauwelaerts), 91 pp. The second edition of 1934.

1956. **Intuitionism. An introduction.** Amsterdam (North-Holland Pub. Co.), VIII+133 pp.

1959. (editor) **Constructivity in mathematics. Proceedings of the Colloquium held at Amsterdam,** [August 26-31] **1957.** Amsterdam (North-Holland Pub. Co.), 1959, VIII+297 pp.

KABAKOV, F. A. (Kabakow, F. A.)

1963. *Vyvodimost' nekotoryh realizuemyh formul isčislenija vyskazyvanii* (*Über die Ableitbarkeit einiger realisierbare Formeln des Aussagenkalküls*). **Zeitschrift für mathematische Logik und Grundlagen der Mathematik,** vol. 9, pp. 97-104.

KLAUA, DIETER

1961. **Konstruktive Analysis.** Mathematische Forschungsberichte, XI, VEB Deutsches Verlag der Wissenschaften, Berlin, VIII+160 pp.

KLEENE, STEPHEN COLE

1952a. *Finite axiomatizability of theories in the predicate calculus using ad-*

ditional predicate symbols. **Memoirs of the American Mathematical Society,** no. 10, pp. 27-68. Errata in **Jour. Symbolic Logic,** vol. 19, p. 63.

1952b. Introduction to metamathematics. Amsterdam (North-Holland Pub. Co.), Groningen (Noordhoff), New York and Toronto (Van Nostrand), X+550 pp. The most serious errata in early printings are: p. 120, second line of Remark 1, for "same" read "intuitionistic"; p. 161, in lines 13-15 let (ii) read "it contains exactly the variables $a_1, \ldots, a_n$", and replace lines 17-18 by "$A_i(t_1, \ldots, t_{n_i})$ and $A_j(u_1, \ldots, u_{n_j})$ are not the same formula for any terms $t_1, \ldots, t_{n_i}, u_1, \ldots, u_{n_j}$."; p. 338, after "SECOND METHOD" insert ", for $Q_1, \ldots, Q_m$ simultaneously defined"; p. 342, last display, after "$*$" insert "[2 exp", and at the end insert "]"; p. 418, in (Ia) replace "E′ is E." by "$\vdash_1$ E′ $\sim$ E." Also cf. Footnote 1 in Chapter I above.

1955. *Arithmetical predicates and function quantifiers.* **Trans. Amer. Math. Soc.,** vol. 79, pp. 312-340. Errata, ibid. vol. 80 (1955) p. 386 and vol. 81 (1956) p. 524, and **Proceedings of the American Mathematical Society** vol. 8 (1957) p. 1006.

1955a. *On the forms of the predicates in the theory of constructive ordinals (second paper).* **American journal of mathematics,** vol. 77, pp. 405-428. Errata in the bibliography of **1959.**

1955b. *Hierarchies of number-theoretic predicates.* **Bull. Amer. Math. Soc.,** vol. 61, pp. 193-213.

1956. *A note on computable functionals.* **Kon. Ned. Akad. Wet. (Amsterdam), Proc., Ser. A,** vol. 59 (or **Indag. math.,** vol. 18), pp. 275-280.

1957. *Realizability.* **Summaries of talks presented at the Summer Institute of Symbolic Logic in 1957 at Cornell University,** vol. 1, pp. 100-104 (2nd ed., Princeton N.J. (Communications Research Division, Institute for Defense Analyses) 1960), reprinted in **Constructivity in mathematics,** Amsterdam (North-Holland Pub. Co.) 1959, pp. 285-289. Errata: p. 101 line 5 from below (or p. 287 line 2), after "variable" read "free"; 1957 p. 101 (4), for "β" read "γ" (in 1959, 1960, (4) is correct).

1957a. Review of Beth **1956. Jour. symbolic logic,** vol. 22, pp. 363-365.

1958. *Mathematical logic: constructive and non-constructive operations.* **Proceedings of the International Congress of Mathematicians, Edinburgh, 14-21 August 1958,** Cambridge (Cambridge Univ. Press) 1960, pp. 137-153. Erratum: p. 149 line 8, for "$H_{n+1}(a)$" read "$H_{(n+1)}(a)$".

1959. *Recursive functionals and quantifiers of finite types I.* **Trans. Amer. Math. Soc.,** vol. 91, pp. 1-52. Errata in the bibliography of **1963.**

1959a. *Countable functionals.* **Constructivity in mathematics,** Amsterdam (North-Holland Pub. Co.), pp. 81-100. Erratum: p. 83, definition of $\alpha^{(2)}(s)$, middle line, for first "$(s)_1$" read "$(s)_1+1$". A full list of errata is in the bibliography of **1963.**

1959b. *Quantification of number-theoretic functions.* **Compositio mathematica,** vol. 14, pp. 23-40. Clarification: p. 26, line 2 from end of text, before "recursive" insert "suitable".

1960. *Realizability and Shanin's algorithm for the constructive deciphering of* *mathematical sentences.* **Logique et analyse,** 3e Année, Oct. 1960, 11-12, pp. 154-165. Erratum: p. 164 line 4, before "A_3" insert "γ".

1962. *Herbrand-Gödel-style recursive functionals of finite types.* **Recursive function theory,** Proceedings of the Symposia in Pure Mathematics, vol. 5, American Mathematical Society, Providence R. I., pp. 49-75.

1962a. *Disjunction and existence under implication in elementary intuitionistic formalisms.* **Jour. symbolic logic,** vol. 27, pp. 11-18. Errata: p. 16, lines 11 and 16, for "$\vdash$" read "|". *An addendum,* ibid., vol. 28 (1963), pp. 154-156 (cf. (b) of Footnote 1 in Chapter I above). Cf. T. T. Robinson **1963.**

1963. *Recursive functionals and quantifiers of finite types II.* **Trans. Amer. Math. Soc.,** vol. 108, pp. 106-142. Principal errata: p. 129, 4th line of LXIX, before second "q_1" insert ";"; p. 135 Footnote 24, in 4th line for "E_2'" read "E_1'", and in 3rd line from end for "Togué" read "Tugué"; p. 142 line 6, for "τ^{j-1}" read "(τ^{j-1})".

See Kleene and Post.

KLEENE, S. C. and POST, EMIL L.

1954. *The upper semi-lattice of degrees of recursive unsolvability.* **Annals of mathematics,** 2 s., vol. 59, pp. 379-407. Erratum: p. 404, in (56) for next to last "=" read "$\neq$".

KOLMOGOROV, A. N. (Kolmogoroff, A.)

1924-5. *O principe tertium non datur* (*Sur le principe de tertium non datur*). **Matematičeskiĭ sbornik (Recueil mathématique de la Société Mathématique de Moscou),** vol. 32, pp. 646-667 (with brief French abstract).

1932. *Zur Deutung der intuitionistischen Logik.* **Mathematische Zeitschrift,** vol. 35, pp. 58-65.

KÖNIG, DÉNES

1926. *Sur les correspondences multivoques des ensembles.* **Fundamenta mathematicae,** vol. 8, pp. 114-134.

KREISEL, GEORG

1958. *Mathematical significance of consistency proofs.* **Jour. symbolic logic,** vol. 23, pp. 155-182. The result in Remark 6.1 p. 172 is strengthened in **1959b.**

1958a. *Elementary completeness properties of intuitionistic logic with a note on negations of prenex formulae.* Ibid., vol. 23, pp. 317-330.

1958b. *A remark on free choice sequences and the topological completeness proofs.* Ibid., pp. 369-388.

1959. *Interpretation of analysis by means of constructive functionals of finite types.* **Constructivity in mathematics,** Amsterdam (North-Holland Pub. Co.), pp. 101-128.

1959a. *The non-derivability of* $\neg(x)A(x) \rightarrow (Ex)\neg A(x)$, *A primitive recursive, in intuitionistic formal systems* (abstract). **Jour. symbolic logic,** vol. 23 (1958), no. 4 (publ. 1959), pp. 456-457.

1959b. *Inessential extensions of Heyting's arithmetic by means of functionals of finite type* (abstract). Ibid., vol. 24, p. 284.

1959c. *Inessential extensions of intuitionistic analysis by functionals of finite type* (abstract). Ibid., pp. 284-285.

1959d. *Reflection principle for subsystems of Heyting's (first order) arithmetic (H)* (abstract). Ibid., p. 322.

1960. Review of Beth **1956.** **Zentralblatt für Mathematik und ihre Grenzgebiete,** vol. 73, pp. 249-250.

1961. *Explicit definability in intuitionistic logic* (abstract). **Jour. symbolic logic,** vol. 25 (1960), no. 4 (publ. 1962), pp. 389-390.

1962. *On weak completeness of intuitionistic predicate logic.* Ibid., vol. 27, pp. 139-158.

1962a. *Foundations of intuitionistic logic.* **Logic, methodology and philosophy of science,** Stanford Calif. (Stanford Univ. Press), pp. 198-210.

1962b. Review of Beth **1959a.** **Zentralbl. Math. Grentzgeb.,** vol. 91, pp. 9-11.

1962c. Review of van Rootselaar **1960.** Ibid., vol. 95, pp. 242-243.

1962d. *Proof theoretic results on intuitionistic first order arithmetic (HA)* (abstract). **Jour. symbolic logic,** vol. 27, pp. 379-380.

1962e. *Proof theoretic results on intuitionistic higher order arithmetic* (abstract). Ibid., p. 380.

1962f. *Consequences of Brouwer's bar theorem* (abstract). Ibid., pp. 380-381.

See Dyson and Kreisel, Kreisel and Putnam.

KREISEL, G. and PUTNAM, H.

1957. *Eine Unableitbarkeitsbeweismethode für den intuitionistischen Aussagenkalkül.* **Archiv für mathematische Logik und Grundlagenforschung,** vol. 3, nos. 3-4, pp. 74-78.

KURODA, SIGEKATU

1951. *Intuitionistische Untersuchungen der formalistischen Logik.* **Nagoya mathematical journal,** vol. 2, pp. 35-47.

LEBLANC, HUGUES and BELNAP, NUEL D., JR.

1962. *Intuitionism reconsidered.* **Notre Dame journal of formal logic,** vol. 3, pp. 79-82.

LOOR, B. DE

1925. **Die hoofstelling van die algebra van intuisionistiese standpunt** (The fundamental theorem of algebra from the intuitionistic standpoint). Thesis Amsterdam, 63 pp.

ŁUKASIEWICZ, JAN

1952. *On the intuitionistic theory of deduction.* **Kon. Ned. Akad. Wet. (Amsterdam), Proc., ser. A,** vol. 55 (or **Indag. math.,** vol. 14), pp. 202-212.

LUSIN, NICHOLAS N.

1927. *Sur les ensembles analytiques.* **Fund. math.,** vol. 10, pp. 1-95.

1930. **Leçons sur les ensembles analytiques et leurs applications.** Paris (Gauthier-Villars), xv+328 pp.

MAEHARA, SHÔJI

1954. *Eine Darstellung der intuitionistischen Logik in der klassischen.* **Nagoya mathematical journal,** vol. 7, pp. 45-64. Cf. **Mathematical reviews,** vol. 16, p. 325.

Markov, A. A.

1951c. *Teorija algorifmov* (Theory of algorithms). **Trudy Matematičeskogo Instituta imeni V. A. Steklova,** vol. 38, pp. 176-179. Eng. tr. by Edwin Hewitt, **American Mathematical Society translations,** ser. 2, vol. 15, 1960, pp. 1-14.

1954. Teorija algorifmov (Theory of algorithms). **Trudy Matem. Inst. im. V. A. Steklova,** vol. 42; Izdatel'stvo Akademii Nauk SSSR, Moscow-Leningrad, 375 pp. Eng. tr. by Jacques J. Schorr-Kon and PST staff, pub. for the U. S. National Science Foundation and Department of Commerce by the Israel Program for Scientific Translations, Jerusalem, 1961, vi+444 pp.

1954a. *O nepreryvnosti konstruktivnyh funkciĭ* (On the continuity of constructive functions). **Uspehi matematičeskih nauk,** vol. 9, no. 3 (61), pp. 226-230.

McCall, Storrs

1962. *A simple decision procedure for one-variable implication/negation formulae in intuitionist logic.* **Notre Dame jour. formal logic,** vol. 3, pp. 102-122.

McKinsey, J. C. C. and Tarski, Alfred

1946. *On closed elements in closure algebras.* **Ann. Math.,** vol. 47, pp. 122-162.

Medvedev, Ju. T.

1962. *Finitnye zadači* (Finite problems). **Doklady Akademii Nauk SSSR,** vol. 142, pp. 1015-1018. Eng. tr. by Elliott Mendelson, **Soviet mathematics** (Amer. Math. Soc.), vol. 3, pp. 227-230.

Mostowski, Andrzej

1961. (an organizer) **Infinitistic methods. Proceedings of the Symposium on Foundations of Mathematics, Warsaw, 2-9 September 1959.** Oxford, London, New York, Paris (Pergamon Press), Warszawa (Państwowe Wydawnictwo Naukowe), 1961, 362 pp. Lacking the names of the Editors, we use that of one of the Organizing Committee of the Symposium.

Nagel, Ernest; Suppes, Patrick; Tarski, Alfred

1962. (editors) **Logic, methodology and philosophy of science. Proceedings of the 1960 International Congress** [Stanford University, August 24-Sept. 2], Stanford Calif. (Stanford Univ. Press), 1962, ix+661 pp.

Nishimura, Iwao

1960. *On formulas of one variable in intuitionistic propositional calculus.* **Jour. symbolic logic,** vol. 25, pp. 327-331.

Ohnishi, Masao

1953. *On intuitionistic functional calculus.* **Osaka mathematical journal,** vol. 5, pp. 203-209. Cf. de Iongh 1948.

Péter, Rósza

1951. Rekursive Funktionen. Akadémiai Kiadó (Akademischer Verlag) Budapest, 206 pp. 2nd ed., Budapest 1957, 278 pp.

PIL'ČAK, B. JU.

1950. *O probleme razrešimosti dlja isčislenija zadač* (On the decision problem for the calculus of problems). **Doklady Akad. Nauk SSSR,** vol. 75, pp. 773-776.

1952. *Ob isčislenii zadač* (On the calculus of problems). **Ukrainskiĭ matematičeskiĭ žurnal,** vol. 4, pp. 174-194. Cf. Kreisel-Putnam **1957** p, 78.

PORTE, JEAN

1958. *Une propriété du calcul propositionnel intuitionniste.* **Kon. Ned. Akad. Wet. (Amsterdam), Proc., ser. A,** vol. 61 (or **Indag. math.,** vol. 20), pp. 362-365.

RASIOWA, HELENA

1951. *Algebraic treatment of the functional calculi of Heyting and Lewis.* **Fund. math.,** vol. 38, pp. 101-126.

1954. *Constructive theories.* **Bulletin de l'Académie Polonaise des Sciences, Classe III,** vol. 2, pp. 121-124.

1954a. *Algebraic models of axiomatic theories.* **Fund. math.,** vol. 41, pp. 291-310. See Rasiowa and Sikorski.

RASIOWA, HELENA and SIKORSKI, ROMAN

1953. *Algebraic treatment of the notion of satisfiability.* **Fund. math.,** vol. 40, pp. 62-95.

1954. *On existential theorems in non-classical functional calculi.* Ibid., vol. 41, pp. 21-28.

1955. *An application of lattices to logic.* Ibid., vol. 42, pp. 83-100.

1959. *Formalisierte intuitionistische elementare Theorien.* **Constructivity in mathematics,** Amsterdam (North-Holland Pub. Co.), pp. 241-249.

RIDDER, J.

1950-1. *Formalistische Betrachtungen über intuitionistische und verwandte logische Systeme, I-VII.* **Kon. Ned. Akad. Wet. (Amsterdam), Proc., ser. A,** vol. 53 (1950), pp. 327-336, 446-455, 787-799, 1375-1389 (= **Indag. math.,** vol. 12, pp. 75-84, 98-107, 231-243, 445-459) and vol. 54 (1951) (or **Indag. math.,** vol. 13), pp. 94-105, 169-177, 226-236.

RIEGER, LADISLAV

1949. **On the lattice theory of Brouwerian propositional logic.** Acta Facultatis Rerum Naturalium Universitatis Carolinae, no. 189, Prague (F. Řivnáč), 40 pp. (with Czech summary).

ROBINSON, T. THACHER

1963. **Interpretations of Kleene's metamathematical predicate Γ|A in intuitionistic arithmetic N.** Dissertation Princeton (mimeographed), iii+75 pp.

ROGERS, HARTLEY, JR.

1964. **Recursive functions and effective computability.** McGraw-Hill, New York.

ROOTSELAAR, B. VAN

1952. *Un problème de M. Dijkman.* **Kon. Ned. Akad. Wet. (Amsterdam), Proc., ser. A,** vol. 55 (or **Indag. math.,** vol. 14), pp. 405-407.

1960. *On intuitionistic difference relations.* Ibid., vol. 63 (or **Indag. math.,** vol. 22), pp. 316-322. *Corrections to the paper "On intuitionistic difference relations"*, ibid., vol. 66 (or **Indag. math.,** vol. 25) 1963, pp. 132-133. Cf. Kreisel **1962c.**

ROSE, GENE F.

1953. *Propositional calculus and realizability.* **Trans. Amer. Math. Soc.,** vol. 75, pp. 1-19. The publ. version of 1952. Cf. Jaśkowski 1936, Pil'čak **1952,** Gál-Rosser-Scott **1958,** Medvedev **1962.**

ROSSER, J. BARKLEY

1957. (anonymous editor) **Summaries of talks presented at the Summer Institute of Symbolic Logic in 1957 at Cornell University** (mimeographed). 3 vols., 432 pp. 2nd ed., Princeton N. J. (Communications Research Division, Institute for Defense Analyses) 1960, xvi+427 pp.

See Church and Rosser; Gál, Rosser and Scott.

SACKS, GERALD E.

1963. **Degrees of unsolvability.** Annals of Mathematics studies, no. 55, Princeton Univ. Press, Princeton N. J., xi+174 pp.

ŠANIN, N. A. (Schanin, N. A.; Shanin, N. A.)

1953. *O nekotoryh operacijah nad logiko-arifmetičeskimi formulami* (Some operations on logico-arithmetical formulas). **Doklady Akademii Nauk SSSR,** vol. 93, pp. 779-782.

1954. *O pogruženijah klassičeskogo logiko-arifmetičeskogo isčislenija v konstruktivnoe logiko-arifmetičeskoe isčislenie* (On imbeddings of the classical logico-arithmetical calculus into the constructive logico-arithmetical calculus). Ibid., vol. 94, pp. 193-196.

1955. *O nekotoryh logičeskih problemah arifmetiki* (On some logical problems of arithmetic). **Trudy Matem. Inst. im. V. A. Steklova,** vol. 43; Izdat. Akad. Nauk SSSR, Moscow, 112 pp.

1958. *O konstruktivnom ponimanii matematičeskih suždeniĭ* (On the constructive interpretation of mathematical judgements). Ibid., vol. 52, pp. 226-311. Eng. tr. by Elliott Mendelson, **Amer. Math. Soc. translations,** ser. 2, vol. 23 (1963), pp. 109-189.

1958a. *Ob algorifme konstruktivnoĭ rasšifrovki matematičeskih suždeniĭ (Über einen Algorithmus zur konstruktiven Dechiffrierung mathematischer Urteile).* **Zeitschrift für mathematische Logik und Grundlagen der Mathematik,** vol. 4, pp. 293-303.

SCHMIDT, H. ARNOLD

1958. *Un procédé maniable de décision pour la logique propositionnelle intuitionniste.* **Le raisonnement en mathématiques et en sciences expérimentales,** pp. 57-66. Colloques Internationaux du Centre National de la Recherche Scientifique, LXX. Editions du Centre National de la Recherche Scientifique, Paris.

SCHRÖTER, KARL

1956. *Über den Zusammenhang der in den Implikationsaxiomen vollständigen*

Axiomensysteme des zweiwertigen mit denen des intuitionistischen Aussagenkalküls. **Zeitsch. math. Logik Grundlagen Math.,** vol. 2, pp. 173-176.
1957. *Eine Umformung des Heytingschen Axiomensystems für den intuitionistischen Aussagenkalkül.* Ibid., vol. 3, pp. 18-29.

SCHÜTTE, KURT
1962. *Der Interpolationssatz der intuitionistische Prädikatenlogik.* **Math. Ann.,** vol. 148, pp. 192-200.

SCOTT, DANA
1957. *Completeness proofs for the intuitionistic sentential calculus.* **Summaries of talks presented at the Summer Institute of Symbolic Logic in 1957 at Cornell University** (mimeographed), pp. 231-242. 2nd ed., Princeton N. J. (Communications Research Division, Institute for Defense Analyses) 1960.
1960. Review of Skolem **1958. Math. reviews,** vol. 21, pp. 1031-1032.
See Gál, Rosser and Scott.

SIKORSKI, ROMAN
1959. *Der Heytingsche Prädikatenkalkul und metrische Räume.* **Constructivity in mathematics,** Amsterdam (North-Holland Pub. Co.), pp. 250-253.
See Rasiowa and Sikorski.

SKOLEM, TH.
1958. *Remarks on the connection between intuitionistic logic and a certain class of lattices.* **Mathematica Scandinavica,** vol. 6, pp. 231-236. Cf. Scott **1960.**

SMULLYAN, RAYMOND
1959. Theory of formal systems. Massachusetts Institute of Technology, Lincoln Laboratory, Group Report 54-5, Lexington, Mass. Also Ann. Math. studies, no. 47, Princeton N. J. (Princeton Univ. Press) 1961, x+142 pp.

SPECTOR, CLIFFORD
1962. *Provably recursive functionals of analysis: a consistency proof of analysis by an extension of principles formulated in current intuitionistic mathematics.* **Recursive function theory,** Proc. Symposia Pure Math., vol. 5, Amer. Math. Soc., Providence R. I., pp. 1-27.

STONE, M. H.
1937-8. *Topological representations of distributive lattices and Brouwerian logics.* **Časopis pro pěstování matematiky a fysiky,** vol. 67, pp. 1-25.

TARSKI, ALFRED
1938. *Der Aussagenkalkül und die Topologie.* **Fund. math.,** vol. 31, pp. 103-134.
See McKinsey and Tarski; Tarski, Mostowski, Robinson; Nagel, Suppes, Tarski.

TARSKI, ALFRED; MOSTOWSKI, ANDRZEJ; ROBINSON, RAPHAEL M.
1953. Undecidable theories. Amsterdam (North-Holland Pub. Co.), XI+98 pp.

TRAHTENBROT, B. A. (Trakhtenbrot, B. A.)
1960. Algoritmy i mašinnoe rešenie zalač (Algorithms and machine solution of problems). 2nd ed., ed. by S. V. Jablonskiĭ, Gosudarstv. Izdat. Fiz.-

Mat. Lit., Moscow, 119 pp. **Algorithms and automatic computing machines,** Boston (D. C. Heath) 1963, viii+101 pp. is an Eng. tr. *and adaptation* (in which notes and some additional references were supplied, and the text was cut, rearranged and paraphrased in places) by Jerome Kristian, James D. McCawley and Samuel A. Schmitt (ed. by Alfred L. Putnam and Izaak Wirzup, under a grant from the U. S. National Science Foundation) of the 2nd ed.

UMEZAWA, TOSHIO

1955. *Über die Zwischensysteme der Aussagenlogik.* **Nagoya math. journal,** vol. 9, pp. 181-189.

1959. *On intermediate propositional logics.* **Jour. symbolic logic,** vol. 24, pp. 20-36.

1959a. *On logics intermediate between intuitionistic and classical predicate logic.* Ibid., pp. 141-153.

USPENSKIĬ, V. A.

1960. **Lekcii o vyčislimyh funkcijah** (Lectures on computable functions). Matematičeskaya logika i osnoviniya matematiki, Gosudarstv. Izdat. Fiz.-Mat. Lit., Moscow, 492 pp.

VAN DALEN, DIRK

See Dalen, Dirk van.

VAN DANTZIG, DAVID

See Dantzig, David van.

VAN ROOTSELAAR, B.

See Rootselaar, B. van.

VESLEY, RICHARD E.

1963. *On strengthening intuitionistic logic.* **Notre Dame jour. formal logic,** vol. 4, p. 80.

ŽEGALKIN, I. I.

1936. *O probleme razrešimosti v Brouwer'ovskoĭ logike predloženiĭ* (On the decision problem in Brouwer's propositional logic). **Trudy 2-go Vsesojuznogo Matematičeskogo S'ezda, Leningrad 24-30 ijunja 1934,** vol. 2, Moscow-Leningrad 1936, p. 437. An abstract, stating the existence of a decision procedure for Brouwer's propositional calculus. (Gentzen 1934-5 was received 21 July 1933.)

SYMBOLS AND NOTATIONS

[The notations "used formally" include ones used also informally (possibly changing] Roman to italic type). A bold-faced page reference indicates that the notation is also indexed in IM p. 538. Notations of IM taken over tacitly are indexed only in IM p. 538.

USED FORMALLY

USED ONLY INFORMALLY

INDEX

Names mentioned only in Footnotes 2 and 3 are not indexed.